NUCLEAR ENERGY
Technology from Hell

Neeraj Jain

"Man's dearest possession is life. It is given to him but once, and he must live it so as to feel no torturing regrets for wasted years, never know the burning shame of a mean and petty past; so live that dying, he might say: all my life, all my strength were given to the finest cause in all the world—the fight for liberation of Mankind."

Nikolai Ostrovosky

Nuclear Energy: Technology from Hell
Neeraj Jain

First Published, 2012

ISBN 978-93-5002-180-4 (Hb)

Published by
AAKAR BOOKS
28 E Pocket IV, Mayur Vihar Phase I, Delhi 110 091
Phone : 011 2279 5505 Telefax : 011 2279 5641
info@aakarbooks.com; www.aakarbooks.com

in association with
LOKAYAT
129 B/2, Erandwane,
Pune 411 004
lokayat.india@gmail.com; www.lokayat.org.in

Printed at
Mudrak, 30 A, Patparganj, Delhi 110 091

Dedicated to the fighting people of Madban, Nate, Mithgavane and other villages of the Jaitapur region.

Contents

Acknowledgements

I have much gratitude for the invaluable help, knowledge and advice provided by Dr Surendra Gadekar. I am no expert on nuclear energy and, whenever I had difficulties in understanding any issue, he patiently answered all my queries. He also took out time to read the manuscript, pointed out numerous mistakes and gave invaluable advice, without which this book would never have taken the shape it has finally taken. I am also very grateful to Dr Sulabha Brahme, who initiated me into writing this book, went through the draft and gave many important suggestions, and to Soumya Dutta and Shanker Sharma, who also went through some parts of the draft and provided invaluable comments and criticisms. Then of course, I would like to thank Karen Haydock for making all those wonderful sketches which have been included in the book, and Sumesh Chandran who allowed me to use his cartoon which has been included in the Epilogue of the book. I would also like to thank Milind Joshi who provided help in the layout and design.

Finally, my deepest gratitude to all my colleagues in Lokayat, who provided invaluable support during the long period it took me to write this book.

Aknowledgements for Photographs

We thankfully acknowledge the following links for the photographs included in the book:

P1 – Child Victims of Chernobyl

- *Sasha:* http://picasaweb.google.com/BlacksmithInstitute/WorldSWorstPolluted Places2007#5109019001241573154
- *Annya:* http://www.chernobyl-international.com/Libraries/blog/Annya_2005_ Knoth_1.sflb.ashx
- *Natasha and Vadim*: http://www.flickr.com/photos/greenpeacebrasil/5536720339/lightbox/
- *Alexandra:* http://www.flickr.com/photos/greenpeacebrasil/5536720365/in/photostream/
- *Nastya:* http://www.greenpeace.org/international/en/campaigns/nuclear/

P2 – Child Victims of Jaduguda Uranium Mines

- http://network.earthday.net/profiles/blogs/how-much-safe-is-nuclear-power
- http://network.earthday.net/profiles/blog/list?tag=uranium
- http://www.siliconeer.com/past_issues/2007/siliconeer_march_2007.html
- http://nitishpriyadarshi.blogspot.com/2008/10/growing-sunflower-in-jadugoda-of.html
- http://www.dianuke.org/urgent-an-appeal-for-solidarity-with-anti-uranium-mine-movement-in-jadugoda/

Remaining photographs are courtesy Dr Surendra Gadekar

P3 – Rawatbhata Nuclear Plant, Health Costs:

- All photographs, courtesy Dr Surendra Gadekar

P4 – The Growing Worldwide Anti-nuclear Movement:

- French protest: http://www.beyondnuclear.org/home/2009/10/3/french-anti-nuclear-rally-goes-on-despite-city-under-siege.html
- Japan protest: http://shineyourlight-shineyourlight.blogspot.com/2011/04/7-japans-nuke-crisis-on-same-level-of.html

P5 – The Growing Worldwide Anti-nuclear Movement

- Hongkong protest: http://world.pressenza.org/npermalink/anti-nuclear-demonstrations-in-hong-kong-june-11-2011
- Germany protest: http://www.indymedia.org.au/2011/03/28/biggest-ever-anti-nuclear-protests-in-germany

P6 – The Growing Worldwide Anti-nuclear Movement

- Taiwan protest: http://mirrorsignalmove.blogspot.com/2011/05/anti-nuke-protest-in-kaohsiung.html
- South Korea protest: http://makanaka.wordpress.com/2011/05/05/radiation-truth-there-is-no-threshold-below-which-risk-is-zero/

P7 – Anti-nuclear Protests, India

- Kudankulam protest: http://www.dianuke.org/thirteen-reasons-against-the-koodankulam-nuclear-power-project/
- March to Parliament: http://toxicswatch.blogspot.com/2011/03/anti-nuclear-march-to-parliament.html

P8 – Anti-nuclear Protests, India

- Tarapur Jaitapur Yatra: http://www.thehindu.com/news/national/article1761935.ece
- Jaitapur protest: http://www.deccanchronicle.com/channels/nation/north/one-killed-seven-injured-anti-jaitapur-protests-turn-violent-319

Foreword

The most unfortunate thing about this book is that, even after sixty years of the nuclear age, there is such a need for it. In a fairer and a more just world, nuclear power would have died a natural death long before this. The fact that it has no relevance to peoples' real energy needs—besides being too expensive, extremely dirty and prone to catastrophic accidents—would have led to its rejection in any sane society. Its continued existence, even after Fukushima, is a demonstration of the power of moneyed vested interest to steer large parts of the world towards suicide.

The next most unfortunate thing about this book is the language it is written in. The people who need this book the most, read and understand languages such as Marathi, Hindi, Gujarati, Telugu, Tamil, Bangla, etc. Neeraj Jain has assured me that this book is being translated into Marathi. I hope it also gets translated into other Indian languages and can become a source-book for activists involved in the struggle against the nuclear monster.

The Indian government, its atomic energy establishment and many amongst the ruling elites in the country have turned a blind eye towards the warning posed by the Fukushima disaster in Japan and the many other near misses that have already occurred in India. Chernobyl is, of course, too distant a memory for these worthies to recollect. It is the ordinary people of this country, the *aam adami*, who have not forgotten these disasters or overlooked their relevance to their existence and opposed the setting up of any more of these death-dealing machines. Hopefully, this book shall help in these efforts.

July 20, 2011

Dr Surendra Gadekar
Sampoorna Kranti Vidyalaya
Vedchhi, Dist. Surat
Gujarat 394 641

Abbreviations Used in Text

AEC	Atomic Energy Commission (India)
AERB	Atomic Energy Regulatory Board
BARC	Bhabha Atomic Research Centre
BWR	Boiling Water Reactor
DAE	Department of Atomic Energy (India)
DOE	Department of Energy (USA)
DU	Depleted Uranium
ECCS	Emergency Core Cooling System
EIA	Environmental Impact Assessment
EPA	Environmental Protection Agency (USA)
EPR	European Pressurised Reactor
FBR	Fast Breeder Reactor
FDA	Food and Drug Administration (USA)
GE	General Electric
HWR	Heavy Water Reactor
IAEA	International Atomic Energy Agency
IEA	International Energy Agency
IEP	Integrated Energy Policy (India)
KAPS	Kakrapar Atomic Power Station
KARP	Kalpakkam Atomic Reprocessing Plant
LWR	Light Water Reactor
MAPS	Madras Atomic Power Station (Kalpakkam)
MNRE	Ministry of New and Renewable Energy (India)
MoEF	Ministry of Environment and Forests
MOX	Mixed Oxide Fuel
NPCIL	Nuclear Power Corporation of India Limited
NRC	Nuclear Regulatory Commission (USA)
PHWR	Pressurised Heavy Water Reactor
RAPS	Rajasthan Atomic Power Station, Rawatbhata
TAPS	Tarapur Atomic Power Station
TEPCO	Tokyo Electric Power Company
UCIL	Uranium Corporation of India Limited
WHO	World Health Organisation

About Us : Lokayat

Santhal revolt

Who has become free?
From whose forehead has slavery's stain been removed?
My heart still pains of oppression ...
Mother India's face is still sad ...
Who has become free?

Ali Sardar Jafri wrote these lines a few years after Independence. Nevertheless, these lines accurately describe the current situation in our country too! 'Who has become free', is indeed the real question. This country now belongs to the rich; development is now only for them. Giant-sized malls, ultra-modern cars, express highways, imported luxury goods, five-star hospitals ... and, on the other hand, the few crumbs given to the poor after Independence are also being snatched away.

In the deceptive name of *globalisation*, giant multinational corporations (MNCs) are being invited into the country—the country is now being run solely for the profit maximisation of big foreign and Indian corporations. In connivance with the politicians-bureaucracy-police-courts, they have launched a ferocious assault to dispossess the poor of their lands, forests, water and resources in order to set up Special Economic Zones (SEZs), huge infrastructural projects, golf courses, residential complexes for the rich, et cetera. Laws are being modified to facilitate their loot. In the name of *privatisation*, public sector corporations, including public sector banks and insurance companies, built out of the savings, sweat and toil of the common people, are being handed over at throwaway prices to these scoundrels. Indian agriculture, on which 60 per cent of the Indian people still depend for their livelihoods, is being deliberately strangulated; so that it can be taken over by giant agribusiness corporations. The consequence is that nearly two lakh farmers have committed suicide in the past fifteen years—something which did not happen even during the colonial era. There are simply no decent jobs for the youth: big corporations are retrenching tens of thousands of workers, while small businesses are downing their shutters by the millions. If one uses the standards of the developed countries, probably nearly half the population is unemployed or underemployed. Even welfare services are being taken over by these corporations and transformed into instruments for naked profiteering: government hospitals and municipal schools are being privatised; medicine prices have zoomed; college fees have gone through the roof; electricity prices are rising; bus fares are rising; the public distribution system designed to check speculation in prices of foodgrains is being eliminated; and now, drinking water supply in cities is also being handed over to these corporations, who will then hike its rates 10-15 times. Today, there is no need for the imperialists to rule us by the force of arms. Our black rulers are themselves handing over control of our wealth, resources, economy to them for their unbridled plunder.

The imperialists are farsighted. They are not satisfied with just controlling the economy. They also want to control what we eat, drink, see, think, read. And so, along with MNC capital, imperialist culture is also flowing in.

As the economic system becomes more and more sick, the

social and political system is also becoming more and more degenerate. All-pervasive corruption, an educational system that makes us think that we are incompetent fools, continuation of the age-old, caste-based social system because of which atrocities on the dalits take place almost daily, a communal political system that divides people in the name of religion and fills them with hatred against each other, a value system that promotes crass selfishness and unconcern and apathy for others, a society where cynicism and moral bankruptcy permeate every nook and cranny—this is the reality of today. In the name of fighting terrorism, the criminals and murderers who dominate the Indian Parliament are passing the most draconian laws which give almost unlimited powers to the police to arrest ordinary people and put them behind bars for years without trial!

The common people have not been silent spectators to this sordid drama being enacted by the MNCs and their Indian collaborators. Like flowers springing up in every nook and corner with the onset of spring, people are coming together all over the country, getting organised, forming groups and raising their voices in protest. Though these struggles are presently small, scattered and without resources, the future lies in these magnificent struggles. As more and more people join them, they will strengthen, join hands and become a powerful force which will transform society.

We must stop being skeptics, dream of a better future, believe that it is possible to change the world. Yes, *Another World is Possible*! But to make it a reality, we must start our own small struggles. These will ultimately unite, like the small rivulets hurtling down the Himalayas which ultimately form the mighty Ganges. And so, we have started this forum, **Lokayat**.

Lokayat is actually a thought process that has existed in India since Vedic times. Lokayat is a vision of life that rejects fatalism, is uncompromisingly rationalist, realist. Life must be lived to the full, in the best possible way. All problems are man-made; one should face them and not run away from them. This is the inspiration that Lokayati thought gives us. We accept this tradition; hence the name of our forum, **Lokayat**.

The aim of Lokayat is to bring together ordinary people who wish to take some initiative, who wish to do their bit for transforming society for the better and to take up various activities with their cooperation. Some of the activities that we have initiated so far are:

- We organise public awareness campaigns on various issues of deep concern to common people, such as: privatisation of essential services like education, health, electricity; rise in petrol and diesel prices; destructive effects of nuclear energy; decaying public transport system; harmful effects of genetically modified foods, et cetera. We are also active in many national campaigns like 'Boycott Coke-Pepsi Campaign', 'No More Bhopals Campaign', 'Campaign for Judicial Accountability and Reforms', 'Anti-Posco Campaign', 'Campaign in Defence of the Right to Dissent', et cetera. We use various forms such as street campaigns, poster exhibitions and street plays in these campaigns; likewise we also organise protest programs like rallies, dharnas, et cetera on these issues.
- We organise film shows, seminars and talks on issues like displacement and destruction of livelihoods of common people in the name of development, US invasion of Iraq, targeting of minorities in the name of fighting terrorism, gender inequality and the caste question, global warming, etc. We especially focus on reaching out to the youth in colleges.
- While working on all these fronts, we publish booklets-pamphlets that discuss and analyse current questions—in order to solve a problem, we must first thoroughly understand it.

Dear friends, if you would like to know more about us, you may contact us at :

Contact address:

Lokayat,
Opp. Syndicate Bank, Law College Road
Near Nal Stop, Pune - 411 004

Contact phones:

Neeraj Jain 94222 20311
Abhijit A. M. 94223 08125

E-mail: neerajj61@gmail.com
Website : www.lokayat.org.in

INTRODUCTION

Part I: Government Plans for Nuclear Energy

The government of India is promoting nuclear energy as a solution to the country's future energy needs and is embarking on a massive nuclear energy expansion program. It expects to have 20,000 MW nuclear power capacity online by 2020[1] and 63,000 MW by 2032[2]. The Department of Atomic Energy (DAE) has projected that India would have an astounding 275,000 MW of nuclear power capacity by 2050, which is expected to be 20 per cent of India's total projected electricity generation capacity by then.[3] The signing of the Indo-US Nuclear Deal having opening up the possibility of uranium and nuclear reactor imports, the Prime Minister stated, in September 2009, that India could have an even more amazing 470,000 MW of nuclear capacity by 2050.[4] Dr Anil Kakodkar, then Chairman of India's Atomic Energy Commission (AEC), is even more optimistic. He has predicted that

India's nuclear energy capacity could reach 600-700 thousand MW and account for 40 per cent of the estimated total power generation by 2050.[5]

This would be a quantum leap from the present scenario. As of March 31, 2010, the total installed power generation capacity in the country was 159,400 MW, of which the contribution of nuclear power —more than sixty years after the atomic energy program was established and forty years after the first nuclear reactor started feeding electricity to the grid—was just 4560 MW,[6] or 2.86 per cent of the total. Thus, the projected capacity in 2050 would represent an increase by a factor of over a hundred.

New Projects

The government has taken rapid steps to implement this plan. Following the Indo-US Nuclear Deal, it has given 'in principle' approval to setting up a string of giant size nuclear parks all along India's coastline, each having six to eight reactors of between 1000 to 1650 MW—at Mithivirdi (Gujarat), Jaitapur (Maharashtra), Kudankulam (Tamil Nadu), Kovvada (Andhra Pradesh) and Haripur (West Bengal). It is also proposing to set up four indigenous reactors of 700 MW each at Gorakhpur in Haryana, and another two similar reactors at Chutka in Madhya Pradesh. To meet the fuel needs of these plants, it is proposing to set up several new uranium mining projects: at Tummalapalle (Kadapa district) and Lambapur-Peddagattu (Nalgonda district) in Andhra Pradesh, Gogi (near Gulbarga) in Karnataka and West Khasi Hills district of Meghalaya.

Government Claims

Justifying this huge push for nuclear energy, India's politicians, nuclear scientists and many prominent intellectuals are claiming that nuclear energy is clean, safe, green and cheap. This propaganda campaign is being led from the front by the Prime Minister himself. Here are a few quotes from some of his recent statements (emphasis ours in all quotes):

- At the inauguration of a new fuel reprocessing plant at the Bhabha Atomic Research Centre, Tarapur on January 7, 2011:

He praised the plant at Tarapur as 'an outstanding example of *clean, economic and safe energy* that our nation requires'.[7]

- At the Nuclear Security Summit, held in Washington, D.C. on April 13, 2010:

> Today, nuclear energy has emerged as a *viable source of energy* to meet the growing needs of the world in a manner that is *environmentally sustainable*. There is a real prospect for nuclear technology to address the developmental challenges of our times ... The *nuclear industry's safety record* over the last few years has been *encouraging*. It has helped to restore public faith in nuclear power.[8]

- Speech after dedicating Tarapur-3 and 4 atomic reactors to the nation on August 31, 2007: A *nuclear renaissance* is taking place in the world, 'and we cannot afford to miss the bus or lag behind these global developments.' Elaborating on the reasons for the growing importance of nuclear energy, he stated: 'Our long-term economic growth is critically dependent on our ability to meet our energy requirements of the future ... [Since] our proven reserves of coal, oil, gas and hydropower are totally insufficient to meet our requirements (and) the energy we generate has to be *affordable, not only in terms of its financial cost, but in terms of the cost to our environment*', this was the reason why 'we place so much importance on nuclear energy'.[9]
- Statement to the Indian Parliament on July 29, 2005, after returning from a visit to the United States where the first steps were taken towards signing what has come to be known as the 'Indo-US Nuclear Deal': 'Energy is a crucial input to propel our economic growth ... it is clear that nuclear power has to play an increasing role in our electricity generation plans ... For this purpose, it would be very useful if we can access nuclear fuel as well as nuclear reactors from the international market ... There is also considerable concern with regard to global climate change arising out of CO_2 emissions. Thus, we need to pursue *clean energy* technologies.

> Nuclear power is very important in this context as well.' Since 'the US understood our position in regard to our securing adequate and *affordable* energy supplies, from all sources' and because President Bush was willing to 'work towards promoting nuclear energy as a means for India to achieve energy security', this was the reason why India has decided to enter into a nuclear cooperation agreement with the USA.[10]

On January 18, 2011, at an 'open house' on the Jaitapur Nuclear Power Project organised by the Chief Minister of Maharashtra in coordination with the Nuclear Power Corporation of India Ltd. (NPCIL), to clear misconceptions about nuclear power, an entire galaxy of scientists and doctors emphasised that nuclear power was safe, clean and green. They stated that the claims made by activists and scientists opposing nuclear energy—that radiation leakage from nuclear plants has a horrendous impact on human health, that it causes cancer and birth deformities in children, that mankind has yet to find a solution to the problem of what to do with the terribly radioactive waste generated by nuclear plants and that nuclear plants are prone to catastrophic accidents—were either an exaggeration, or lies:

- S.K. Jain, NPCIL chairman and managing director, claimed that India already runs 20 nuclear plants without any blemish on its safety record.[11]
- The 'experts' claimed that nuclear plants do not harm the environment. Dr S.P. Dharne from the NPCIL said that nuclear power was clean and green energy, and that it could reduce the impact of global warming since it did not generate carbon dioxide.[12] Dr Srikumar Banerjee, current Chairman of the AEC, in fact, came up with the fantastic claim that flora and fauna had actually increased around India's nuclear plants.[13]
- Dr Anil Kakodkar, former Chairman of the AEC, tried to prove that the atomic waste generated by the Jaitapur nuclear plant would not cause any problems, as 'there is no question of the waste being thrown in the open areas'. He stated that the nuclear waste would be 'taken to reprocessing plant after

use', and therefore '[t]here is no hazard of the waste to the biodiversity of Konkan region.'[14]

- On fears about radiation leakages from nuclear power plants, the government experts came up with another amazing explanation: they stated that the belief that nuclear plants cause impotency and cancer and deformities among children is due to superstitions because of illiteracy![15] Dr Rajendra Badwe, head of the Tata Memorial Cancer Hospital, tritely stated that the plant was safe as, otherwise, it would not have been permitted. Referring to the survey by the anti-nuclear activist-scientist Dr Surendra Gadekar on the incidence of abnormalities in children around the Rawatbhata Atomic Power Station in Rajasthan, which has been published in a leading international journal, he blithely lied that the report was without any foundation since it had not been peer-reviewed and published in reputed scientific journals. On the contrary, he made the bewildering claim that radiation was used to cure cancers.[16] Nuclear scientists Sharad Kale and Shrikumar Apte said there would not be any effect of radiation on agricultural products and marine life in the area.[17]

The propaganda is so intense that most people in the country, at least those who read the newspapers and watch television, believe that nuclear energy is an environmentally friendly solution to India's power shortages.

Part II: People's Resistance

The people's movement against nuclear energy in India dates back to the 1980s. The movement was especially strong in Kerala, where people succeeded in forcing the cancellation of plans to set up nuclear plants at Kothamangalam and Peringome. Tens of thousands of people came out onto the streets to protest government plans to set up nuclear plants at Kakrapar (in Gujarat) and Kaiga (in Karnataka). There were also protests against the decision to site a nuclear plant at Narora in the thickly populated state of Uttar Pradesh.[18]

In continuation of this glorious history, people are rising up in revolt at each and every place where the government is proposing to set up a new uranium mining project or a nuclear power plant. Protests have stalled the uranium mining project in **Nalgonda** district in Andhra Pradesh for the last five years,[19] while a powerful movement led by the Khasi Students Union, together with various tribal organisations, has held up the mining project in the state of **Meghalaya** for over one and a half decades now.[20] Likewise, people everywhere are strongly protesting proposals to set up nuclear plants, be it in Haripur (West Bengal), Gorakhpur (Haryana), Mithivirdi (Gujarat) or Jaitapur (Maharashtra).

Kudankulam

The people of Tirunelveli, Kanyakumari and Tuticorin districts have fought long and hard against the two Russian VVER-1000 reactors being built in Kudankulam village in Tirunelveli district of Tamil Nadu. Plans to build the reactors were first announced during the visit of Prime Minister Morarji Desai to Moscow in 1979; a formal agreement for the project was signed during President Gorbachev's visit to New Delhi in 1988. People's opposition to these plans intensified in the late 1980s, with more than 10,000 people participating in a rally in Kanyakumari called by the National Fishworkers Union to focus national attention on environmental issues, including the Kaiga and Kudankulam atomic power plants. Soon after, the collapse of the Soviet Union in 1991 stalled the project.[21]

This fortuitous reprieve lasted only a few years. In 1997, the Indian Prime Minister, Deve Gowda, and the Russian President, Boris Yeltsin, signed an agreement to revive the Kudankulam project. The people, too, revived their struggle. The struggle has further intensified after the government signed another agreement with Russia to build four additional reactors there. Various people's organisations have come together and formed an umbrella organisation, the People's Movement Against Nuclear Energy (PMANE), to fight the nuclear plant. They have held meetings in practically every village in the area and have organised dozens of demonstrations, cycle yatras and seminars against the project.

Construction of the first two reactors was started in 2001, without any environmental clearance, under the excuse that the proposal for these reactors had been first mooted in 1988, when the law providing for environmental clearance for large projects was not in force (this law was passed in 1994). For the additional four reactors proposed, the public hearing on people's objections to the Environmental Impact Assessment report (necessary before a plant is granted environmental clearance) was held on June 2, 2007. Thousands gathered to file their opposition to the plant, despite an intimidating *bandobast* with 1,200 policemen, nasty riot gear and armoured personnel carriers. Yet, none of this prevented the people from expressing their views. However, after just a few people had voiced their objections, the collector brought the public hearing to an abrupt end after two hours, and declared that the people had given their assent to the plant![22]

Haripur

More than 20,000 people, organised under the banner of 'Haripur Paramanu Bidyut Prakalpa Pratirodh Andolan', prevented a team of experts from the NPCIL from visiting the area on November 17, 2006, even though they were accompanied by battalions of armed police. Thousands of men, women and children from villages around the proposed site blockaded all entry points and vowed to embrace instant death rather than allowing their coming generations to suffer from the nuclear menace. The attempt was repeated on the next day; but again the experts and police were forced to go back.

The stakes for building nuclear plants are very high, and it makes for strange bedfellows. While the CPI(M) was strongly against the Indo-US Nuclear Deal, which was crucial for the construction of the Haripur plant to go ahead, and has also been protesting the Jaitapur nuclear plant probably because it is in the opposition in the state of Maharashtra, the West Bengal Chief Minister has repeatedly expressed his support for building the Haripur plant, and the local goons of CPI(M) have tried to portray the opposition as either Maoists or as being anti-development environmentalists. Yet, repression has not broken the resolve of the people, and they have not allowed a single official of India's atomic energy establishment to visit the area for the last 5 years.[23]

Mithivirdi

A powerful movement of the people of Mithivirdi, Jaspara and nearly 40 surrounding villages in district Bhavnagar of Gujarat has being going on for the last three years against government plans to construct a 6000-8000 MW nuclear power plant there. 7000 people attended a public meeting against the project on April 25, 2010. In June 2010, NPCIL officials together with truck loads of police tried to visit the area to take soil samples for testing, but thousands of people surrounded them and firmly told them to go back. After trying to use force, the officials and police finally retreated.[24]

Gorakhpur

NPCIL is proposing to set up four indigenous reactors in Gorakhpur village, in Fatehabad district of Haryana. Despite efforts by NPCIL scientists to convince the local people about the benefits of nuclear power, the villagers of Gorakhpur and nearby villages have launched a militant protest against the project. They have been sitting on a dharna outside the office of the District Collector since October 2010. The biting cold wave led to one farmer being martyred and many farmers being hospitalised. However, this has not broken the resolve of the people. Support groups for the struggle have been formed in a number of nearby cities, including Chandigarh.

Jaitapur

Amongst the most heroic of these struggles has been the militant struggle of the people of Madban, Nate and other nearby villages against the Jaitapur nuclear plant in Ratnagiri district of Maharashtra. The government has forcibly acquired land from 2275 families, after more than 95 per cent of them refused to accept the hiked compensation offered by the government of Rs.10 lakh per acre and the promise of a job. The few people who have accepted the cheques are mostly absentee landlords. The issue for the people is not displacement, which is why not just the affected people, but people from dozens of nearby villages too, are waging a fantastic struggle despite intense police repression. Farmers, mango growers, rickshaw drivers, transporters, fisherfolk, shopkeepers, everyone has joined the movement. They are refusing to believe assurances given by the top official scientists of the country, media intellectuals and politicians of various parties, that nuclear energy is safe, clean and green. They firmly believe that the plant will destroy not just their livelihoods, but will also affect the very sustainability of life in the entire Konkan region for centuries. When the government issued a directive to school teachers to brainwash students into believing that nuclear energy is green, the children boycotted the schools for a few days!

The government has unleashed savage repression on the people. It has promulgated prohibitory orders disallowing people from holding meetings and demonstrations under Section 144 of the CrPC and Section 37 of the Bombay Police Act. It has resorted to *lathi*-charges, beatings, indiscriminate arrests, registering of false cases against hundreds of men, women and even children, including the atrocious charge of 'attempt to murder' on many of them. Thousands of people have courted arrest, and many have spent several nights in jail on trumped up charges. Leading activists of the area have been issued externment notices from Ratnagiri district. Eminent citizens of the region who have extended support to the struggle, including former Supreme Court Judge P.B. Sawant, retired Chief of Naval Staff Admiral Ramdas and noted economist Dr Sulabha Brahme, have been barred from entering the district! The government is using every trick in the book to divide the people and break their will, by trying to split them along communal lines, labelling activists as Maoists and 'outsiders' with an ideological agenda, setting up police camps in the area to intimidate the people, issuing threats, and so on.

However, the people are standing firm and have refused to be cowed down! They are united in their resolve that, come what may, they will fight, till the plant is cancelled!!

Part III: About this Book

We, on behalf of Lokayat, our activist group in Pune, have been campaigning against government plans to promote nuclear energy for the last three years. We have actively campaigned in the Jaitapur-Madban region against the proposed nuclear plant there, and have also campaigned in Pune to build support for the struggle of the people of that area. We have also participated in efforts to build a united platform of all the various anti-nuclear energy struggles taking place in the country.

To create awareness amongst people regarding the hazards of nuclear energy, we have brought out numerous pamphlets in Marathi and English. There was also a booklet on nuclear energy in Marathi, written by Dr Sulabha Brahme. With the government pressing ahead with its program of setting up a string of nuclear parks all across the country, the anti-nuclear struggle has also gradually picked up strength across the country. So we decided to bring out a comprehensive booklet in English. We started writing it in September 2009; the first draft was ready a year ago. However, so many suggestions on this draft came in from friends across the country that the booklet expanded into a book. Yet, because of severe time constraints due to the many activities of Lokayat, it has taken us more than a year to bring this book to completion.

Outline of this Book

In Chapter 1, we take a brief look at the history of the global nuclear energy scenario, as to how after much initial promise, nuclear energy entered into a long period of stagnation, and then why, over the last decade, everyone, including the Indian Prime Minister, is claiming that a nuclear 'renaissance' is underway in the world.

We take a look at the science of nuclear energy and the various components of the Nuclear Fuel Cycle in Chapter 2. We then critically examine the most important claims made about the benefits of nuclear energy, that it is clean and safe (in Chapter 3), cheap (in Chapter 4), and green and is the answer to global warming (in Chapter 5). In the light of this analysis, we take a close look at the reality of the claims about a 'global nuclear renaissance' in Chapter 6, by examining the present scenario and the likely future prospects for nuclear energy in North America and Western Europe.

We then move on to examining the nuclear energy scenario in India. We first give a brief history of India's nuclear energy program in Chapter 7, including the recent steps taken by the government of India towards massively increasing nuclear electricity generation capacity in the country. In Chapter 8, we take a look at the present cost of nuclear electricity in India and the likely cost of electricity from the proposed new imported reactors, especially the EPR reactors proposed to be installed in Jaitapur. Finally, in Chapter 9, we investigate the claim made by India's nuclear establishment that India has amongst the best safety records in the world, in the light of the known facts about the existing safety situation at India's nuclear installations: from uranium mining to India's nuclear reactors, and also India's much touted fast breeder reactor program. We then examine the safety issues associated with the new generation reactors being imported by NPCIL —the Russian VVER-1000 reactor and Areva's EPR.

In Chapter 10, we take a look at alternate, genuinely sustainable, solutions to India's energy crisis. If such solutions exist, why isn't the Indian government seeking to implement these solutions? We attempt to answer this question in the concluding chapter, along with a call to action. What is the point in simply understanding what is going on in the world? All understanding should lead us into action, to changing the world ...

Neeraj Jain
Pune, February 26, 2011

Postscript

We had just finished writing the book when on March 11, 2011, the TV channels flashed the news of a devastating accident at the Fukushima Nuclear Plant in Japan.

Not very long ago, in 2006, Dr Helen Caldicott, the pioneering Australian anti-nuclear activist, had prophetically warned: 'Statistically speaking, an accidental meltdown is almost a certainty sooner or later in one of the 438 nuclear power plants located in thirty-three countries around the world.' Quoting her, we had written in Chapter 3 of this book: 'In its greed for profits, the world's nuclear industry is pushing to making her grim foreboding come true sooner than later.'[25] When we wrote these lines, little did we know that an accident even bigger than Chernobyl was going to take place so soon ...

Till before March 11, 2011, the global nuclear industry had been claiming that it had been more than two decades since Chernobyl happened and no major accidents had occurred because lessons had been learned from the Chernobyl accident and safety issues adequately addressed. And so, nuclear authorities and political leaders from the US to India were claiming that no more Chernobyls will take place, and that nuclear energy was a safe and environmentally sustainable solution to the energy crisis. The Fukushima accident has blown apart their claims.

With hydrogen explosions ripping off the roofs of reactor buildings, three reactor cores spewing radiation unabated, one spent fuel pool on fire, another spent fuel pool suffering an explosion that scattered its fuel rods for miles, water poured in to cool the reactors flowing out from the bottom due to damaged pressure vessels and containments, millions of litres of radioactive water accumulating in the basement of the plant and draining out into the ocean, radiation levels in the Pacific Ocean spiking to unheard of levels, Reactor 4 of the plant tipping due to softening of the ground and threatening to collapse, groundwater in danger of getting contaminated, extremely dangerous MOX fuel in Reactor 3, fallout from Fukushima detected as far away as North America and Europe within a week of the accident ... the Fukushima accident is in an apocalyptic downward spiral.[26]

The global nuclear industry knows that it is standing on the edge of the precipice of extinction. If the full magnitude of the Fukushima accident becomes known to the people of the world, they are going to rise and demand the closure of nuclear plants everywhere. It is already happening in Germany and Switzerland. So, it has launched a huge propaganda offensive to downplay the Fukushima disaster. Pro-nuclear governments from USA to France and China are claiming that the accident was specific to Japan, and their reactors are safe. The corporate controlled media is claiming that nothing much has happened in Japan, and that radiation from the accident will not affect the world. Such is the power of the nuclear industry that news about Fukushima has disappeared from the media.

India's nuclear authorities have declared that the Fukushima accident will not affect India's nuclear program in any way; the Prime Minister chose the 26th anniversary of the Chernobyl accident to announce the government's resolve to go ahead with the Jaitapur nuclear plant. Our establishment scientists are making the most hare-brained claims: that Fukushima was not a nuclear emergency, only a chemical reaction; that our nuclear safety systems are safer than Japan's; and so on.[27]

So we decided we must write an epilogue on the Fukushima accident, to bring out the truth about what is happening in Fukushima, and its lessons for the world ...

Neeraj Jain
Pune, October 24, 2011

1

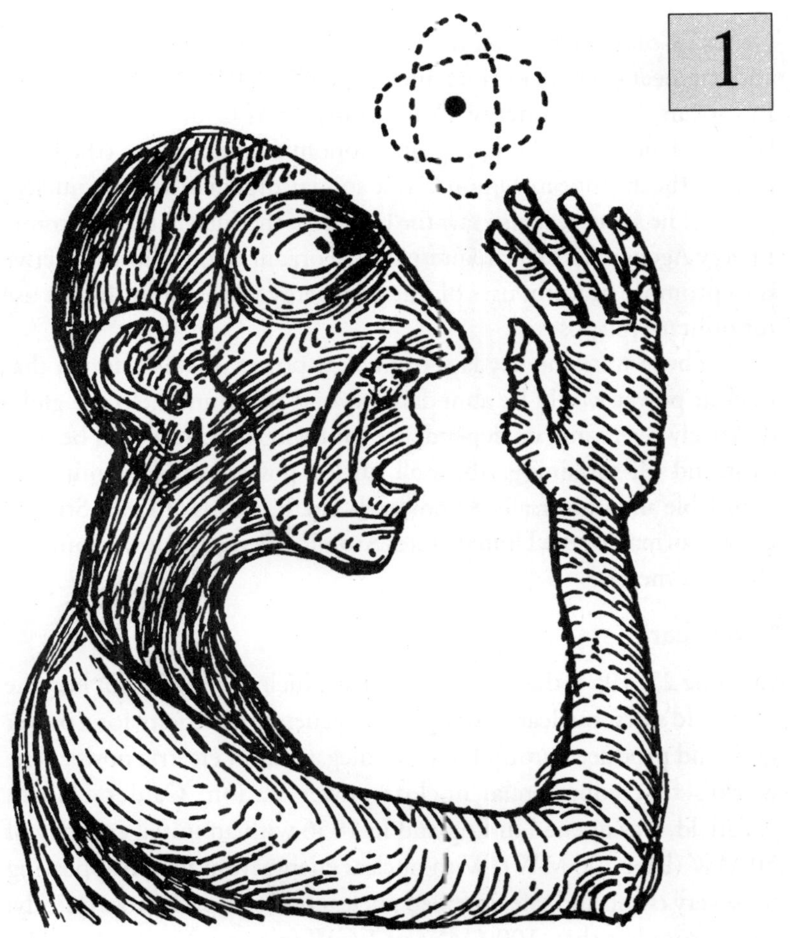

NUCLEAR ENERGY: FROM SLOWDOWN TO 'RENAISSANCE'

It was in December 1953, in his famous 'Atoms for Peace' speech before the UN General Assembly, that President Eisenhower of the United States first spoke of the peaceful uses of the atom, including the generation of electricity from nuclear fission as a solution to the world's growing energy needs.[1] In 1955, the United Nations' 'First

Geneva Conference', then the world's largest gathering of scientists and engineers, met to explore nuclear power technology. In 1957, the European Atomic Energy Community (EAEC or Euratom) was launched alongside the European Economic Community (the latter is now the European Union), as a special organisation for nuclear power. The same year also saw the launch of the International Atomic Energy Agency (IAEA), an international organisation whose objective is to promote peaceful uses of nuclear energy while inhibiting its use for military purposes.

Those were the heydays of nuclear power. It was claimed that nuclear power would be abundant beyond belief and help the globe decisively overcome its dependence on fossil fuels. It would be safe, clean and self-sustaining. Above all, nuclear power would be eminently affordable and universally economical—in the words of Lewis Strauss, then chairman of the United States Atomic Energy Commission, 'too cheap to meter'.[2]

Early Years

On June 27, 1954, the USSR's Obninsk Nuclear Power Plant became the world's first nuclear power plant to generate electricity for a power grid, and produced around 5 MW (megawatt) of electric power. The world's first commercial nuclear power station, Calder Hall in Sellafield, was opened in England in 1956 with an initial capacity of 50 MW (later 200 MW). With nuclear energy from fission appearing to be very cheap and safe, installed nuclear power capacity rose quickly: rising from less than 1000 MW or 1 GW (gigawatt) in 1960 to 100 GW in the late 1970s, and 300 GW in the late 1980s.[3] The IAEA euphorically forecast that global installed nuclear capacity would reach 4,450 GW by the year 2000![4]

Problems and Slowdown

Soon, the problems started becoming evident. As nuclear plant construction costs mounted, the claim that nuclear energy was going to be 'too cheap to meter' went through the roof. It became clear that finding a way of safely disposing of the rising mountains of nuclear waste was going to be very difficult, if not impossible. Several scientists started challenging the prevailing view that the small amounts of radiation released by nuclear power plants during normal operation were not a problem. One of these was John William Gofman, professor emeritus of Medical Physics at UC Berkeley, who emphatically stated in the late 1960s that any amount of radiation, howsoever small, causes damage to human genes and health.[5]

In 1976, four nuclear engineers—three from General Electric (GE) and one from the US Nuclear Regulatory Commission (NRC) —resigned, stating that nuclear power was not as safe as their superiors were claiming. They testified to the Joint Committee on Atomic Energy (a United States congressional committee) that 'the cumulative effect of all design defects and deficiencies in the design, construction and operations of nuclear power plants makes a nuclear power plant accident, in our opinion, a certain event. The only question is when, and where.' The three GE engineers announced that they would now work full time for Project Survival, an organisation which coordinated the 1976 anti-nuclear referendum drive in California. These men were engineers who had spent most of their working life building reactors, and their defection galvanised anti-nuclear sentiment across Europe and America.[6]

Soon after, there occurred the Three Mile Island (in 1979) and Chernobyl (in 1986) disasters. The catastrophic consequences led to an explosion of protests against nuclear power plants the world over. In the US, over the next decade and a half, it resulted in some 47,000 anti-nuclear arrests during at least 1800 actions at more than 250 different sites (these actions were against both nuclear bombs and nuclear power).[7] They succeeded in not only bringing new nuclear plant ordering to a halt, but also forced cancellation of plants whose construction had already begun. In the US, all nuclear plants ordered after 1973 were eventually cancelled.[8] Many West European parliaments (for example, Italy, Germany, Sweden, Belgium) imposed a moratorium on new nuclear reactors and decided to phase out existing ones. Worldwide, more than two-thirds of all nuclear plants ordered after January 1970 were eventually cancelled.[9]

The nuclear industry went into a tailspin. To give a few examples of the multi-billion dollar losses suffered by the nuclear industry (much of which was transferred to the public): in 1983, the Washington Public Power Supply System abandoned three nuclear plants after sinking $24 billion into them; the next year, a new nuclear reactor at Shoreham in Long Island (near New York, USA), completed at a cost of $5.3 billion, could not be licensed and had to be scrapped.[10] By 1985, *Forbes* magazine was calling nuclear power 'the largest managerial disaster in history';[11] while energy expert Amory B Lovins, CEO of the Rocky Mountain Institute, termed it the greatest failure in the industrial history of the world, which has lost more than $1 trillion in subsidies, losses, abandoned projects and other damage to the public.[12]

Consequently, since the late 1980s, worldwide nuclear capacity has risen very slowly: from roughly 320 GW in 1990, it reached just 366 GW in 2005, and has been hovering around that figure ever since.[13]

Funding a 'Nuclear Renaissance'

By the beginning of this century, it was apparent that the nuclear power industry had entered into a long period of stagnation, and nuclear power was becoming a technology without a future. In a desperate attempt to revive its sagging fortunes, the global nuclear industry has launched a massive funding effort and propaganda drive during the last decade. Helping it along has been the rise of deeply conservative currents in the politics of some developed countries, from the USA to Germany and the United Kingdom, due to the deepening economic crisis there. (Discussing the reasons for this shift is beyond the scope of this essay.)

For its propaganda offensive, the nuclear industry has taken advantage of the growing crisis of global warming and the increasing public awareness and concern about it, and launched a multi-billion dollar public relations campaign claiming that nuclear energy is the answer to global warming. It has been so successful in its propaganda campaign that today, every political leader in the world—from the President of the United States to the Prime Minister of India, every Environment Minister in the world—from the Climate Minister of Britain to our own Jairam Ramesh, and even many scientists, are speaking of nuclear energy being green as if it is a self evident truth.

In the USA, the Nuclear Energy Institute (NEI), the propaganda wing and trade arm of the American nuclear industry, has poured out hundreds of millions of dollars not only to bribe politicians and win billions of dollars in new subsidies for the nuclear industry, but also block the implementation of distributed, home-based solar systems that would allow millions of people to break free from having to write big cheques each month to their electricity distribution company.[14] The industry has been the third largest influence peddler in Washington, D.C. over the past decade, spending more than $1 billion lobbying Congress and the Executive Branch since 1998, behind only the pharmaceutical and insurance industries.[15] The larger portion of this money has gone to maintaining its traditional base among Republicans, who are all for building a 100 new nuclear plants: in 2000, it backed the Bush-Cheney ticket with nearly $270,000 in contributions.[16] Simultaneously, it has also spent money in building bridges to Democrats in both houses, from House Majority Whip James Clyburb to White House Chief of Staff Rahm Emanuel and strategist David Axelrod, right up to President Obama—Exelon Corporation (the largest nuclear operator in the United States) contributed nearly $210,000 to his Presidential campaign through its employees.[17]

Its efforts have borne fruits. Within months of coming to power in 2000, the Bush administration announced the 'Nuclear Power 2010 program', whose declared objective was to get a new generation of nuclear reactors up and running by 'early in the next decade.'[18] In 2005, Bush launched the biggest subsidy program since the 1960s to promote nuclear energy, including a huge loan guarantee of $18.5 billion for new nuclear plants (for more details, see Chapter 3). President Barack Obama has not only continued with Bush's loan guarantees, but also nearly tripled it to $54 billion in his budget request to the US Congress in 2010.[19]

Likewise, in Europe too, the nuclear lobby is one of the most influential and well-funded groups—in just one year, 2007, the group spent 1.6 million euros ($2.2 million) on lobbying the various arms of the European Union (EU).[20] Simultaneously, it launched a massive propaganda offensive to convince people that nuclear energy is clean and green. Thus, in Britain, the Nuclear Industry Association, the trade association of the country's civil nuclear industry, fashioned a classy public relations campaign targeting politicians, media and the public beginning in 2004;[21] it even got the national curriculum changed to make it compulsory for schools to teach all 14-16 year olds about the 'benefits' of nuclear power.[22]

The campaign has had the desired results; many European Parliamentarians are promoting the concerns and interests of the nuclear industry.[23] In Western Europe, many newly elected governments have announced that they are reconsidering plans to phase out nuclear plants in their countries. The Energy Ministers of the G-8 countries met in Rome in May 2009 and issued a statement emphasising nuclear power as a means of meeting energy demand and combating climate change. This statement was subsequently endorsed by the heads of state and government leaders of these countries at their annual summit meeting held in L'Aquila, Italy in July 2009.[24]

Along with the governments of the developed countries, various international organisations controlled by them have also begun issuing enthusiastic statements in favour of nuclear power and predicting a positive future for it. The Organisation for Economic Cooperation and Development's *World Energy Outlook* (WEO), the US Department of Energy's *International Energy Outlook* and the IAEA have all given very optimistic projections about the future of nuclear energy in their recent reports. For instance, the IAEA in its 2010 report projects that global nuclear energy generation should rise from 2558 TWh (TWh = tera watt-hour = thousand billion watt-hours or 10^{12} watt-hours) at present (in 2009) to between 4040 TWh (low estimate) and 5938 TWh (high estimate) by 2030—an increase of between 58 per cent and 132 per cent; likewise, it also estimates that global nuclear generating capacity should rise to between 546 GW (low estimate) and 803 GW (high estimate) by 2030, from 372 GW at the end of 2009.[25]

The nuclear industry has also succeeded in winning endorsement from many prominent intellectuals from all over the world. The effect has been an echo chamber of support for nuclear power. Swayed by this offensive, many common people have also started believing that nuclear energy is a solution to meet the world's energy needs and simultaneously tackle the growing crisis of global warming. It is a perfect example of what Prof. Noam Chomsky, the world renowned scholar, has called 'Manufacturing Consent': use of massive propaganda by those in power to control what people think.

In jubilation, the nuclear industry and its apologists have started proclaiming: the 'nuclear renaissance is here' (in the words of the Chairman of the US Nuclear Regulatory Commission, Dale E. Klein).[26]

2

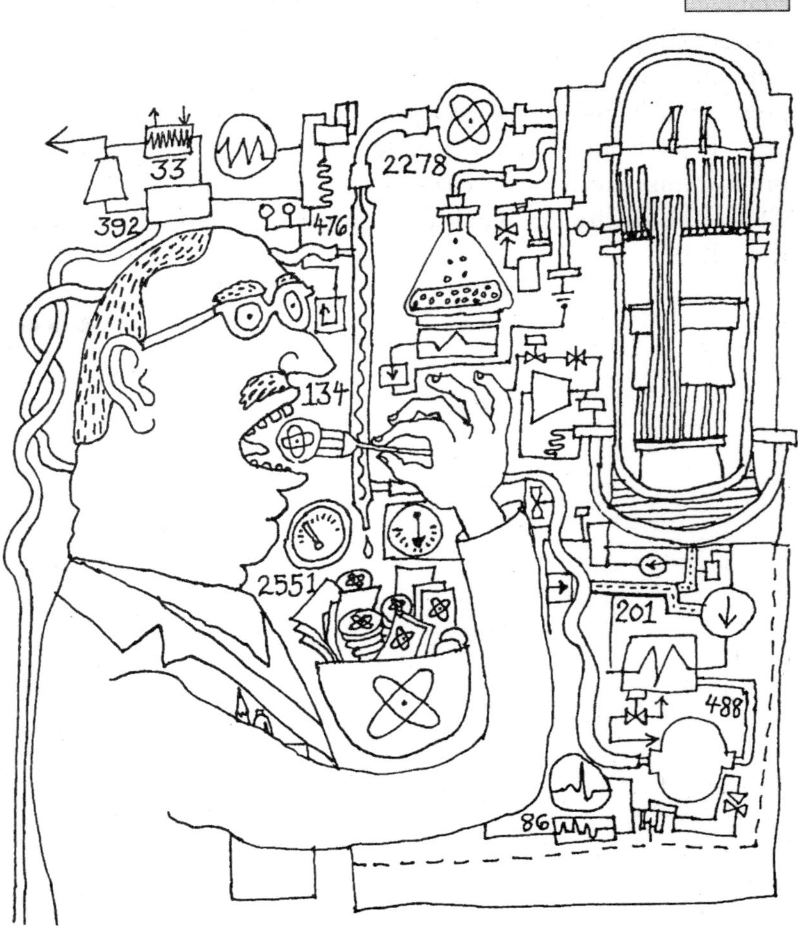

WHAT IS NUCLEAR ENERGY?

Part I: The Basics of Nuclear Power

The basic operation of a nuclear power plant is no different from that of a conventional power plant that burns coal or gas. Both heat water to convert it into pressurised steam, which drives a turbine to generate electricity. The key difference between the two plants lies in the method of heating the water. Conventional power plants burn fossil fuels to heat the water. In a nuclear power plant, this heat is produced by a nuclear fission reaction, wherein energy in the nucleus of an atom is released by splitting the atom into two.

The Atom

Everything is made of atoms. Any atom found in nature will be one of 92 types of atoms, also known as elements (actually, element is a pure chemical substance containing only one kind of atoms). Atoms bind together into molecules. So a water molecule is made from two hydrogen atoms and one oxygen atom bound together into a single unit. Every substance on Earth—metal, plastics, hair, clothing, leaves, glass—is made up of combinations of the 92 atoms that are found in nature. An ordered list of these 92 atoms found in nature, plus a number of man-made elements, is known as the Periodic Table of Elements.

Atoms are made up of three subatomic particles: the positively charged protons, the neutral neutrons and the negatively charged electrons. Protons and neutrons bind together to form the nucleus of the atom, while the electrons surround and orbit the nucleus. The number of protons is equal to the number of electrons, making the atom electrically neutral. The nucleus and electrons are held together by the coulomb force, the same force that produces static electricity and lightning. The nucleus contains most of the mass of the atom. The protons and neutrons in the nucleus are bound together by very strong nuclear forces, much greater than the electrical forces that

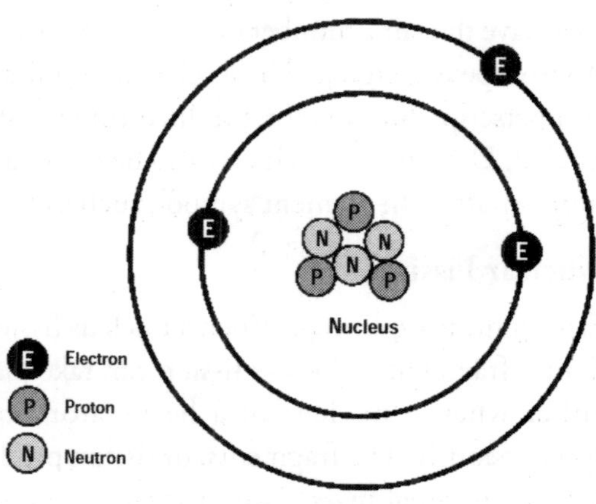

Figure: Lithium Atom

bind the electrons to the nucleus. This is why the electrostatic coulomb repulsion between positively charged protons does not lead to the nucleus falling apart.

Atoms are so small that they can't be seen except with the help of an electron microscope. An atom is roughly 0.1 nanometers, that is, 0.0000000001 meters. In other words, if we make a tiny dot with a pencil, a dot of roughly one mm in size, then this dot would have ten million, or one crore, atoms.

Every element is characterised by its atomic mass number and atomic number. The mass number A of an element is the total number of nucleons, that is, the total number of neutrons and protons contained in its nucleus; the atomic number Z is the number of protons. The atomic number of an element (that is, the number of protons in it) determines its chemical properties and its place in the Periodic Table. Hydrogen has the lowest periodic number (Z =1), whereas uranium has the highest atomic number among the naturally occurring elements (Z = 92). Elements with higher atomic numbers, like Neptunium (Z = 93) and Plutonium (Z = 94) have been created artificially.[1]

The chemical properties of an atom depend upon the number of protons in it, that is, its atomic number. There are atoms whose nuclei have the same number of protons, but different number of neutrons. The chemical properties of these atoms are identical, since they have the same number of protons. Such atoms are called **isotopes.** An isotope is designated by its element symbol with the mass number as superscript; for instance, the three isotopes of uranium are designated as U^{234}, U^{235} and U^{238}. (It can also be designated by writing the mass number after the element symbol, such as U-235).

Nuclear Fission

Fission means splitting. When a nucleus fissions, it splits into several lighter fragments. Nuclear fission can take place in one of two ways: either when a nucleus of a heavy atom captures a neutron, or spontaneously. The fragments, or fission products, are about equal to half the original mass. Two or three neutrons are also emitted. The sum of the masses of these fragments (and emitted neutrons) is less than the original mass. This 'missing' mass (about 0.1 per cent of the original mass) has been converted into energy.

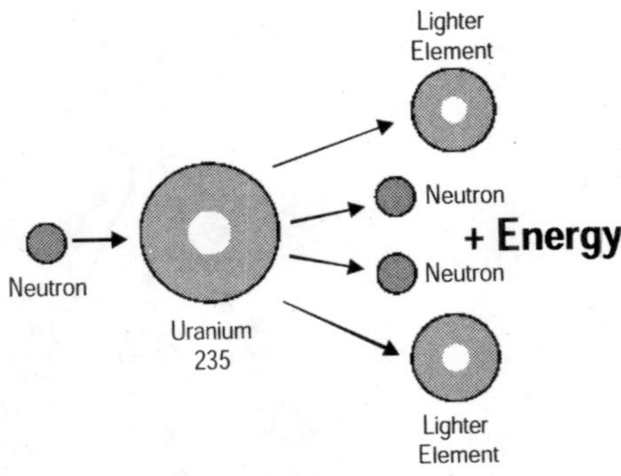

Figure: Nuclear Fission

The amount of energy released in this process can be obtained from Einstein's famous equation $E = mc^2$, where E is energy, m is mass and c is the speed of light (approximately 300,000 kilometers per second). The concept behind this equation is simple: that matter and energy are essentially interchangeable—matter can be converted into energy, and energy can be converted into matter.

Typical fission events release about 200 million eV (electron volts) for each fission event, that is, for the splitting of each atom. In contrast, when a fossil fuel like coal is burnt, it releases only a few eV as energy for each event (that is, for each carbon atom). This is why nuclear fuel contains so much more, millions of times more, energy than fossil fuel. To get an idea of the energy released in a fission reaction: the energy found in half a kilogram of uranium is equivalent to 4.2 million litres of gasoline.

The energy of nuclear fission is released as kinetic energy of the fission products and fragments, and as electromagnetic radiation in the form of gamma rays. In a nuclear reactor, this energy is converted to heat as the particles and gamma rays collide with atoms of the coolant, the moderator, the reactor vessel, et cetera, and give up part of their kinetic energy.

Nuclear Chain Reaction

A chain reaction refers to a process in which neutrons released in a fission reaction produce an additional fission in at least one further nucleus. This nucleus in turn emits neutrons, and the process repeats. The process may be controlled (to generate nuclear power) or uncontrolled (to produce a nuclear explosion, as in nuclear bombs).

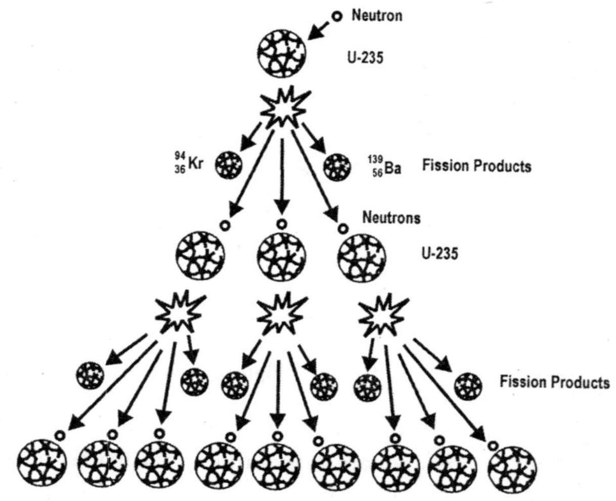

Figure: Nuclear Chain Reaction

Nuclear Fuel

The isotopes that can sustain a fission chain reaction are called nuclear fuels. The most common nuclear fuels are U-235 (an isotope of uranium) and Pu-239 (an isotope of plutonium). We discuss the use of U-235 as nuclear fuel here.

Uranium has many isotopes. Two, uranium-238 primarily, and to a lesser extent, uranium-235, are commonly found in nature. (A third isotope, uranium-234, also exists naturally, but its abundance is only 0.0055 per cent) Both U-235 and U-238 undergo spontaneous fission (that is, spontaneous radioactive decay), but this takes place over periods of millennia: the half-life of uranium-238 (half-life is the amount of time taken by half the atoms to decay) is about 4.47 billion years and that of uranium-235 is 704 million years. (For more on radioactivity and half-life, see Chapter 3, Part I.)

What makes U-235 special and useful for both nuclear power production and nuclear bomb production is that it has an extra property: U-235 is fissile, that is, it undergoes fission when struck by a **slow moving** or **thermal neutron.** U-235 is the only isotope existing in nature (in any appreciable amount) that is fissionable by thermal neutrons, and releases enough neutrons (typically 2 or 3, average 2.5) to sustain a chain reaction. It is this property which makes it possible for U-235 to be used as nuclear fuel.

However, the concentration of U-235 in naturally occurring uranium ore is just around 0.71 per cent, the remainder being mostly the non-fissile isotope U-238. For most types of reactors, this concentration is insufficient for sustaining a chain reaction and needs to be increased to about 3 to 5 per cent in order that it can be used as nuclear fuel. This can be done by separating out some U-238 from the uranium mass. This process is called **enrichment**, and the resulting uranium is called **enriched uranium**. (Not all nuclear reactors need enriched uranium; for example, heavy water reactors use natural (unenriched) uranium.)

As mentioned above, U-235 also undergoes a small amount of spontaneous fission, which releases a few free neutrons into any sample of nuclear fuel. One possibility is that such free neutrons escape rapidly from the fuel mass and decay. That is because free neutrons are unstable, that is, they are radioactive, each decaying spontaneously, with a half-life of about 15 minutes, into a proton, an electron and an electron-antineutrino. However, the greater possibility is that these neutrons collide with other U-235 nuclei in the vicinity, and induce further fissions, releasing yet more neutrons, thus starting a chain reaction.

If exactly one out of the average of roughly 2.5 neutrons released in the fission reaction is captured by another U-235 nucleus to cause another fission, then the chain reaction proceeds in a controlled manner and a steady flow of energy results. If the chain reaction sustains, it is said to be **critical**; and the mass of U-235 required to produce a controlled chain reaction is called a **critical mass**. However, if on the average less than one neutron is captured by another U-235 atom, then the chain reaction gradually dies away. And if more than one neutrons are captured, then an uncontrolled chain reaction results, which

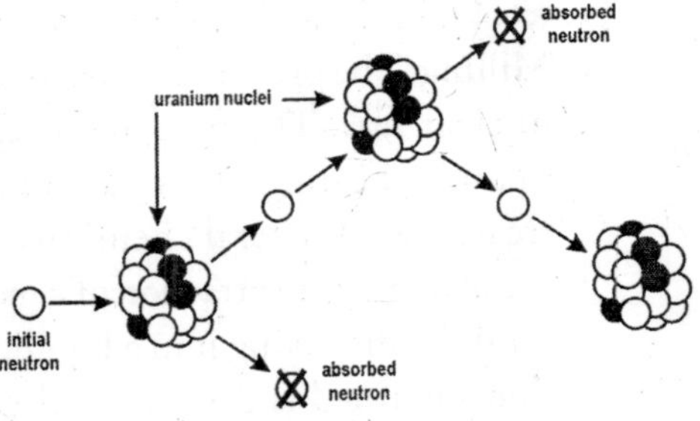

Figure: Controlled Chain Reaction

can cause the nuclear reactor to meltdown; this is also what happens in an atomic bomb. To control the fission reaction in a nuclear reactor, most reactors use **control rods** that are made of a strongly neutron-absorbent material such as boron or cadmium.

The neutrons released in a fission reaction travel extremely fast (energy = 1 MeV or speed ~ 107 m/s). At such speeds, the possibility of their being captured by another U-235 nucleus is very low. If they are slowed down, or moderated, the probability of fission rises dramatically. In that case, a critical condition (that is, a controlled chain reaction) can be achieved with lower concentrations of U-235. In a nuclear reactor, the fast neutrons are slowed down using a **moderator** such as heavy water, graphite or ordinary water.

Part II: The Nuclear Fuel Cycle

The nuclear fission reaction that we have discussed above is only a small part of the entire complex process of generating electricity from uranium. This entire process is known as the nuclear fuel cycle. We now take a brief look at the various stages of this process. The fuel cycle described here includes the phase of uranium enrichment, necessary for obtaining the fuel for light water reactors, which constitute the largest number of the world's nuclear reactors.

- **Mining:** The nuclear fuel cycle starts with mining of uranium. Most uranium mines are open-pit mines. Since 90 per cent of the worldwide uranium ores have uranium content of less than 1 per cent, and more than two-thirds have less than 0.1 per cent,[2] therefore, large amounts of ore have to be mined to obtain the amounts of uranium required.
- **Milling:** The ore is then processed in two stages to obtain the nuclear fuel. The first step is called milling, wherein the uranium bearing ore is ground into fine powder, and then treated with several chemicals to leach out the uranium. (Leaching is the extraction of a material, say a metal, from the solids by dissolving it in a liquid.) The uranium concentrate thus obtained is dried and filtered to yield what is called 'yellowcake', 70-90 per cent of which is uranium oxide.

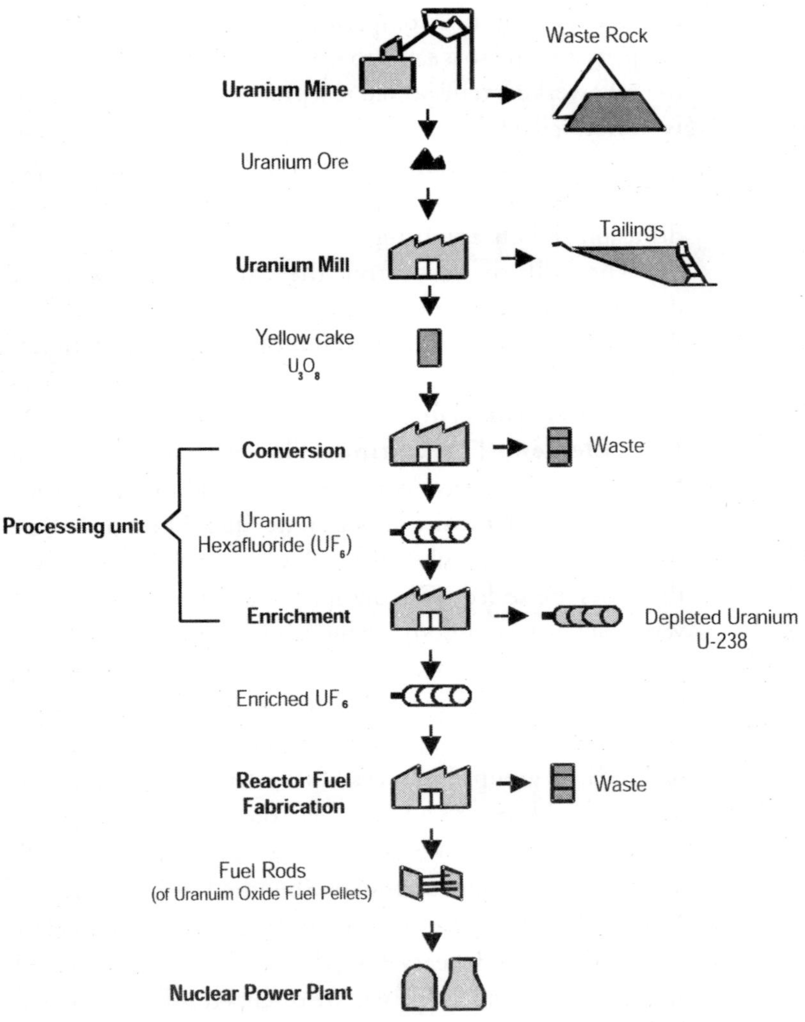

Figure: Nuclear Fuel Cycle

- **Enrichment:** The uranium oxide obtained from milling contains both the usual isotopes of uranium, the fissile uranium-235 and non-fissile uranium-238. It must now be enriched, that is, the proportion of fissile uranium-235 in it must be increased. For this, uranium oxide must first be converted to uranium hexafluoride. Uranium hexafluoride is the only uranium compound which is gaseous at low temperatures and so is easier to work with. For this conversion, the yellowcake is transported to a processing facility, where it is converted to uranium hexafluoride. The subsequent enrichment of uranium-235 from 0.7 per cent to 3-5 per cent can be done using one of two basic methods—gaseous diffusion and ultracentrifuge.

The bulk of waste from the enrichment process is **depleted uranium**, so-called because most of the uranium-235 has been extracted from it. It is thus primarily uranium-238, and contains less than one-third uranium-235 as compared to natural uranium.

- **Fuel Element Fabrication:** The enriched uranium hexafluoride gas is now converted into solid uranium oxide fuel pellets, each the size of a cigarette filter. These pellets are packed in very thin tubes of an alloy of zirconium, and then the tubes are sealed. (Zirconium is chosen because it has a very small neutron absorption probability, is a metal and hence can be worked into thin tubes, and has a fairly high melting point.) These tubes are called fuel rods. Each fuel rod is normally twelve feet long and half-an-inch thick. The finished fuel rods are grouped in special fuel assemblies that are then used to build up the nuclear fuel core of a nuclear power reactor. A typical 1,000 megawatt reactor contains 50,000 fuel rods, amounting to about one hundred tons of uranium.
- **Nuclear Reactor:** The nuclear reactor is where the nuclear fuel is fissioned and the resulting chain reactions are controlled and sustained at a steady rate. We discuss this in more detail in Part III below.
- **Decommissioning and Dismantling:** Nuclear power plants were originally designed for an operating life of 30 years. However, nowadays, the nuclear industry believes that it can safely operate nuclear plants for around 40 to 60 years. When the reactor completes its working life, it is dismantled. Unlike conventional coal and gas power plants, the dismantling of a nuclear power plant is a very long-term, complicated and

costly operation, because the entire nuclear power plant has become contaminated; all of its parts including the concrete reactor building have become radioactive. The long-term management and clean up of these closed reactors is known as 'decommissioning'. It involves managing and cleaning up the highly radioactive fuel, residues, massive quantities of radioactive equipment and components, mixed hazardous wastes and the mountain of contaminated concrete and debris (making up the reactor building) that have now accumulated at the plant site.

Only a very small number of nuclear power plants have so far been completely dismantled. Generally, the most common decommissioning method is that the intensely radioactive products, especially the deadly cobalt-60 and iron-55 formed inside the reactor vessel from neutron bombardment, are first allowed to decay considerably. During this period, which can be anywhere from 5 to 100 years, depending upon the decommissioning plan, these huge, intensely radioactive mausoleums must be guarded and protected from damage or unwarranted intrusion. After this, the actual process of dismantling begins. The reactor is now cut apart into small pieces either by humans or by remote control, and the still-radioactive pieces packed into containers for transportation and final disposal at some 'low-level'[3] nuclear waste disposal site.[4]

- **Disposal of Radioactive Nuclear Fuel Waste:** Every year, one-third of the nuclear fuel rods must be removed from the reactor, because they are so contaminated with fission products that they hinder the efficiency of electricity production. The uranium fuel after being subjected to the fission reaction in the reactor core becomes one billion times more radioactive.[5] A person standing near a single spent fuel rod can acquire a lethal dose within seconds. This spent nuclear fuel is going to be intensely radioactive for tens of thousands of years; therefore it needs to be safely stored for centuries to come.

However, it is also thermally extremely hot when removed from the reactor. Therefore, the spent fuel is first stored for many years in on-site storage ponds and continually cooled by air or water. If it is not continually cooled, the zirconium cladding of the rod could become so hot that it would spontaneously burn, releasing its radioactive inventory. After an adequate cooling period (generally five years in the US[6]), there are two options for the waste—either it is reprocessed, or it is moved to dry cask storage.

In the latter case, the spent fuel rods are packed by remote control into highly specialised containers made of metal or concrete designed to shield the radiation. These casks are to be stored for thousands of years, till the radiation diminishes to a point where it is no longer hazardous. Presently, in most countries having nuclear plants, these casks are 'temporarily' stored near the spent fuel cooling ponds.[7]

Countries having nuclear plants are hoping to build a long-term nuclear waste dump site where this waste can be safely stored for centuries. However, no country, including the USA has succeeded in building such a site so far.

- **Reprocessing Spent Fuel:** Reprocessing is a chemical process for separating out the uranium and plutonium contained in the spent fuel, which can then be used as fuel for what are known as fast breeder reactors. The technology generally used to extract the plutonium (and uranium) from the spent fuel is called Plutonium-Uranium Redox Extraction (PUREX).

  Reprocessing also segregates the waste into high-level, intermediate-level[8] and low-level wastes (HLW, ILW and LLW respectively). Because HLW, which contains the bulk of the radioactivity present in the original spent fuel, occupies only a fraction of the volume of spent nuclear fuel, this has been used as an argument by countries like France and India that reprocess their spent fuel waste to claim that reprocessing is a superior way of managing the radioactive spent fuel, when compared to the direct disposal of spent fuel in geological repositories.

Part III: The Nuclear Reactor

The nuclear fuel is fissioned in the nuclear reactor. The energy released in the fission reaction is harnessed as heat to convert water to steam. This steam is used to drive a turbine and produce electricity. Most nuclear reactors work on the same basic principles. The basic components common to most types of nuclear reactors are as below:

Reactor core: The part of the nuclear reactor where the nuclear fuel assembly is located.

Moderator: The material in the core which slows down the neutrons released during fission, so that can they cause more fission. It is usually ordinary water (used in light water reactors), but it may also be heavy water (used in heavy water reactors) or even graphite (used in certain types of reactors, for example the RBMK).

Control rods: These are made with neutron-absorbing material such as cadmium, hafnium or boron, and are inserted or withdrawn from the core to control the rate of reaction, or to halt it.

Coolant: A liquid or gas circulating through the core so as to transfer the heat from it. In light water reactors, the moderator which is water also serves as the primary coolant. Except in boiling water reactors (BWRs), this primary coolant passes through another heat exchanger, to convert another loop of water into steam. This steam drives the turbine. The advantage of this design is that the radioactive water (that is, the primary coolant) does not come into contact with the turbine.

Pressure vessel: Usually a robust steel vessel containing the reactor core and moderator/coolant.

Steam generator (not in BWR): Here, the primary coolant bringing heat from the reactor is used to convert another loop of water into steam to drive the turbine.

Containment: The structure around the reactor core which is designed to protect it from outside intrusion and to protect those outside from the effects of radiation in case of any malfunction inside. It is typically a metre-thick concrete and steel structure.

Refuelling the reactors: In most reactors, the refuelling is done at intervals of one to two years, when a quarter to one-third of the fuel assemblies are replaced with fresh ones. For this, the reactor needs to be shut down. CANDU type reactors are an exception to this: these have pressure tubes (instead of a pressure vessel enclosing the reactor core) which can be refuelled even when the reactor is operating by disconnecting individual pressure tubes.

Safety Systems in Nuclear Reactors

Nuclear reactors have a number of safety systems to quickly shut down the reactor under accident conditions and prevent the release of radioactivity into the atmosphere. All reactors have some form of the following safety systems:

- **Scram system:** This is a system designed for emergency termination of the fission chain reaction.
- **Emergency core cooling system (ECCS):** During an accident, even if the reactor is shut down by the Scram system, the reactor needs to be cooled continuously as the core continues to produce heat due to the decay of radioactive fission products. If there is a loss of coolant, then alternate methods of cooling are required to prevent damage to the nuclear fuel. This is achieved by the ECCS, which has its own separate water and power supply.
- **Containment:** After the zirconium fuel cladding and the reactor pressure vessel, this is the last barrier against a catastrophic release of radioactivity into the atmosphere. Apart from a primary containment, many reactors, especially boiling water reactors, have a secondary containment too—which is normally a concrete dome enveloping the entire steam generating unit.

Types of Nuclear Power Reactors

At a basic level, reactors may be classified into two categories: Light Water Reactors (LWRs) and Heavy Water Reactors (HWRs). LWRs are largely of two types, Pressurised Water Reactors (PWRs) and Boiling Water Reactors (BWRs); each come in multiple variations. Heavy Water Reactors can also be of different types, one of the most well known being the CANDU (acronym for 'CANada Deuterium

Uranium') reactors developed by Canada which are a type of Pressurised Heavy Water Reactors (PHWRs).

LWRs are the most widespread type of reactors in operation today. Of the 437 reactors in operation at the end of 2009, 357 were LWRs, of which 265 were PWRs and 92 BWRs. Apart from these, the other reactor types in operation were: 45 Pressurised Heavy Water Reactors (PHWRs), 18 Gas-cooled Graphite-moderated Reactors (GCRs), 15 Light-water-cooled Graphite-moderated Reactors (LWGR) and two Fast Breeder Reactors (FBRs).[9]

Below, we discuss the most well-known type of nuclear power reactor, the PWR, and also the reactor design of most of India's reactors, the PHWR or CANDU reactor. We also discuss the Fast Breeder Reactor, which India has been trying to build for many decades and of which there are only two reactors in operation today, a prototype unit in Japan and the BN-600 reactor in Russia (which, technically speaking, is actually not a fast breeder—see Chapter 9, Part VI).

Pressurised Water Reactor

This is the most common type, with 265 in use for power generation and several hundred more employed for naval propulsion. The PWR uses ordinary water as both coolant and moderator.

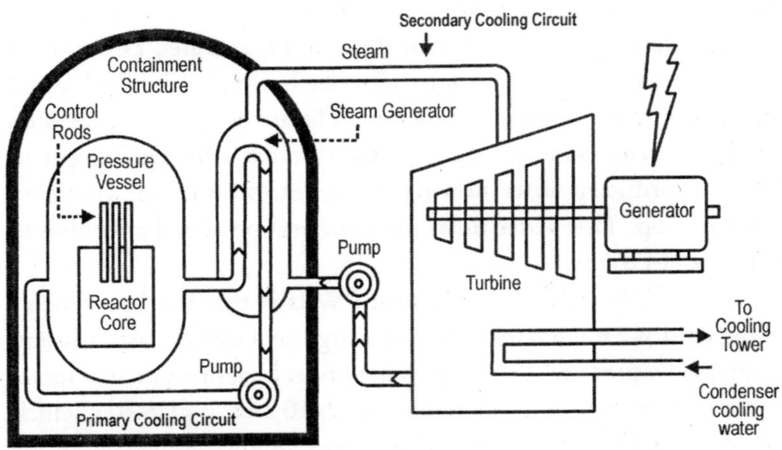

Figure: Pressurised Water Reactor

It has a primary cooling circuit which flows through the core of the reactor under very high pressure, a secondary circuit in which steam is generated to drive the turbine, and also a tertiary circuit which condenses this steam back into water. Water in the primary circuit which flows through the reactor core reaches about 325°C; hence it must be kept under about 150 times atmospheric pressure to prevent it from boiling. Water in the primary circuit is also the moderator, and if it starts turning into steam, the fission reaction would slow down. This negative feedback effect is one of the safety features of this type of reactors.

The pressurised hot water in the primary cooling circuit heats the water in the secondary circuit, which is under less pressure and therefore gets converted into steam. The steam drives the turbine to produce electricity. The steam is then condensed by water flowing in the tertiary circuit and returned to the steam generator.

Pressurised Heavy Water Reactor (PHWR or CANDU)

This design was originally developed in Canada in the 1950s. In this design, unenriched uranium, that is natural uranium (0.7 per cent U-235) oxide, is used as fuel, along with heavy water as moderator, which is a more efficient moderator than ordinary water.

Conceptually, this reactor is similar to PWRs discussed above. Instead of water, heavy water is used as the coolant and moderator. Fission reactions in the reactor core heat the heavy water. This coolant is kept under high pressure to raise its boiling point and avoid significant steam formation in the core. The hot heavy water generated in this primary cooling loop is passed through a heat exchanger to heat the ordinary water flowing in the less-pressurised secondary cooling loop. This water turns to steam and powers the turbine to generate electricity.

The difference in design with PWRs is that the heavy water being used as a moderator is kept in a large tank called Calandria and is under low pressure. The heavy water under high pressure that serves as the coolant is kept in small tubes, each 10 cms in diameter, which also contain the fuel bundles. These tubes are then immersed in the moderator tank, the Calandria.

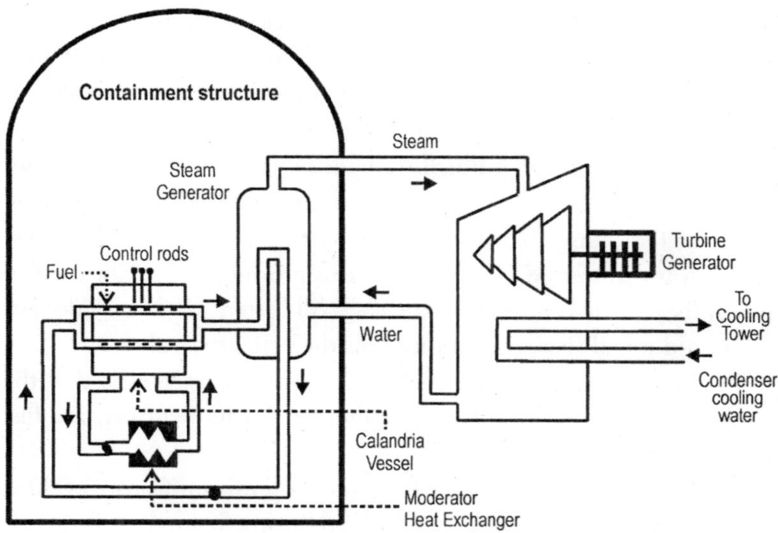

Figure: Pressurised Heavy Water Reactor

Heavy water is a more efficient moderator than ordinary water as it absorbs 600 times fewer neutrons than the latter.[10] Further, the design of the PHWR is such that most of the moderator which is in the Calandria is at a lower temperature than the moderator in the PWRs. Therefore, the neutrons in the PHWRs are at optimum speeds to cause fission, implying that the PHWR is more efficient in fissioning U-235 nuclei. Both these advantages mean that the PHWR can sustain a chain reaction with lesser number of U-235 nuclei in uranium as compared to the PWRs, which is why it uses unenriched uranium as nuclear fuel.

Thus, in the PHWRs, enrichment costs are saved, but the disadvantage is that heavy water is very costly, amounting to hundreds of dollars per kilogram.

Fast Breeder Reactor

In the uranium fuelled reactors discussed above, U-235 is fissioned by slow moving neutrons, and the energy released in the fission reaction is used to generate steam to run a turbine and generate electricity. Neutrons produced by fission have high energies and move extremely quickly. These so-called fast neutrons do not cause fission as efficiently as slower-moving ones, so they are slowed down using a moderator.

In contrast, a Fast Breeder Reactor uses a mix of oxides of plutonium-239 (or Pu-239) and uranium-238 as the fuel—also called MOX fuel. (Plutonium is usually obtained by reprocessing waste from the uranium fuelled reactor.) Pu-239 is the fissile material. It is even better at fissioning than U-235. The energy released in the fission reaction is transferred via the coolant to produce the steam used to power the electricity generating turbines. All current fast reactor designs use liquid metal (generally sodium) as the primary coolant.

The second difference with U-235 fuelled reactors is that the coolant used in FBRs is not a moderator. So its neutrons are fast moving.

This brings us to the third and most significant feature of these reactors. While the fast neutrons released during fission in this reactor are not good at causing fission, they are readily captured by U-238 to transform it into U-239 which then beta decays (that is, emits electrons from its nucleus—see Chapter 3 for more on this) to form Pu-239. Thus, this reactor breeds fuel (Pu-239) as it operates, hence its name. In many FBR designs, the reactor core is surrounded by a blanket of tubes containing non-fissile U-238. This too captures fast neutrons from the reaction in the core and is partially converted to fissile Pu-239.

The advantage of this design is that it uses uranium-238 as fuel, which makes up 99.27 per cent of naturally occurring uranium, whereas conventional uranium fuelled reactors use uranium-235—this makes up just 0.72 per cent of naturally occurring uranium.

This is the reason why FBRs have been an attractive option for countries like India which possess limited amounts of uranium deposits, and that too of low-grade.

3

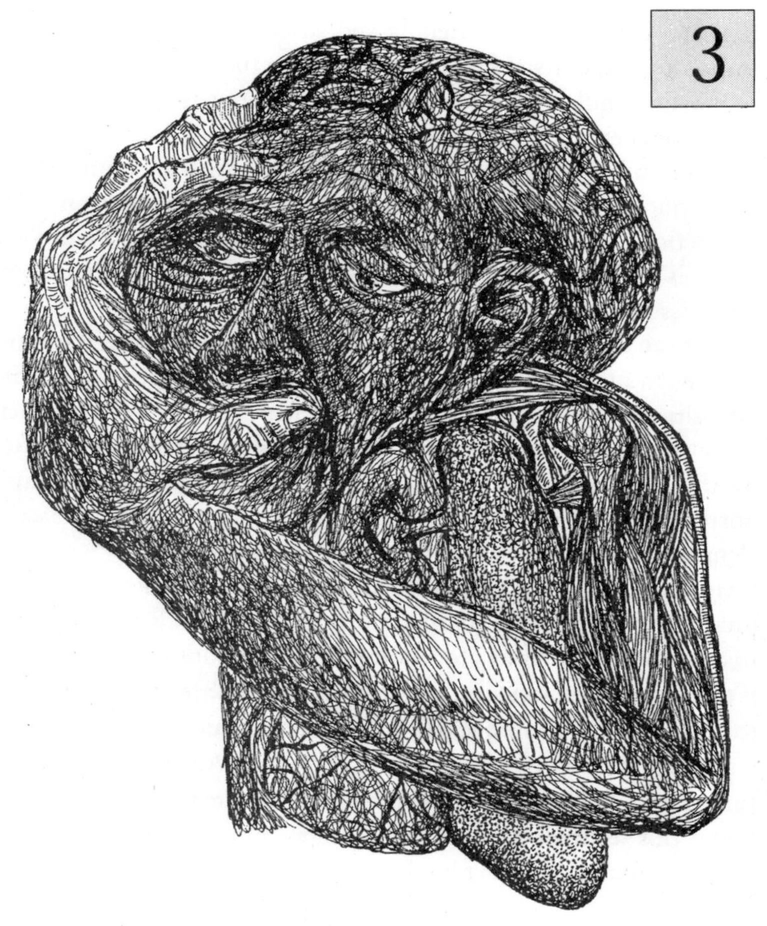

IS NUCLEAR ENERGY SAFE?

According to President Bush, who very strongly voiced support for nuclear energy in various international meetings, nuclear energy was a 'safe and clean' energy source for the future.[1] His envoy to India, David Mulford, made the same claim at a function to celebrate World Environment Day in 2008.[2] Echoing the Master's Voice, Prime Minister Manmohan Singh has repeatedly claimed that nuclear energy is a safe, environmental friendly and sustainable source of energy.[3] India's top scientists and leading 'intellectuals' are also strongly supporting India's push towards nuclear energy as a safe and clean source of energy.[4]

From US to India, politicians and intellectuals are blithely lying about nuclear energy. They believe that if you lie frequently and with conviction, people will believe you. We discuss how much 'clean and green' is nuclear energy in Chapter 5. In this chapter, we discuss the safety issues associated with nuclear energy.

Even if nuclear power plants are operating normally, the entire cycle from uranium mining to nuclear reactors routinely emits huge quantities of extremely toxic radioactive elements into the atmosphere every year. The environmental costs of the deadly radiation emitted by these elements and its impact on human health are simply horrendous. What is infinitely more worse, since these radioactive elements will continue to emit radiation for tens of thousands of years, their effects will continue to plague the human race not just for the present, not just during our lifetime, but for thousands of generations to come—yes, we repeat, for thousands of generations to come! And if there is a major accident, and nuclear reactors are inherently prone to accidents, the consequences will be cataclysmic!!

As Dr Helen Caldicott, the renowned Australian physician turned anti-nuclear activist who has worked tirelessly to expose the threat this technology from hell poses to human survival, wrote in *Nuclear Madness*, her first book:

> As a physician, I contend that nuclear technology threatens life on our planet with extinction. If present trends continue, the air we breathe, the food we eat, and the water we drink will soon be contaminated with enough radioactive pollutants to pose a potential health hazard far greater than any plague humanity has ever experienced.[5]

In this chapter, we discuss the radiation emitted at each stage of the nuclear fuel cycle and its consequences for the human race. We also discuss the possibility of a major accident occurring in nuclear reactors, and its probable impact in the light of past experience. However, first, let us discuss what is radiation and how it affects human health.

PART I: WHAT IS RADIATION?

Ionising and Non-ionising Radiation

Radiation is energy in the form of particles or electromagnetic waves that moves through space. It can have a wide range of energies.

Based on its energy levels, radiation can broadly be divided into two categories:

- Non-ionising radiation
- Ionising radiation

Radiation that is of low energies, which at most can only move or vibrate electrons or atoms, but cannot remove electrons from atoms, is referred to as **non-ionising radiation**. Examples of this kind of radiation are sound waves, visible light, and microwaves. **Ionising radiation** is high energy radiation. It has enough energy to remove electrons from atoms, thus creating ions, hence the name. In case of very high energy radiation, it can even break apart the nucleus of atoms, and break apart molecules.

Various types of ionising radiation are produced in a variety of ways. One of these is what is known as **radioactive decay**. This is the mechanism by which energy in the form of ionising radiation is released during the various stages of the nuclear fuel cycle, to cause terrible health effects.

Radioactive Decay: Stable and Unstable Atoms

Most atoms found in nature are **stable**; that is, they do not undergo changes on their own. For instance, if we put an atom of Aluminium (Z = 27)[6] in a bottle, seal it and open it after a million years, it would still be an atom of aluminium. Aluminium is therefore called a stable atom.

There also exist many stable atoms which have **unstable** isotopes. An unstable atom is one whose nucleus undergoes some internal

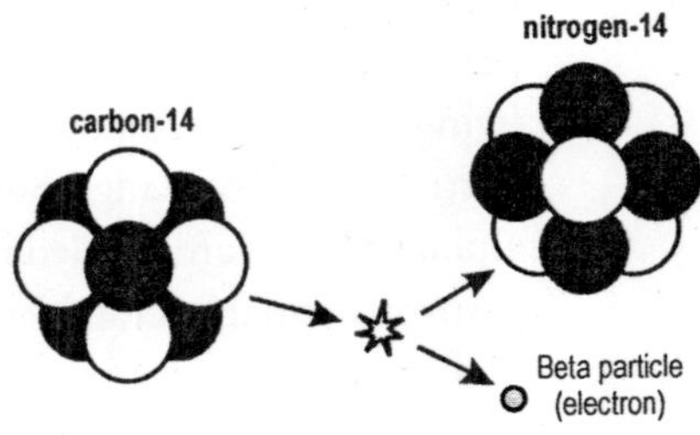

Figure: Unstable Atom

change spontaneously. In this change, the nucleus emits **ionising radiation** in the form of subatomic particles, or a burst of energy, or both. This emission of radiation by the unstable nucleus is called **radioactivity**, and the nucleus is said to have undergone **radioactive decay**, or just decay. In this process, the nucleus changes its composition and may actually become a different nucleus entirely. The process continues till the nucleus achieves stability.

For example, most carbon (Z = 6, A=12)[7] atoms are stable, with the nucleus having six protons and six neutrons. Carbon has an isotope, C-14, whose nucleus consists of six protons and eight neutrons, which is unstable. In its attempt to achieve stability, a C-14 nucleus gives off a beta particle (that is, an electron emitted by the atomic nucleus). After the C-14 nucleus has emitted the beta particle, it now consists of seven protons and seven neutrons. But a nucleus consisting of seven protons and seven neutrons is no longer a carbon nucleus. It is now the nucleus of a nitrogen atom. By giving off a beta particle, the C-14 atom has changed into a N-14 atom.

In the periodic table, elements with atomic number 83 and above are unstable, meaning all their isotopes emit radioactivity.[8] While elements with atomic number from 1 to 82 are mostly stable [with the exceptions of technetium (Z=43) and promethium (Z=61)], many have unstable isotopes.

Types of Radiation Emitted by Radioactive Elements

Radioactive isotopes emit three types of ionising radiation:

(i) **Alpha radiation:** Alpha particles are composed of two protons and two neutrons. Being heavy (as compared to beta particles), these particles do not travel very far, and are not able to penetrate dead cells in the skin to damage the underlying living cells. Therefore, when outside the human body, alpha particles are not dangerous to human life. However, when alpha particles are inhaled into the lungs or ingested into the gastrointestinal tract, they come into contact with living cells and severely damage them. The biological damage can have serious consequences for human health, including the possibility of causing cancer. For instance, plutonium is an alpha emitter, and no quantity inhaled has been found to be too small to induce lung cancer in animals.

(ii) **Beta radiation:** This is composed of electrons. How does a nucleus emit a electron? The answer: a neutron breaks up into a proton and electron, and the latter is emitted. Beta particles are lighter than alpha particles, and so while they travel farther than alpha particles in body tissues, the biological damage caused by them is less, like a bullet compared to a cannon ball. They can penetrate the outer layer of dead skin and damage the underlying living cèlls. If they are inhaled or consumed or absorbed into the blood stream, then they can damage tissues and cause cancer. Usually, health effects of beta particles develop relatively slowly, typically over five to 30 years. Typically, how they act is by accumulating in the human body, causing low level exposure over a prolonged period of time, and the main health effect of this prolonged exposure is cancer. For instance, iodine-131, a beta emitter, concentrates heavily in the thyroid gland, increasing the risk of thyroid cancer and other disorders.

(iii) **Gamma radiation:** This is akin to X-rays. It is composed of photons, that is, high energy light waves. It has great penetrating power and can travel large distances. Gamma radiation goes straight through human bodies. As gamma rays pass through the body, their energy can be transferred to the body cells and can cause damage.

The danger with all these three types of ionising radiation is that when people are exposed to this radiation, the energy they contain hits body tissue, releasing the energy contained in the radiation, and this energy always causes some damage to the tissue. This can damage cellular DNA leading to cancer. If damage is caused to cells in the reproductive organs, mutations and malformations may occur. Of course, there are cellular mechanisms that can also repair damage to the DNA. The mechanisms of damage and repair are very complicated and not completely understood.

However, what is very well understood today is that there is no lower limit of exposure below which there is no damage, in other words, there is no minimum safe dose of radiation. Any amount of radiation will damage cells and it is the delicate balance of repair mechanisms that determines the ultimate outcome of health or disease. (We will discuss this in greater detail in a later section.)

In order to make sense of how much radiation exposure will cause how much damage, it is necessary to understand the various units of radiation.

Units of Radiation

Becquerel and Curie: This unit applies to the strength of the source of ionising radiation, that is, the radioactive isotope. In the International System of units (SI), it is measured in Becquerel (Bq). One Bq is defined as one disintegration per second. Becquerel is a very small unit. An older, non-SI, and much larger unit of radioactivity is Curie, defined as: Curie (Ci) = 3.7×10^{10} disintegrations per second.

Rad and Gray: The radiation emitted by a radioactive element is not the same as the radiation absorbed by the body. The difference between the two is like a boxer who hits at his opponent, but he may or may not strike him. The radiation dose absorbed by the body is measured in a unit called Rad, or Radiation Absorbed Dose. In the SI system of units, the unit is Gray. A dose of one gray means the absorption of one joule of radiation energy per kilogram of absorbing material. The conversion factor is: 1 gray = 100 rad.

Rem and Sievert: Not all radiation has the same biological effect, even for the same amount of absorbed dose. Thus, the same Rad of alpha particles when absorbed cause much more damage than beta particles. This difference is measured by a unit called Rem, or Roentgen Equivalent Man. It relates the absorbed dose of radiation in human tissue to the effective biological damage caused by it. To determine Rem, the absorbed dose in Rad is multiplied by a quality factor (Q) that is unique to the type of incident radiation. For gamma rays and beta particles, one rad of exposure results in one rem of dose, while for alpha particles, one rad of exposure is equivalent to 20 rems of dose. Another unit for measuring biological impact of absorbed radiation is Sievert or Sv: 1Sv = 100 rem.

Thus, becquerel and sievert are two different kinds of units of radiation. The former represents exposure to radiation; it is what is outside, in the environment. The latter measures the dose, how much is received or absorbed. In the case of ionising radiation, what ultimately matters is the dose. However, in the media and in many informational articles, exposure and dose are often used interchangeably. Though it is not very accurate, in a way exposure and dose are related in that exposure is a good approximation of how high doses can be.

Table 3.1: Some Examples of Doses

Radiation dose	Source
0.1 mSv	X-ray (chest)
0.4 mSv	Mammography
1.5 mSv	X-ray (spine)
2 mSv	CT scan (head)
15 mSv	CT scan (abdomen and pelvis)
250 mSv	US limit for fire-fighters, police officers and other emergency workers engaged in life-saving activity

Radiation is often measured in dose rate, such as millisievert per hour. Dose rates matter because faster delivery of radiation can have a relatively stronger impact; getting the same dose in 1 hour is usually worse than getting the same dose stretched out over the course of a year.

Table 3.2: Some Examples of Dose Rates

Radiation dose rate	Source
1 mSv / year	Maximum exposure limit for members of the public in the US by a facility licensed by the NRC
2-3 mSv / year	Average background radiation from natural sources
6.2 mSv / year	Average exposure of people in the US from natural and human caused sources, according to the NRC
20 mSv / year	Limit of radiation exposure for adult employees in nuclear installations in most countries, including India
50 mSv / year	Limit of radiation exposure for adult employees in nuclear installations in the US, as set by the NRC
1000 mSv / hour	Causes radiation sickness after short exposure; for 3 hours of exposure—50% fatality rate, for 6 hours exposure—essentially 100% fatality rate
350 mSv/lifetime	Criterion for relocating people after Chernobyl accident

Half-life

Each radioactive isotope has a specific half-life. Half-life of an isotope is the amount of time it takes for half the number of atoms of that isotope to decay. For example, radioactive iodine-131 has a half-life of eight days, so that in eight days it loses half its radioactive energy, in another eight days it further decays to one quarter of the original radiation, ad infinitum. The amount of time taken by a radioactive isotope to decay to a harmless level can be obtained by a simple thumb rule: multiply the half-life by 20. (There is of course no unanimity on this, with many experts saying that radiation becomes harmless in 10 half-lives.) Thus, in the case of iodine-131, its radioactive life is 8 x

20 = 160 days. Some isotopes created during the fission reaction in a nuclear reactor have very short half-lives (less than a second), and some extremely long (millions of years).

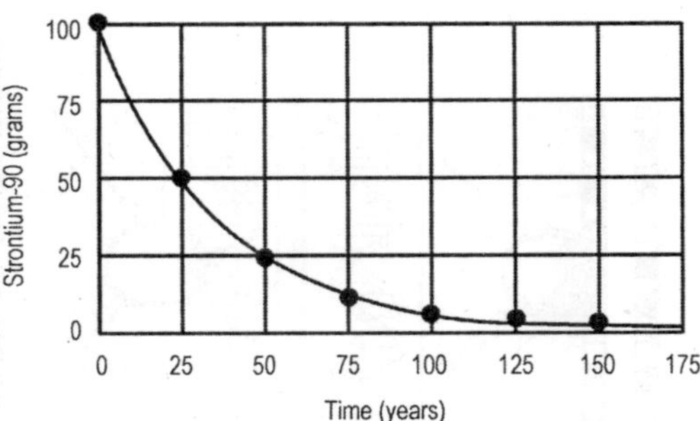

Figure: Radioactive Decay Curve

PART II: RADIATION AND HUMAN HEALTH[9]

Radiation and Reproduction

Instructions providing all the information necessary for a living organism to grow and live reside in every cell of the body of the organism. These instructions are stored in a molecule called the DNA, or Deoxyribonucleic acid, whose shape is like a twisted ladder, called a 'double-helix'. The DNA molecules are stranded together like letters in a sentence, and these strands are called genes. Genes are packed into thread like structures, called chromosomes. All genes come in pairs, one inherited from each parent. In the human cell, they are organised in two sets of 23 chromosomes, one coming from the mother, the other from the father. The human cell thus has a total of 46 chromosomes.

Genes are the very building blocks of life, responsible for every inherited characteristic in all species—plants, animals and humans. Every person inherits half of his/her genes from the mother, and half from the father. While every human cell has 46 chromosomes, the egg and sperm have 23. At the time of conception, the mother's egg cell unites with the father's sperm, to form the zygote, which has a full complement of 46 genes. This cell then duplicates itself, and develops into the child. Most genes are the same in all human beings, which is why all human beings are similar. A small number of genes are different, and it is these which are responsible for each human being's unique features.

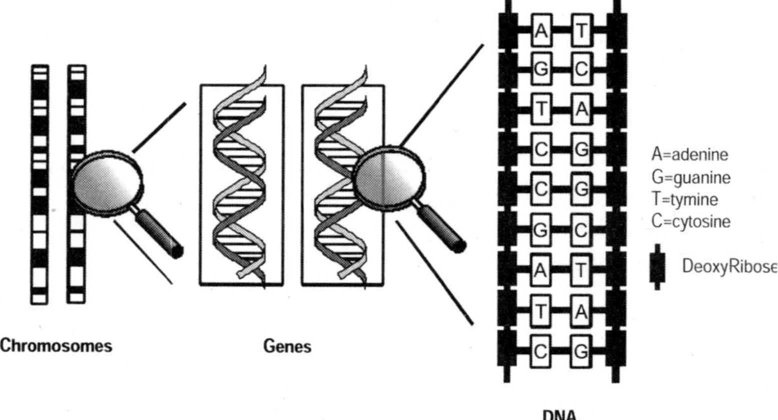

Figure: DNA, Genes and Chromosomes

Radiation can induce mutation, that is, a chemical change, in the DNA molecule, thereby causing a change in the gene. (Apart from radiation, mutation can be caused by other reasons too.) If this mutation takes place in the reproductive gene, then it can cause the most unexpected changes in the offspring. This can be understood from the fact that according to evolution theory, radiation from the atmosphere and earth's crust (called background radiation) is one of the important causal factors of evolution. While most mutations caused by this radiation were 'harmful', causing disease and death in the offspring, some were 'advantageous': that is, they produced changes in the offspring that enabled it to better survive and multiply in the hostile environment. Thus, it is because of such mutations that fish developed lungs and climbed out of water to become land-dwelling amphibians, dinosaur-like creatures developed wings and became the earliest form of birds, and humans evolved from early primates. But this also means that there must have been incomparably more mutations which led to the birth of monstrous offspring which were unfit to survive and died. Mutations are also responsible for thousands of genetically inherited diseases, like heart diseases, cystic fibrosis and sickle cell anaemia—medical literature describes 19,000 genetically inherited diseases!

Apart from causing mutations in genes, radiation can also cause a break in chromosomes. This can cause a baby to be born with serious mental and physical genetic disorders.

If a pregnant woman is exposed to radiation, then it may so happen that the radiation kills a cell of the foetus that was going to become a leg or a valve of the heart or some other part of some other important organ. Such a mutation, which is not passed on to the offspring as it did not take place in the reproductive gene, is called teratogenic mutation.

Radiation and Cancer

All non-reproductive cells of the body have regulatory genes that control the rate of cell division. If a regulatory gene is exposed to radiation, and it mutates, then the cell may become carcinogenic. However, cancer does not develop right away; there is a long incubation period which can be from two to 20 years, and even up to 40 years and more. Then one day, instead of the cell dividing into two daughter cells in a regulated fashion, it will begin to divide in a random, uncontrolled fashion into millions and trillions of daughter cells, creating a cancer. In many cases it is difficult if not impossible to stop this random growth of abnormal cells. All kinds of cancers can be caused by exposure to radiation: from cancer of the upper digestive tract and lungs to bone cancer and leukaemia.

Other Impacts of Radiation on Human Health

Radiation exposure causes cancer, but the incubation period can be many years. Apart from cancer, non-cancerous health effects of radiation exposure are also there. It can cause radiation sickness, whose symptoms include: nausea, weakness, hair loss, skin burns and diminished organ function. If the dose is high, it can also cause premature aging and death. Exposure to radiation can damage body cells, causing a wide variety of effects. Thus, it can damage the reproductive system, causing infertility and spontaneous abortion. It deforms red blood cells, inhibiting their passage into the tiny capillaries and depriving the muscles and brain of adequate oxygen and nutrients. This can lead to impairment of many organs, especially the kidneys, liver, lungs and cardiovascular system, and can also damage the system that causes formation of blood in the body (known as haematopoietic system). Radiation can also cause disorders of protein and carbohydrate metabolism, leading to symptoms ranging from severe headache to brain dysfunction.

No Safe Dose of Radiation

Many pioneers of radiation physics died of radiation induced illnesses. The dangerous effects of ionising radiation were not known then, and so no one took any safety measures. Thus, the discoverer of X-rays, Wilhelm Roentgen, died of bone cancer in 1923, while the two pioneers in its medical use, Madame Marie Curie and her daughter, Irene, died of leukaemia. Isaac Asimov writes that at least one hundred of the early workers with x-rays and radioactive materials died of cancer; many others had to have their fingers and arms amputated. Over the years, as the painful understanding of harmful effects of radiation grew, there was a gradual lowering of the level of radiation exposures permitted for workers in radiation-related occupations. Thus, permissible occupational exposure to ionising radiation in the United States was set at 52 roentgen (or rem) per year in 1925, 36 roentgen (or rem) per year in 1934, 15 rem per year in 1949 and five to 12 rem per year from 1959 (depending on average per year over age 18) to the present.[10]

During the past few decades, many eminent scientists came to realise that these so-called 'permissible' or 'safe' levels of radiation were actually a permit to commit murder! Among the many scientists who did pioneering work on the dangers of low level radiation, one of the most brilliant was the late Dr John W. Gofman, who began work in the field of nuclear energy in the 1940s. He pioneered the process that enabled plutonium to be separated from the uranium and fission products of irradiated nuclear fuel. His work in this field eventually led to him joining the US Atomic Energy Commission (US AEC). At the US AEC's request, at the beginning of the 1960s, Gofman established the Biomedical Research Division at the Lawrence Livermore National Laboratory in order to study the effects of nuclear energy on health. He now began to focus his attention on studying cancer and chromosomes, and the effect that radiation had on chromosomal mutations and gene stability.

By 1969, Dr Gofman, once a strong proponent of the nuclear energy industry, had concluded that human exposure to ionising radiation was far more dangerous than scientists, including himself, had previously realised. His research led him to conclude that there is no safe threshold of ionising radiation for any human being. Gofman now became a leading anti-nuclear activist, calling for a moratorium on the expansion of nuclear power plants.[11] In an interview published in the 1982 book *Nuclear Witnesses, Insiders Speak Out* he stated:

> Licensing a nuclear power plant is in my view, licensing random premeditated murder ... [T]he evidence on radiation producing cancer is beyond doubt. I've worked fifteen years on it, and so have many others. It is not a question any more: radiation produces cancer, and the evidence is good all the way down to the lowest doses.
>
> The only way you could license nuclear power plants and not have murder is if you could guarantee perfect containment. But they admit that they're not going to contain it perfectly. They allow workers to get irradiated, and they have an allowable dose for the population. So in essence I can figure out from their allowable amounts how many they are willing to kill per year.
>
> I view this as a disgrace, as a public health disgrace ...
>
> People like myself and a lot of the atomic energy scientists in the late fifties deserve Nuremberg trials. At Nuremberg we said those who participate in human experimentation are committing a crime. Scientists like myself ... were experimenting on humans, weren't we? But once you know that your nuclear power plants are going to release radioactivity and kill a certain number of people, you are no longer committing the crime of experimentation—you are committing a higher crime. Scientists who support these nuclear plants—knowing the effects of radiation—don't deserve trials for experimentation; they deserve trials for murder.[12]

While Dr Gofman was initially much riled by the nuclear industry for his work, and the US nuclear authorities defunded his research on chromosomes and cancer, soon, over the next few decades, scientific evidence mounted in support of his conclusions. Many eminent scientists emphatically stated that there is nothing like a safe dose of radiation, that even the smallest amount has the possibility of a lethal outcome. To give an example, one of these was the American physicist Dr Karl Morgan, the legendary founder of the field of radiation health physics. Like Gofman, he too began his career with the US nuclear establishment. After a long career in the Manhattan Project and at the Oak Ridge National Laboratory, he too came out against the nuclear industry when he understood the danger of low levels of ionising radiation. In an article in the *Bulletin of Atomic Scientists* in 1978, he wrote: 'There is no safe level of exposure and there is no dose of radiation so low that the risk of a malignancy is zero.'[13]

Today, the evidence is so overwhelming that many official studies too have come to the same conclusion. In 2005, a panel of the US National Academy of Sciences charged to investigate the dangers of low-energy, low-dose ionising radiation came to the conclusion that there is no safe threshold dose. The panel came to this conclusion despite having a number of pro-nuclear individuals; the evidence was simply too overwhelming for them to ignore.[14] Every US agency that regulates radiation exposure, from the US Environmental Protection Agency to the US Nuclear Regulatory Commission, agrees that there is no safe dose no matter how small.[15]

Yet, nuclear regulatory authorities prescribe what are loosely called 'safe' radiation doses, but are, in actuality, 'allowable' or 'legally permissible' radiation doses. In the US, according to standards set by the Nuclear Regulatory Commission (NRC), nuclear plant operators cannot legally expose the general public to more than 100 millirems per person annually. Rules are more lenient for nuclear workers: they are allowed an yearly exposure of 5,000 millirems (5 rems or .05 Sv).[16] In India, the standards set by the Atomic Energy Regulatory Board are that workers must not be exposed to more than 2000 millirems a year averaged over five consecutive years (and not more than 3000 millirems in any single year), and pregnant women 200 millirems for the entire duration of pregnancy.[17]

These standards and those set by other governments worldwide are based on risk coefficients for ionising radiation exposure promulgated by the International Commission on Radiological Protection (ICRP). Recently, on May 6, 2009, a very important conference took place in Lesvos, Greece: the European Committee on Radiation Risk Conference. Sixteen world experts who participated in the conference issued *The Lesvos Declaration* which unequivocally stated that existing methods to determine safe radiation doses are clearly outdated. The declaration states that the ICRP risk model was developed before the DNA structure was discovered and before new discoveries such as that 'certain radionuclides have chemical affinities for DNA'. Therefore, the experts asserted, 'the ICRP risk coefficients are out of date', that 'employing the ICRP risk model to predict the health effects of radiation leads to errors which are at minimum 10 fold', that 'damage to the cardio-vascular, immune, central nervous and reproductive systems' due to radiation exposure is significant but as yet unquantified, and called for more research into the health effects of radiation.[18]

This means that all the health effects of radiation outlined above are an underestimate, the full extent of effects of radiation on health of human beings is yet to be understood ...

Background Radiation and Man-made Radiation

Cancers have always plagued the human race. It is generally accepted that most cancers in the past and present are due to what is known as 'background radiation', that is, radiation that is constantly present in the environment and is emitted from a variety of natural sources, like cosmic rays coming from space and radioactive materials present in the Earth's crust. Nuclear authorities argue that there is nothing to fear from routine radioactive releases from nuclear plants, as it is much less than the naturally occurring background radiation.[19] Even if this fact is correct, it is a strange argument. While we cannot do anything about background radiation, and therefore cannot prevent a certain number of people from developing cancer due to this, should we not try and ensure that this number does not increase by preventing man-made radiation from adding to background radiation!

We are exposed to a background radiation of around 100 millirems per year from the earth and sun.[20] The US NRC has decided that it is acceptable for the public to receive an additional 100 millirems per year from man-made radiation created through generation of nuclear energy. This means that for the NRC, it is acceptable that the number of cancer patients double (as compared to the number of cancer patients that would have occurred due to naturally occurring radiation)!

The US National Academy of Sciences estimates that man-made radiation in the United States accounts for 18 per cent of human exposure.[21] However, what is not realised is that as more and more of the huge quantity of radioactive waste accumulating near nuclear power plants leaks and contaminates the environment and enters the water and food chains around the world, the percentage of radiation exposure from these sources is going to increase. And since this radiation remains potent for tens of thousands of years, by using nuclear electricity today, we are bequeathing our descendants a radioactive legacy tomorrow.

Internal and External Radiation[22]

In recent times, as the debate over effect of radiation leakages from nuclear reactors on human health has intensified, especially after the Fukushima accident in Japan, many prominent pro-nuclear intellectuals have questioned the theory that 'there is no safe dose of radiation'. They claim that the anti-nuclear lobby has been exaggerating the impact of low-level radiation emitted from nuclear reactors on human health.

They counter the argument given above about 'background radiation versus radiation from nuclear reactors' by saying that since radiation is ubiquitous, why worry about it? Thus Patricia Hansen, senior scientist at the US FDA, argues: 'Radiation is all around us in our daily lives ... A person would be exposed to low levels of radiation on a round-trip cross-country flight, watching television and even from construction materials.'[23]

A similar argument is the following statement by the nuclear engineering department at the University of California, Berkeley, which reported that radioactive isotopes of iodine-131, cesium-134 and cesium-137 have been found in milk from a local organic dairy after the Fukushima accident in Japan. To assure the people of

California that the levels of these isotopes were not very high and that the milk was safe to drink, the nuclear engineering department stated: 'Please note ... I-131 ... levels are still very low—one would have to consume at least 1,900 litres of milk to receive the same radiation dose as a cross-country airplane trip.'[24]

In essence, these experts are saying that if we are getting exposed to so much radiation daily in our lives, why worry about some additional radiation from nuclear reactors.

After the Fukushima accident, George Monbiot, British journalist and author, argued in his much publicised debate with Dr Helen Caldicott, one of the world's foremost anti-nuclear activists, that while it 'may cause health effects for some people' and 'that it will cause mass evacuation' for the present, 'I would disagree, though, that it will devastate a large part of Japan forever, which I think was a term that she (Helen Caldicott) used. I think that's an overstatement of the impacts of the radiation.'[25]

The most important mistake made in the above arguments is that these intellectuals confuse external radiation with internal radiation. Dr Helen Caldicott explains the difference:

> The former is what populations were exposed to when the atomic bombs were detonated over Hiroshima and Nagasaki in 1945; their profound and on-going medical effects are well documented.
>
> Internal radiation, on the other hand, emanates from radioactive elements which enter the body by inhalation, ingestion, or skin absorption. Hazardous radionuclides such as iodine-131, caesium-137, and other isotopes currently being released in the sea and air around Fukushima ... [after] they enter the body, these elements—called internal emitters—migrate to specific organs such as the thyroid, liver, bone, and brain, where they continuously irradiate small volumes of cells with high doses of alpha, beta and/or gamma radiation, and over many years, can induce uncontrolled cell replication—that is, cancer. Further, many of the nuclides remain radioactive in the environment for generations, and ultimately will cause

> increased incidences of cancer and genetic diseases over time.
>
> The grave effects of internal emitters are of the most profound concern at Fukushima. It is inaccurate and misleading to use the term 'acceptable levels of external radiation' in assessing internal radiation exposures.[26]

Howsoever small be their quantity, when these radionuclides enter the human body, they are going to irradiate the body cells in whichever part they deposit, continuously, for years, to ultimately cause cancer and other diseases. Obviously, the more the nuclides in the body, the more the harm. But there is no minimum safe dose of radiation.

Hence, exposure to radiation from space during airline flights is not even remotely comparable to ingesting or inhaling radioactive isotopes into the human body (as made in the statement by the department of nuclear engineering of the University of Berkeley quoted above). Dr Bill Deagle, MD, a Board Certified family physician and a teacher for the College of Occupational and Environmental Medicine and the American Academy of Environmental Medicine, USA, recently commented on such statements:

> When you're flying in an aircraft, you're being exposed to cosmic background radiation and just some very weak gamma rays, etc, and x-rays from the cosmos. It has a momentary and transitory effect that ... doesn't cause the extreme danger. When you have internal contamination it's ... like having a Fukushima nuclear reactor at the cellular level. This is not the same, and anyone who says that is obviously proving that, no matter what credentials they have, they're either a liar or they are not really a scientist.[27]

PART III: RADIATION EMISSION IN NUCLEAR FUEL CYCLE, AND ITS IMPACT ON LIFE

Man-made radiation is released during all stages of the nuclear fuel cycle.

1. Uranium Mining

Uranium miners are at great risk because they are exposed to high concentrations of a radioactive gas called radon-222. Radon-222 is a decay product of uranium and is a highly carcinogenic alpha emitter. If inhaled, it can deposit in the air passages of the lung, irradiating cells that then become malignant.

Uranium miners are also exposed to radium-226, another lethal uranium daughter, which is an alpha and gamma emitter with a half-life of 1,600 years. Radium-226 is an integral component of uranium dust in the mine. When this is swallowed, radium is absorbed from the stomach into the body and deposits in the bones. It causes osteogenic sarcoma, a highly malignant bone cancer, and leukaemia, because white blood cells are manufactured in the bone marrow.

Uranium daughters present in the ore emit gamma radiation too, which emanates from the surface of the uranium mine. So, miners are exposed to a constant, whole-body radiation (like X-rays), which irradiates their bodies and continuously exposes their reproductive organs.[28]

As a result, uranium miners suffer from a very high incidence of cancer. One-fifth to one-half of the uranium miners in North America, many of whom were Native Americans, have died and are continuing to die of lung cancer. Records reveal that uranium miners in other countries, including Germany, Namibia and Russia, suffer a similar fate.[29]

Waste Rock

The waste produced during mining, called waste rock or mine tailings, is in huge quantities—it is several times larger than the amount of ore mined. This is left lying in huge heaps adjacent to the mine, exposed to the air and the rain. The waste rock contains uranium ore of too low grade for processing in the mill, and decay products of uranium.

So long as the uranium deposit was undisturbed, the radiation was trapped underground. But now that the ore is mined, the waste rock piles present hazards to residents and the environment, even after the shutdown of the mines: radon gas can escape into the air; ore dust can be blown by the wind; and uranium and its decay products can seep into surface water bodies and groundwater. Being radioactive and toxic, they contaminate the environment.

Most uranium mines in the USA are situated on or are adjacent to indigenous tribal lands of the Navajo nation, located in the Four Corners area (the intersection of Arizona, Colorado, New Mexico and Utah) in the American Southwest. The more than 250 million tons of uranium mine tailings lying here in the open constantly leak radon-222 into the air, exposing the indigenous populations who live nearby. As they inhale the radon gas, many of these people have developed or are developing lung cancer. The radioactive debris has also polluted the underground water and the Colorado River; water from this river is used for agriculture and drinking by 30 million people downstream in Arizona, Nevada and California. Because it is tasteless and odourless, people in these contaminated populations cannot tell whether they are drinking radioactive water, breathing radioactive air, or eating fish or food that will induce bone cancer or leukaemia. These wastes are taking a terrible toll: thousands of Navajos are suffering and dying from uranium-induced cancers. No one knows how many exactly, because the authorities do not keep a track. Epidemiological studies reveal that Navajo children living near the mines and mills suffer 5 times the rate of bone cancer and 15 times the rate of testicular and ovarian cancers as other Americans.

The people living on these lands will continue to pay this price in the future too, for thousands of years, unless these wastes are cleaned up. But that is a very costly operation, would cost billions of dollars. The US government and the nuclear industry have made no attempts to clean up this massive radioactive pollution, as it is tribals, and not the well-heeled of America, who are affected![30]

2. Uranium Milling and Mill Tailings[31]

Uranium mills are normally located near the mines to save transportation costs. The wastes generated from the milling process are in the form of sludge and are called uranium mill tailings. Uranium mill tailings are normally pumped to settling ponds, where they are abandoned.

Since uranium represents only a minor fraction of the ore

(around 0.1 per cent or less), the amount of sludge or mill tailings is nearly identical to that of the ore mined. In the US, over the last 40 years, apart from the mine waste, over 100 million tons of mill waste has also accumulated on Navajo lands. In Europe, the largest settling ponds are in Germany: the Culmitzsch tailings dam contains 90 million tons and the Helmsdorf tailings dam 50 million tons of solids.

Since only the uranium is removed, the sludge contains all the remaining constituents of the ore, including the long-lived decay products of uranium—thorium-230 and radium-226. Further, due to technical limitations, all of the uranium present in the ore cannot be extracted. Therefore, the sludge also contains 5 per cent to 10 per cent of the uranium initially present in the ore.

The sludge thus contains around 85 per cent of the initial radioactivity of the ore. One of its deadly radioactive constituents is the uranium decay product, thorium-230. Thorium-230 is the uranium decay product with the longest lifetime, decaying at a half-life of 80,000 years. This means that it emits radioactivity for 8-16 lakh years—in human terms, forever. Thorium-230 is especially toxic to the liver and the spleen. It has been known to cause leukaemia and other blood diseases. It decays to produce radium-226, which in turn produces radon gas (discussed in the previous section on uranium mining), a very powerful cancer-causing agent. Even small doses inhaled repeatedly over a long time can cause lung cancer. Even though radon-222 has a comparatively short half-life of 3.8 days, its quantity will not diminish for a long time, because it is constantly being replenished by the decay of the very long-lived thorium-230.

Hence, the radioactivity emitted from the tailing ponds will continue to be in significant amounts for hundreds of thousands of years! This will severely affect residents living near these enclosures:

(i) Radon gas can affect people very far away from the tailings pond, as it can travel thousands of kilometres with a light breeze in just a few days.

(ii) Heavy rainfall or floods can cause spillover of the sludge into nearby areas; it may also cause a failure of the tailings dam. This has been occurring around the world with frightening regularity. These failures can be huge. One of the biggest of these accidents was the collapse of the Church Rock tailings dam in New Mexico on July 16, 1979, spilling ninety million gallons of liquid radioactive waste and eleven hundred tons of solid mill wastes into the Rio Puerco River. It is the largest release of radioactive waste ever in the US, and second only to the Chernobyl meltdown globally. Few people have heard of this disaster, because it took place in tribal lands, so the media simply ignored it. The Navajos, of course, continue to suffer its consequences.[32]

(iii) Seepage from the tailing ponds can contaminate the ground and surface water. For example, seepage is known to be occurring at the uranium mill tailings pond in the city of Pecs in Hungary and Stráz pod Ralskem mill tailings pond in North Bohemia. Sooner or later, it is going to contaminate the drinking water sources of both these places.[33] Seepage from tailings pond of the Atlas uranium mill in Moab, Utah, USA, had been contaminating the Colorado River for decades. The river is the drinking water supply for millions of Americans downstream. After an intense struggle waged by local people for nearly a decade, finally, in 2005, the authorities agreed to relocate the tailings pond.[34]

The tailings therefore need to be safeguarded for tens of thousands of years. In practice, the settling ponds are simply abandoned. Only when there is a major seepage from the pond, or the dam breaks, do governments move to take some damage control measures.

3. Uranium Enrichment

The uranium-235 isotope is enriched from a low concentration of 0.7 per cent to 3 per cent for fuel in nuclear power plants (except in PHWRs). Workers at all stages of the enrichment process are exposed to whole-body gamma radiation from by-products of uranium decay. But the most serious aspect of enrichment is the material that is discarded: uranium-238. This is called 'depleted uranium' (DU) because it has been depleted of its uranium-235. But it is not depleted radioactively.[35]

In March 2009, bowing to pressure from the nuclear industry, the US Nuclear Regulatory Commission (NRC) voted to declare that depleted uranium from enrichment plants is a Class A low-level radioactive waste—the least dangerous kind that supposedly consists mainly of short-lived radionuclides. In actual fact, the depleted uranium becomes more radioactive with time, and hence more dangerous, because of the growth of decay products of uranium like thorium-230 and radium-226. The uranium-238 and its decay products will continue to pollute the environment for thousands of years.

The NRC decision will make disposal of DU cheap for the US nuclear enrichment industry. Currently some 740,000 tons of depleted uranium in unstable hexafluoride form are stockpiled at US Department of Energy sites at Paducah (Kentucky), Portsmouth (Ohio) and Oak Ridge (Tennessee). Even before this ruling, barrels in which this depleted uranium had been stored at these sites had been leaking and disintegrating and had polluted the groundwater. Now the industry can store this even more callously.[36]

The situation is much the same in Europe. In France, despite objections of residents, the court ruled that depleted uranium is no waste, but a 'directly usable raw material that is effectively used for multiple uses' and allowed the French nuclear fuel company Cogéma to 'store' 2 lakh metric tons of DU at the site of the former uranium mill of Bessines-sur-Gartempe (Haute Vienne) near Limoges. Cogema claims that it is storing this depleted uranium at this site for possible future use.[37]

Governments led by the US and UK have now found a new way of disposing of at least some of this DU—they are using it in bombs and have used hundreds of tons of depleted uranium in Gulf War I, the Balkan wars, the Afghanistan war and now the most recent invasion of Iraq that began in 2002. Israel also dropped US-made DU bombs on Lebanon in 2006, and then again on the hapless residents of Gaza during its invasion of December 2008.[38] Uranium-238 has a half-life of 4.5 billion years. It effectively means these lands are contaminated till the end of time.

Hundreds of thousands of US and UK troops who served in Gulf Wars I and II are sick and slowly dying from what their governments say is a 'mystery disease', which they have labelled Gulf War Syndrome. Many have given birth to deformed children. The reality is, these diseases have been caused by exposure to depleted uranium! The governments of these killer countries are refusing to admit the truth, because it will mean paying out billions of dollars of compensation to their troops!![39]

The impact of these bombs on the people of Iraq (and also other countries where these bombs have been dropped) is of course far more terrible. Their lands have become polluted with nuclear material, and they are condemned to die of malignancy and congenital disease till eternity. Because of the extremely long half-life of uranium-238, the food, the air, and the water in the cradle of civilisation has been forever contaminated. Already, the effects are visible: leukaemia rates have shot up several hundred times, and babies are being born with deformities, including heart defects, cleft lip or palate, Down's syndrome and limb defects on a scale never seen before.[40]

4. Routine Releases from Operation of Nuclear Plants

The process of splitting uranium in nuclear reactors creates more than 200 new, radioactive elements that didn't exist till uranium was fissioned by man. The resulting uranium fuel is a billion times more radioactive than its original radioactive inventory. A regular 1,000 megawatt nuclear power plant contains an amount of long-lived radiation equivalent to that released by the explosion of 1,000 Hiroshima-sized nuclear bombs.[41]

The diabolical elements created in the fission reaction leak out through cracks in the zirconium fuel rods: over time the uranium in the fuel rods swells, due to which pinhole breaks appear in the zirconium cladding; some faulty welds in the zirconium fuel rods also rupture. They now find their way into the environment through a number of ways:

(i) The radioactive isotopes leaking from the fuel rods mix with the primary coolant, that is, the water that cools the reactor core, making it radioactive. The primary coolant in nuclear reactors is constantly taken out for chemical treatment, volume control and to reduce its radioactivity. Most of it is then returned to the primary coolant circuit, and the remaining is kept in holding tanks and then after further treatment is periodically released into the environment.[42] In the USA, nuclear reactors intentionally release about 4000 gallons (15,000 litres) of primary coolant water into the environment every day, while some just leaks out unplanned.[43]

(ii) The thermally hot primary coolant is piped through a steam generator to heat the secondary cooling system. The primary coolant is not supposed to mix with the secondary coolant, but it routinely does (through cracks in the piping). Nuclear utilities in the US admit that about 12 gallons (45 litres) of intensely radioactive primary coolant leaks daily into the secondary coolant via the steam generator through breaks in the pipes. The secondary coolant is converted to steam to drive the turbines. Being at very high pressure, some radioactive steam routinely escapes into the environment from the reactor.[44]

(iii) Apart from mixing with the coolant water, radioactive gases that leak from fuel rods are also routinely released into the atmosphere at every nuclear reactor. This is known as 'venting'. The gases are temporarily stored to allow the short-lived isotopes to decay and then released to the atmosphere through engineered holes in the reactor roof and from the steam generators. The nuclear industry claims that filters are used to remove the most radioactive isotopes, but in reality not all dangerous isotopes are removed and some escape into the environment.[45]

(iv) As we discuss later in greater detail, nuclear plants are inherently prone to accidents. Even if a major accident does not take place, accidental releases of large quantities of radioactive water or gases take place very frequently.

Radioactive Elements Contained in Routine and Accidental Emissions

While the nuclear industry admits that it vents gases from nuclear reactors into the atmosphere, it claims that these gases are harmless as they are noble gases, while the dangerous gases like iodine-131 are removed by filters. In reality, noble gases are high energy gamma emitters and on inhalation, they are readily absorbed into the blood stream from the lungs. Because they are very fat soluble, they tend to locate in the abdominal fat pad and upper thighs, adjacent to the testicles and ovaries. There, they can induce significant mutations in the eggs and sperm of the people living adjacent to a reactor.[46]

More damaging is the fact that many noble gases decay to other more dangerous isotopes, all of which have different actions on the human body. Some of the more dangerous of these isotopes are:[47]

- Xenon-137 with a half-life of 3.9 minutes converts almost immediately to the notoriously dangerous cesium-137 with a half-life of 30 years.
- Krypton-90, half-life of 33 seconds, decays to rubidium-90, half-life of 2.9 minutes, and then to the medically toxic strontium-90, half-life of 28 years.
- Xenon-135 decays to cesium-135 with an incredibly long half-life of 3 million years.
- Large amounts of xenon-133 are released at operating reactors; it has a relatively short half-life of 5.3 days, and so remains radioactive for 106 days.
- Krypton-85 which has a half-life of 10.4 years is a powerful gamma emitter.

Apart from noble gases, small amounts of the deadly radioactive elements created during the fission reaction also escape into the atmosphere fairly frequently during routine emissions from reactors, as they are not entirely trapped by filters. Some of these are:[48]

- Cesium-137 with a half-life of thirty years: it mimics potassium and tends to concentrate in the muscle cells in the body, causing cancer.
- Strontium-90 (half-life of twenty eight years, meaning it remains radioactive for 560 years): the body treats it like calcium and so it concentrates in breast milk and bones, to cause breast cancer and bone cancer years later.
- Iodine-131, half-life of 8 days: it is both a beta and gamma emitter, and hence very carcinogenic; on entering the body,

it circulates in the blood stream and is readily absorbed by the thyroid, to cause the rare thyroid cancer.

Radioactive releases from the Indian Point nuclear power plant in the United States have been polluting the Hudson River. In 2007, strontium-90 was detected in four out of twelve Hudson River fish.[49] A fish taken from the Connecticut River, near the Vermont Yankee nuclear power facility, has also tested positive for strontium-90.[50]

An important toxic isotope that is routinely emitted in large quantities into the air and nearby water bodies from nuclear power plants is tritium (H-3), a radioactive isotope of hydrogen, composed of one proton and two neutrons. Tritium is produced in the fuel rods of nuclear reactors as a fission byproduct. It is also produced in the primary coolant due to interaction of neutrons emitted from the fuel rods with water molecules. Tritium has a half-life of 12.4 years and as such is radioactive for 248 years. H-3 combines readily with oxygen to form tritiated water (H_3O). When the primary coolant water is filtered to remove some of its radioactivity, tritium is not removed, as it is chemically the same as water. Similarly, tritium water vapour is also not trapped by gas filters, and is discharged into the atmosphere during venting. The tritium produced in the reactors thus continuously finds its way into the atmosphere.[51]

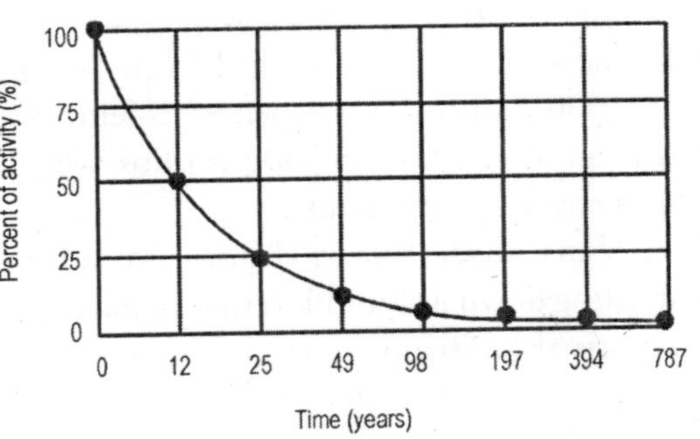

Figure: Decay Curve for Tritium

Tritium is a particularly scary material, as it is a beta emitter and is biologically very mutagenic, being readily absorbed through the skin, lungs and the GI tract. On absorption, it behaves like a water molecule and becomes part of the cell. Tritium causes tumours and cancer in the lungs and GI tract. In animal experiments, even at low doses, it has been shown to shrink the testicles and ovaries, and cause birth defects, ovarian tumours, mental retardation, brain tumours, decreased brain weight, perinatal mortality, and stunted, deformed foetuses.[52]

In the US, tritium releases have occurred quite frequently at nuclear reactors, due to leaks caused by malfunctions. In a recent report released in September 2010, the US NRC has acknowledged that more than half of America's 65 nuclear sites housing its 104 aging atomic reactors are leaking radioactive tritium. The US Environmental Protection Agency's 'allowable' standard ('allowable' does not necessarily mean 'safe') for radioactive tritium in drinking water is 740 becquerels per litre of water. According to the NRC, since January 2009, that level has been met or exceeded by releases into groundwater (not necessarily drinking water) at 37 reactor sites (out of 65). Radiation levels have ranged from 740 Bq/litre to an astonishing 555,000 Bq/litre (at New Jersey's Salem reactor complex). Radioactive tritium levels above 37,000 Bq/litre were measured at nine nuclear sites covering 18 reactors.[53]

In Canada, Ontario Hydro revealed in July 1997, a month before shutting down its four Pickering 'A' nuclear reactors, that it had hidden reports about tritium contamination at the Pickering site for the last twenty years. It revealed that it had found tritium levels in the groundwater to be as high as 700,000 Bq/litre in 1994.[54]

Leakages Due to Radioactive Corrosion

Apart from being created during the fission reaction, radioactive products are also created in another way in the nuclear reactor: due to bombardment of the metal piping and the reactor containment by neutrons. This is known as radioactive corrosion. The elements thus created, which are powerfully radioactive, include cobalt-60, iron-55, nickel-63, radioactive manganese, niobium, zinc and chromium. These materials slough off from the pipes into the primary coolant. Officially called CRUD, it is so intensely radioactive that it poses a severe hazard to maintenance workers and inspectors. During shutdowns of nuclear reactors for maintenance or refuelling, pipes, heat exchangers, et cetera are routinely flushed to remove the highly radioactive CRUD build-up. Some of the CRUD is sent to radioactive waste dumps while some is released into the river, lake or sea near the reactor.[55]

To Sum-up...

Although the nuclear industry claims it is 'emission' free, in reality it is collectively releasing millions of curies annually. The total gaseous and liquid radioactive releases from nuclear reactors vary enormously depending upon accidental and larger-than-normal routine releases. For instance, in 1974, the total release from all reactors in the United States was 6.48 million curies, while in 1993 it ranged between 96,600 to 214,000 curies.[56]

Even these astounding figures are an underestimate, because not all the emissions are monitored by utilities. These figures also do not include the emissions due to the CRUD removed from reactors or the emissions due to the radioactive elements trapped in the primary coolant filters and gas filters—these filters are sent to waste dumps, from where the carcinogenic radioactive isotopes will inevitably leak and contaminate water supplies and food chains.

Impact on Human Life

The routine emission and accidental leakages of radiation from nuclear plants obviously means that there must be an increased incidence of cancer and other diseases in the people living around nuclear power plants. However, there have been very few studies on this issue; the general stand of the environmental authorities, like for instance the US Environmental Protection Agency (EPA) and the US Food and Drug Administration (FDA), has been that these leakages are no cause for panic as radioactive substances pose a threat to our health only when consumed in large quantities![57]

The few studies that have been done on this issue have however come up with alarming findings. A study by researchers at the prestigious Medical University of South Carolina, USA, who carried out a sophisticated meta-analysis of 17 research papers covering 136 nuclear sites in the United Kingdom, Canada, France, United States, Germany, Japan and Spain found evidence of elevated leukaemia rates among children and young people living near nuclear facilities.[58] Elevated leukaemia rates among children were also found in a recent study that examined areas around all 16 major nuclear power plants in Germany.[59] A Canadian federal government study found high rates of Down's Syndrome in Pickering and Ajax, two communities near the Pickering nuclear generating station.[60]

5. Radioactive Waste: Leaking Everywhere

Probably the most monstrous problem created by nuclear power is that of spent fuel, which is intensely radioactive, and is going to remain so for more than two lakh years! Each regular 1,000 MW nuclear power plant generates 30 tons of extremely potent radioactive waste annually.[61] Even though nuclear power plants have been in operation for more than fifty years now, mankind has not yet found a way of safely disposing of the massive amounts of lethal waste that these plants have continued to produce at an ever increasing pace and, scientifically speaking, considering the intensely radioactive and chemically corrosive nature of this waste, it is not possible to find such a permanent storage system.

Since there is no way of removing the radioactive nature of these wastes, therefore, these wastes will have to be stored in temporary storage sites and constantly monitored to see that they don't leak. For tens of thousands of years! This is clearly iniquitous to future generations since they would have to take care of these wastes and bear the consequences for any leakage, while we use the electricity generated by these reactors. Ethical dilemmas aside, no technology that generates such long lived radioactive wastes can be considered environmentally sustainable.

Radioactive waste from nuclear power plants is currently stored in huge cooling pools, euphemistically called 'swimming pools', at reactor sites, or in dry storage casks beside the reactor. Many countries have other temporary storage sites too. In the US, nuclear waste is currently stored at 121 temporary locations in 39 states across the country. Everywhere, this exceedingly toxic waste is leaking, leaching, seeping through the soil into aquifers, rivers, lakes and seas, to ultimately enter the bodies of plants, fish, animals and humans.[62] Its consequences are going to be with us for the rest of time.

To give an idea of the deathly nature of this radioactive waste, we briefly discuss the health impact on human beings of plutonium, just one of the more than 200 elements created directly or indirectly during the fission process and which are present in this waste. Roughly one per cent of the spent fuel created in a uranium-fuelled reactor consists of plutonium.[63]

Plutonium[64]

Named after Pluto, the Greek God of Hell, it is supposed to be one of the most dangerous substances on earth. An alpha emitter, it is so toxic and carcinogenic that less than one-millionth of a gram if inhaled will cause lung cancer. The half ton of plutonium released from the Chernobyl meltdown is theoretically enough to kill everyone on earth with lung cancer 1,100 times if it were to be uniformly distributed into the lung of every human being.

On being transported to the liver, plutonium causes liver cancer. Because it is like iron, on inhalation, it eventually finds its way to the bone marrow to be incorporated into the haemoglobin molecule in the red blood cells. Here it irradiates bone cells to cause bone cancer and white blood cells made in the bone marrow to cause leukaemia. Plutonium is also stored in the testicles, and causes mutations in reproductive genes and increases the incidence of genetic disease in future generations.

The half-life of plutonium-239 is 24,400 years, so it remains radioactive for half a million years. Therefore, once created, it is going to live on and cause cancer and genetic mutations for the rest of time.

Storing the Waste

Spent fuel is highly radioactive and chemically active, and intensely hot. If stored in a repository, the temperature inside the repository will be above boiling point for 1250 years, with temperatures inside the canister holding the waste touching 662 degrees Fahrenheit.[65] There is no way of guaranteeing that any storage system designed for such waste will not corrode and leak a few hundred years from now. All attempts to build even medium term storage systems for this waste have ended in complete disasters. We look at the attempts in the US and Germany below.

The Yucca Mountain Disaster

All told, the nuclear reactors in the United States produce more than

2,000 metric tons of radioactive waste a year—and most of it ends up sitting on-site because there is nowhere else to put it. As of 2008, more than 64,000 tons of dangerously radioactive waste from nuclear power reactors had accumulated in the United States.[66]

The private nuclear industry in the US has never taken the responsibility for disposing of the massive quantities of radioactive waste produced by it. Obligingly, the US government passed the Nuclear Waste Policy Act in 1982, promising to take responsibility for it, and in 1987, designated the Yucca Mountain in Nevada as the primary repository for this waste. The project was envisaged as a complex of tunnels deep inside Yucca Mountain, 100 miles northwest of Las Vegas, where at least 77,000 tons of spent fuel from commercial plants, and government-generated nuclear waste, would be stored and ultimately buried. The site soon ran into huge problems; it also became clear that the site's geology was inappropriate to contain the waste. However, the US Department of Energy (DOE) was desperate to somehow make the site qualify for storing the waste; it even went to the extent of relaxing its norms. Despite its best efforts, including incurring costs to the tune of $13.5 billion, the project fell more than a decade beyond schedule because of a series of management missteps, legal challenges and budget cuts engineered by opponents in Congress. Finally, in 2010, President Obama decided to cancel the project, and set up a 15-member panel of experts to chart new paths to manage highly radioactive nuclear waste. Let's see what solutions they come up with.[67]

Germany's Model Storage System

The German government has invested several hundred million euros in research at the Asse nuclear storage facility in Lower Saxony, Germany in an attempt to solve the long-term waste storage dilemma of the nuclear energy industry. Asse is actually an abandoned salt mine; more than a hundred thousand barrels of low-level and medium-level waste has been stored in this facility since the 1960s. There are suspicions that highly radioactive waste has also been dumped in this facility without authorisation. Now, it turns out that water has been seeping into the mines for decades, rendering the facility in a precarious condition and in danger of collapsing. Authorities are now making an unprecedented attempt to retrieve and relocate hundreds of tons of waste from the site, something that has never been attempted anywhere in the world.

Asse was to serve as a model project for Germany's plans to build a permanent nuclear waste storage facility at Gorleben, also in Lower Saxony. Not only has the plan collapsed, the German Environment Minister Sigmar Gabriel has admitted that the Asse site was 'the most problematic nuclear facility in all of Europe.'[68]

A sobering thought …

Bittu Sahgal, editor of *Sanctuary Magazine*, Mumbai writes: 'If the centralised bureaucracy of Maurya Kings two thousand years ago had discovered nuclear power, we in India and Pakistan would probably be spending half our current national budget storing and caring for or repairing the damage done by atomic wastes.' Indeed!

6. Reprocessing Nuclear Waste: Worsening the Problem

Currently, six countries with nuclear reactors, China, France, India, Japan, Russia and the United Kingdom, reprocess at least some of their spent fuel. Some of the other countries with a nuclear power program have also been shipping their spent fuel to France, Russia or the UK for reprocessing, though most of them have now ended this practice.[69]

Supporters of reprocessing argue that it reduces the nuclear waste problem by segregating out the high level radioactive waste—only this reduced volume now needs to be stored for thousands of years. However, decades of experience from reprocessing plants wherever they have been built in the world provides overwhelming evidence that not only is this argument illogical, reprocessing actually worsens the problems created by nuclear energy.

Firstly, reprocessing does not reduce the total amount of radioactivity to be dealt with. On the contrary, it increases the total volume of waste to be dealt with because reprocessing additionally creates a large amount of low-level and intermediate-level radioactive waste as all the equipment used in reprocessing becomes radioactive.

According to US DOE data, reprocessing increases the total volume of radioactive waste by a factor of seven![70]

Neither does reprocessing reduce the waste disposal costs. The general consensus based on cost data from Western countries is that reprocessing as a waste management technique is far more expensive than direct disposal. A study done for the French Prime Minister in 2000 estimated that reprocessing and plutonium recycle increases the cost of nuclear power by about 0.2 US cents/kWh (assuming 5 francs/dollar).[71] This is primarily because of the enormous capital cost of the reprocessing facility.

Another major problem with reprocessing is that since it segregates plutonium from the spent fuel, and this pure plutonium can be used for making nuclear weapons, reprocessing increases nuclear proliferation and nuclear terrorism risks. In 1976, US stopped the reprocessing of spent fuel because of proliferation concerns. In 2006, the Bush administration announced a new plan to reprocess spent nuclear fuel in a way that renders the plutonium in it usable for nuclear fuel but not for nuclear weapons, but the Obama administration scrapped these plans in July 2009.[72]

Finally, reprocessing plants discharge huge quantities of radioactive waste into the sea and air. The reprocessing plants in Sellafield in UK and La Hague in France are the biggest source of radioactive pollution in Europe. Radioactive contamination of the seas from these plants can be traced to as far away as the Arctic and eastern Canada.

Radioactive Discharges from Sellafield

This nuclear complex on the coast of north-west England has reprocessing facilities, fuel fabrication and other installations. It has one of the highest concentrations of radioactive waste on the planet as well as a disastrous safety record with hundreds of accidents involving the release of radioactive substances into the environment and their radiation of workers.[73]

The reprocessing plants at Sellafield discharge some eight million litres of nuclear waste into the sea each day.[74] It is estimated that over 40,000 TBq (trillion becquerels) of cesium-137, 113,000 Tbq of beta emitters and 1,600 TBq of alpha emitters have been discharged into the Irish Sea since the inception of reprocessing at Sellafield. Between 250 and 500 kilograms of plutonium from Sellafield is now adsorbed on sediments on the bed of the Irish Sea. Discharges of the noxious technetium-99 (half-life 214,000 years) into the sea have also been very high.[75] Radioactive contamination from the Sellafield plant is now reported to have extended through the Arctic Ocean into the waters of northern Canada.[76]

Radioactive pollution from the Sellafield plant has made the Irish Sea one of the most radioactively contaminated seas in the world. Marine life, in particular algae, plankton, and crustaceans including lobsters, have absorbed significant amounts of radionuclides, in many cases exceeding seafood safety levels by many times. In the vicinity of the complex, groundwater, estuaries and soil are contaminated, with levels in the area around Sellafield exceeding contamination levels inside the Chernobyl exclusion zone.[77]

The effects are visible in the local population. Compared to the British average, there has been a ten-fold increase of childhood leukaemia and non-Hodgkin's lymphoma around Sellafield. Plutonium dust has been found in the houses of residents living along the Irish Sea coast.[78]

7. France: A Radioactive Mess

Let us discuss the situation in France in greater detail, because it is the most nuclear powered country in the world. It derives over 75 per cent of its electricity from nuclear energy, a result of a political decision taken by the government in 1974 after the first oil shock to rapidly expand the country's nuclear power capacity.

France's monopolistic dependence on splitting the atom to turn on the lights has come with a huge price: the country has become a radioactive mess, for which the people of France will pay with their health for many centuries to come. As awareness has spread, polls in the last few years have consistently found more than 60 per cent of French people favouring a phase out of nuclear energy.[79]

Uranium Mining Waste Dumpsite

For decades, France was the largest uranium producer in Western Europe; the last uranium mine in the country was shut down in May 2001. 300 million tons of radioactive waste is lying at 210 abandoned uranium mining sites in France without any protective measures or surveillance.[80] This radioactive waste has been used in school playgrounds and ski-resort parking lots. Water contamination from uranium waste threatens French agriculture. Efforts to force France's state run uranium mining corporation Areva to clean up this mess have been met with resistance from the company.[81]

Areva's mining waste footprint extends beyond the borders of France, to Niger. Four decades of uranium mining by Areva's various subsidiaries have contaminated the entire region around the uranium mines of Niger with radioactive dust—dust that is going to remain radioactive for centuries. It has poisoned the air, contaminated the soil, and has even polluted the already scarce water sources in this desert region. This is having catastrophic consequences on the health and livelihoods of the local people.[82]

Radioactive Leakages

The French nuclear reactors have been plagued by thousands of nuclear accidents, euphemistically dubbed as 'events' by the IAEA—between 1986 and 2006, more than 10,000 events were reported. In recent years, the number of safety related 'events' at French nuclear reactors have increased: they have increased steadily from 7.1 per reactor per year in 2000 to 10.93 in 2009. The French nuclear power plant operator, EDF, used to stress that the number of more serious events was on the decline; this too is no longer true. The number of relatively more serious events, which are serious enough to be given a rating on the INES scale, has increased during the period 2005-9.[83]

Several of these accidents have led to radioactive leakages contaminating the environment, including the soil, air and nearby water bodies. For instance, recently, in a major accident, a massive uranium spill occurred at the Tricastin nuclear complex in 2008, contaminating the nearby Gaffière and Lauzon Rivers. While Areva, France's nuclear corporation, denied the spill endangered human health, nevertheless, French authorities banned the use of water from the Gaffière and Lauzon for drinking and watering of crops for two weeks. Swimming, water sports and fishing were also banned. Tricastin wine growers have struggled to market their products since the accident.[84] A week later, nuclear safety officials discovered a burst underground pipe at a plant in Romans-sur-Isere, southeastern France, run by an Areva subsidiary; the pipe had been broken and leaking uranium for several years and didn't meet safety standards.[85]

'No Leakage' Storage Systems[86]

The Centre de Stockage de La Manche (CSM), in Normandy, France, which contains a massive 520,000 cubic meters of nuclear waste, is one of the largest dump sites of nuclear waste in the world. Dumping here started way back in 1969 and continued for 25 years till its closure in 1994. Even though the site was designed to contain low level waste, a government appointed commission found that the site contains unknown quantities of high level waste too. And now, it has been found that the radioactive waste from the storage facility is leaking into the groundwater used by local farmers. In 2005, scientists from a French laboratory found that radioactivity levels in underground aquifers close to the dumpsite averaged 9000 Bq/l, or 90 times above the European safety limit of 100 Bq/l ! The situation is bound to get worse in the coming years, as more hazardous radionuclides, including plutonium and strontium, gradually leak out.

After closure of CSM in 1994, the so-called low and intermediate level waste was transported to another dumpsite, the Centre de Stockage de l'Aube (CSA) in the Champagne-Ardenne region. This was claimed to be a state-of-the-art facility, and when the site was being built in the 1980s, people of the area were assured that there would be no radioactive leakage from the site. By 2002, the site had already received over 100,000 cubic meters of nuclear waste. Soon after, in 2006, it was found that the storage site had started leaking. Tritium leaking from CSA was found in underground water! And this is just the beginning; gradually, more deadly radionuclides are bound to find their way into the groundwater in the coming years. Moreover, in an attempt to make it a high quality nuclear dump site, the storage site has been designed in such a way that no corrective measures can be taken! The people of this region and all their future generations are condemned to live with this radioactivity and all its consequences till eternity.

La Hague Reprocessing Plant: Polluting the Planet

The dirtiest French nuclear site—with the cleanest of reputations—is the vast reprocessing plant at La Hague on the Normandy coast. It is the world's number one plant, reprocessing more than 1700 tons of highly irradiated nuclear fuel rods annually. The plant produces huge volumes of liquid radioactive waste and radioactive gases. These are simply dispersed into the sea and air.

The plant has been pumping as much as 370 million litres of liquid radioactive waste every year into the English Channel! That's unbelievable, but even more shocking is that it has been going on since the 1980s! It has radioactively contaminated the seas as far as the Arctic Circle. These liquid wastes have been measured at 17 million times more radioactive than normal sea water according to an analysis by a French laboratory at the University of Breme.[87] The dumping of this low-level waste into the sea clearly violates the 1992 OSPAR (Oslo-Paris) Convention for the Protection of the Marine Environment of the North-East Atlantic, signed by 15 European countries and also the European Community. Despite concerns voiced by many European governments, Areva has continued to dump the waste into the sea, taking advantage of a technical loophole in the Convention.[88]

The site is also one of the world's worst radioactive air polluters. Aerial discharges from La Hague have been found to contain radioactive krypton-85 at 90,000 times higher values than natural levels. Krypton gas released from La Hague has been traced across the globe.

An analysis of samples of leaves and grasses from around La Hague has revealed that radioactive carbon-14, which had been taken up by the plants in gaseous form, is two to seven times higher than normal background levels.[89]

It is therefore no surprise that health impacts of these radiation releases are showing up in the local population. The *British Medical*

Journal published findings of increased levels of leukaemia around La Hague in 1997, findings that were confirmed by the French government's own studies made later in the same year. Subsequently, another study published in the July 2001 issue of the *Journal of Epidemiology and Community Health* also showed an increased incidence of leukaemia among under-25 year olds living within 10 kilometres of the plant.[90]

Part IV: Impact of Nuclear Reactors on Marine Life

Fifty nine out of 103 nuclear plants in the US rely on what are known as 'once through cooling systems' to remove waste heat. In a nuclear plant with a 'once through cooling system', there are a series of three heat exchanging loops of water:

i) The water in the first loop carries the heat from the reactor core to the steam generator, where it is transferred to the secondary loop.
ii) The water in the secondary loop is at lower pressure than the first loop, and is allowed to get converted to steam, which drives the turbine to generate electricity.
iii) After the steam passes through the turbine, it flows over pipes containing cold water from the river/sea. These pipes constitute the third loop. The contact of steam with this third loop causes the steam to condense into water. It is then pumped back to the steam generator to repeat the process. The water in the third loop, which was sucked from the river or the sea, is dumped back into the same source. In a typical nuclear plant in the US, the water circulating in the third loop can be many billions of litres a day!

Nuclear plant authorities have always claimed that their intake and discharge of billions of litres of water a day did very little harm to the surrounding marine life.

Some years ago, a major report, *Licensed To Kill: How the nuclear power industry destroys endangered marine wildlife and ocean habitat to save money*, released by the well-respected Nuclear Information and Resource Service on February 22, 2001,[91] brought out in devastating detail the impact of these 'once through cooling systems' on marine life. These cooling systems suck in and discharge as much as four million litres of water per minute. This water is sucked in at such a high velocity that along with the water, marine life is also sucked in; it is unable to resist the velocity. The bigger marine animals like the endangered sea turtles impinge on 'prevention devices' such as screens and barrier nets, and either drown or suffocate. While billions of smaller organisms, including small fish, fish larvae, and spawn, all very essential to the food chain, pass through these screens, and are drawn into the reactor cooling system where 95 per cent of them are scalded and discharged back into the water body as lifeless sediment. These high destruction rates can overtake recovery rates, resulting in extensive depletion of the affected species. In this way, entire marine communities can lose their capacity to sustain themselves.

With millions of litres of hot water being discharged into the waterway every minute, the total heat dumped into the waterway is tremendous. How much? Roger Witherspoon, the well-known US journalist, author and editor, in a recent article has given some figures. Citing company records, he points out that the nuclear power plants at Salem, New Jersey, USA, dump about 30 billion BTUs of heat hourly into Delaware Bay. That is the equivalent of the heat which would be generated by exploding a nuclear bomb, the size of the bomb which destroyed Hiroshima, in the waters of Delaware Bay every two hours, all day, every day.[92]

Such a huge hot water discharge damages and destroys fish and other marine life and dramatically alters the immediate marine environment. In the immediate discharge areas, the ocean floor is scoured clean of sediment by the force of the thermal discharge, resulting in bare rock and creating a virtual marine desert. Areas farther away from the discharge become coated in heavy, life-stifling sediment. The water is discharged into the waterway at a temperature around 10-13 degrees Celsius higher than before. The increased temperature drives away indigenous species and attracts others, thus causing huge changes in the marine environment. Warmer waters also cause fatal disease, and disruption of normal behaviour patterns, of some species of fish.

An analysis by the US Environmental Protection Agency has confirmed these findings. It has concluded that power plants with 'once through cooling systems' are collectively killing more than one trillion fish annually and disrupting the local aquatic ecosystems with their hot water discharges. More significantly, the National Marine Fisheries Service (NMFS) has also now publicly acknowledged that 'once through cooling systems' are vacuuming up trillions of newly hatched fish—those under half an inch in length—and destroying them in their heat exchangers. The NMFS has in fact gone so far as to state that there is 'strong evidence' that the decline in fish stocks along the entire northeast Atlantic seaboard is due more to the destruction of baby fish than to overfishing of adults.

Despite this overwhelming evidence, such is the clout of the nuclear power companies that environmental and nuclear regulators have not moved to force nuclear plants using 'once through cooling systems' to retrofit their plants with modern cooling systems which won't kill billions of fish annually. A variety of such systems are there, ranging from the mechanical draft system to cooling towers. The mechanical draft is cheaper and reduces mortality by 95 per cent, while cooling towers are more expensive but eliminate 98-100 per cent of fish mortality. Cooling towers also eliminate the need to discharge large volumes of heated water into the water source and the resulting damage to the marine environment in the discharge area. It costs just about $1.5 billion to build a cooling tower, which though large in absolute terms represents only a small percentage of the profits of these utilities.

Last year (2010), for the first time in the US, environmental authorities of two states, New Jersey and New York, acknowledged that nuclear plants (and also coal power plants) with 'once through cooling systems' are killing billions of juvenile and mature fish annually, and that this was negatively impacting a wide variety of fish. They are now mounting pressure on such utilities in their states to retrofit their plants with modern cooling systems which won't kill fish, or cease operations.[93]

Part V: The Inevitability of Nuclear Accidents

The fission reaction produces such a deadly concoction of radioactive elements that long-lived radiation contained within the reactor of a 1000 MW nuclear power plant is equivalent to that of 1000 Hiroshima bombs! As discussed above, some of this will inevitably find its way into the environment and contaminate it, with all its terrible consequences. However, what if an accident in the nuclear reactor releases a significant part of these deadly radioactive elements into the environment in one go? It has happened before. Not once, but quite a few times. We discuss the two biggest such accidents (till 2010) below, the Three Mile Island meltdown of 1979 and the Chernobyl disaster of 1986. Then we go on to discuss the performance of the nuclear industry in the 24 years since Chernobyl (that is, till before the Fukushima accident), to see if it has become any safer.

Three Mile Island

Beginning at 4 am on March 28, 1979, a relatively minor problem in the Unit 2 reactor of the Three Mile Island (TMI) nuclear power plant in Pennsylvania quickly cascaded into a series of automated events that led to the meltdown of the reactor core.

America's private nuclear power industry and government authorities continue to maintain, to this day, that despite the meltdown of almost half of the uranium fuel in the reactor core, there were only minimal releases of radiation to the environment. According to the NRC, in the month after the TMI accident, a total of around 13 million curies of radioactive gases were released into the atmosphere, most of which were noble gases—which the government claims are harmless—and only 13 to 17 curies of it was the dangerous iodine-131. Strangely, it makes this claim despite the fact that government authorities (including the NRC) and the industry made no serious attempt to monitor the radioactive releases after the meltdown. To date, no survey has ever been made by them of the air and soil around the plant for long-lived highly radioactive elements like strontium, americium and plutonium, despite the fact that these must have most definitely escaped from the reactor core due to the meltdown.[94] A fact sheet distributed by the NRC says: TMI 'led to no deaths or injuries to plant workers or members of the nearby community.'[95]

Child Victims of Chernobyl

Natasha (12) was born with microcephaly, Vadim (8) has a bone disease and is mentally retarded.

Nastya, from Belarus, has cancer of the uterus and lungs.

Sasha, one of the five million children living in the Chernobyl fallout areas of Belarus, Russia and Ukraine.

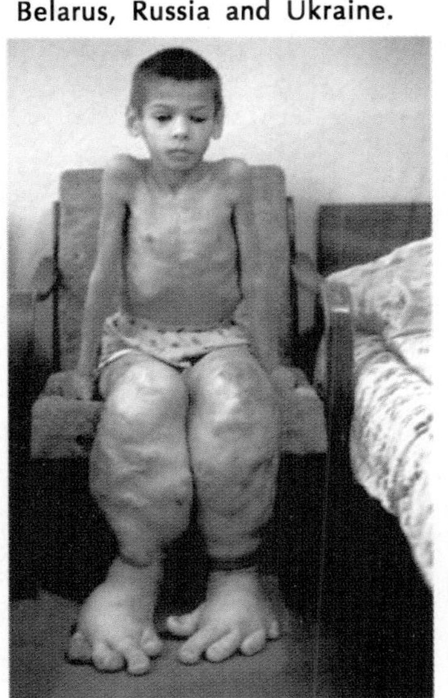

Alexandra (9), from Belarus, has a birth defect called hydrocephalus.

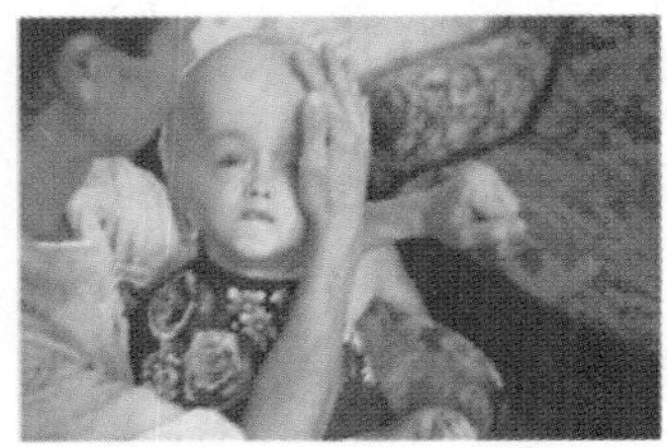

Annya has a cancerous brain tumour since the age of 4.

CHILD VICTIMS OF JADUGUDA URANIUM MINES

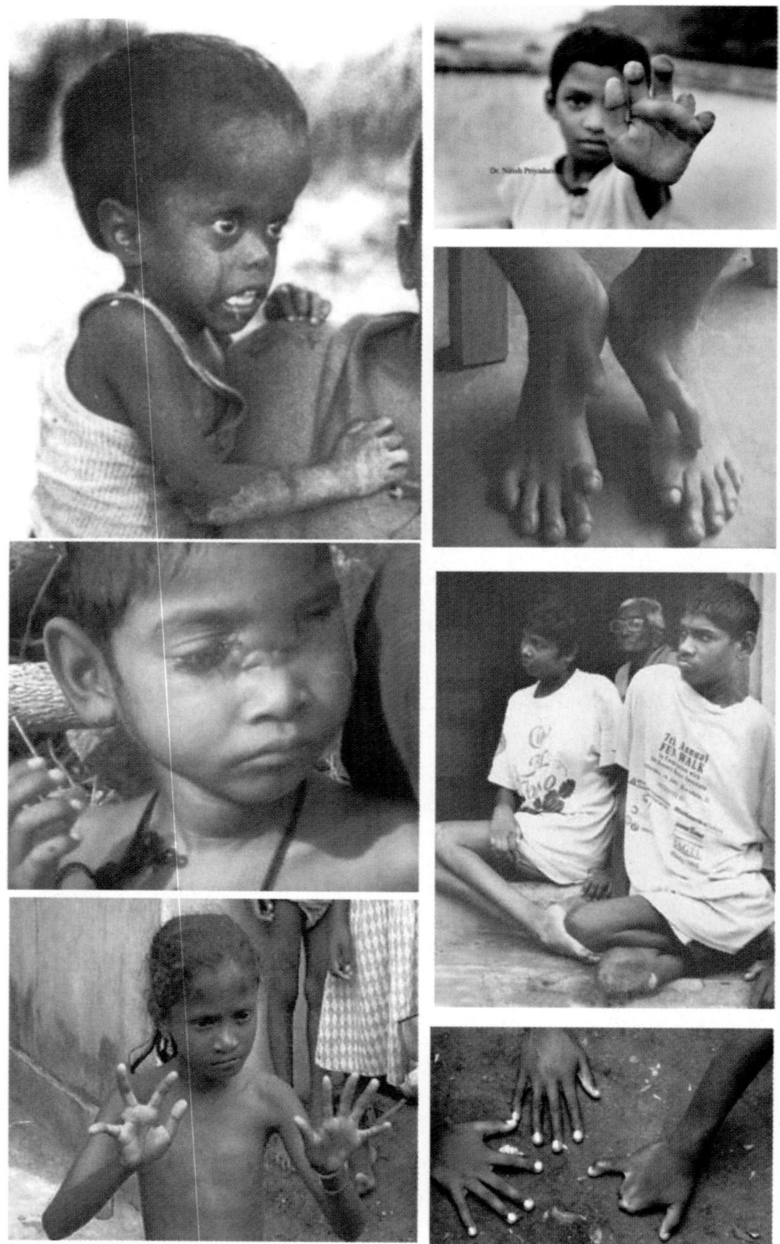

Rawatbhata Nuclear Plant, Health Costs

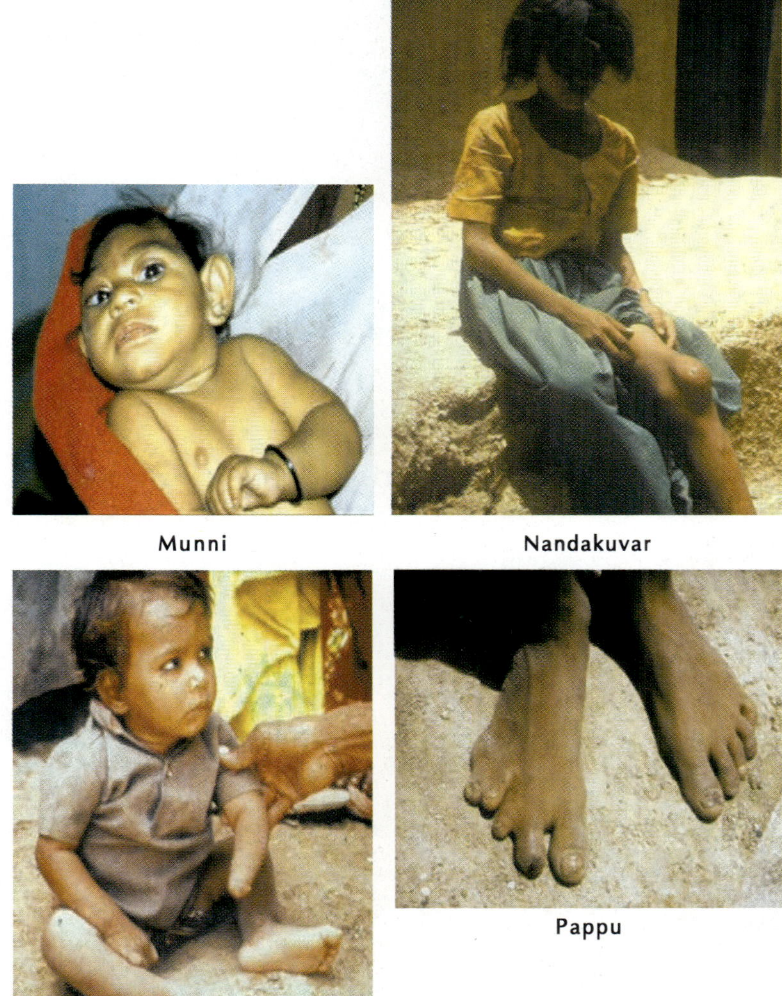

Munni

Nandakuvar

Pappu

Banwari

The Growing Worldwide Anti-nuclear Movement

October 3, 2009: Protesters from All Over Europe at an Anti-nuclear Demonstration in Colmar, France

April 10, 2011 : Anti-nuclear Protestors March in the Streets of Tokyo, Japan

THE GROWING WORLDWIDE ANTI-NUCLEAR MOVEMENT

June 11, 2011 : 30 Anti-Nuclear Groups Demonstrate in Hong Kong Against Daya Bay Nuclear Power Plant

March 26, 2011: Anti-nuclear Demonstration, Cologne; 200,000 People March in Germany Against Nuclear Power

The Growing Worldwide Anti-nuclear Movement

April 30, 2011: Thousands Protest in Taiwan, Demanding Closure of New Nuclear Plants and a More Sustainable Energy Policy

April 19, 2011 : South Korean Mothers Stage Anti-nuclear Energy Rally in Seoul, South Korea

ANTI-NUCLEAR PROTESTS, INDIA

March 26, 2011 : Kudankulam

March 25, 2011 : Anti-Nuclear March to Parliament

Anti-nuclear Protests, India

April 23, 2011 : Tarapur-Jaitapur Yatra

December 4, 2010 : Jaitapur

We know today for a fact that the nuclear industry and the government are lying outright. There is a growing body of scientific evidence presented by numerous independent experts which suggests that radiation releases from the TMI plant due to the meltdown were much higher than what the government and industry have acknowledged.

According to estimates by Dr Karl Morgan (published in 1982), the Three Mile Island accident released at least 45 million curies of noble gases and 64,000 curies of radioactive iodine. Dr Karl Morgan is acclaimed as the founder of Health Physics, the science of human health and radiation exposure. It is also known that on day three of the accident, 172,000 cubic feet of high-level radioactive water was released into the Susquehanna River by the nuclear plant company without NRC permission, an event unheard of in the history of the US nuclear industry. The Susquehanna River drains into Chesapeake Bay, a major fishing location. Then, in June 1980, large quantities of radioactive krypton-85 were purposefully vented from the damaged reactor. Again in November 1990, 8.7 million litres of radioactive water containing tritium was purposefully evaporated from the damaged reactor building.[96]

In March 2009, at a symposium in Harrisburg (capital of the state of Pennsylvania) to mark the 30th anniversary of the TMI disaster, independent experts presented yet more evidence to prove the official story—that minimal radiation escaped from the TMI meltdown—to be a big lie. For instance, Arnie Gundersen, a nuclear engineer and long-time nuclear industry executive, stated that it could be deduced from industry data that at least one billion curies of radiation were released—more than 75 times the official figure. More serious is the finding by these independent experts that hidden in a government commission's technical report is the admission that at least one million curies of iodine-131 was released into the atmosphere—that is way off from the official admission of just 13-17 curies![97]

The experiences of people living near Three Mile Island also do not match with official claims that the meltdown did not have any health effects on the people living near the plant. On the contrary, their acute sufferings further prove that they were exposed to high levels of radiation. Surveys of nearby residents have revealed a huge increase in cancers, including lung cancer and leukaemia, skin and reproductive problems, organ collapse and chromosomal damage—all associated with high levels of radiation exposure.[98]

The Three Mile Island case eventually went to court, with approximately 2,000 residents claiming that the radioactive releases from the meltdown were much larger than admitted by the nuclear industry and government. After several dismissals and appeals, the sick plaintiffs gave up and decided to settle for out-of-court settlements. They got hardly anything, the largest settlement being $1.1 million for a child born with Down's syndrome.[99]

This is the real reason why the nuclear industry and its concubine governments refuse to admit to the release of radiation from nuclear power plants, even when there is a major accident: so that compensation payments to the affected people are minimised!

Chernobyl

On April 26, 1986, Unit 4 of the Chernobyl nuclear power plant exploded, spewing almost a quarter of the deadly radioactive fission products in its reactor core into the environment.[100] This medical catastrophe will continue to plague much of Russia, Belarus, the Ukraine and Europe for the rest of time.

To this day, international institutions dealing with nuclear energy and the various agencies of United Nations maintain a conspiracy of silence over the true effects of Chernobyl on human life on Earth. The World Health Organisation does not independently research the health consequences emanating from nuclear accidents. It signed an agreement with the International Atomic Energy Agency (IAEA) in 1959 subordinating itself to the latter in matters of interest to the IAEA. Dr Michael Fernex, formerly on the faculty of the University of Basel, who worked with the WHO, said in 2004, 'Six years ago we tried to have a conference. The proceedings were never published. This is because in this matter the organisations at the UN are subordinate to the IAEA ... The IAEA blocked the proceedings; the truth would have been a disaster for the nuclear industry.'[101]

The IAEA is a cheerleader for the global nuclear industry, and has been actively promoting the construction of nuclear reactors worldwide. Obviously then, the IAEA would seek to obfuscate the true magnitude of the Chernobyl disaster. In September 2005, the IAEA and the WHO released the draft of a study by the UN Chernobyl Forum. The most important figures of this study were: just under 50 dead; 4,000 curable cases of thyroid cancer; no proof for an increase in miscarriages and sterility or leukaemia and other forms of cancer in relation to the reactor accident; total number of future deaths as a result of the disaster could possibly reach a maximum of 4000 people. The IAEA declared: the Chernobyl case is closed.[102]

Let us compare these 'official' figures with some of the medical and ecological consequences of Chernobyl known today[103]:

- Nearly nine million people living in Belarus, Ukraine and Russia were heavily exposed to radiation. In all, 20 per cent of the land area of Belarus, 8 per cent of the Ukraine, and 0.5 to 1 per cent of Russia—100,000 square miles—was heavily contaminated, an area slightly less than the area of the state of Maharashtra. It will remain so for thousands of years. Agricultural areas covering nearly 52,000 square kilometers, slightly less than the area of the state of Himachal Pradesh, were ruined.
- While 400,000 people living in the most contaminated areas near the plant, in a perimeter of 30 km around the plant, were evacuated and resettled elsewhere, more than 5 million people continue to live in the affected regions, over 1 million of whom are children, who are inordinately sensitive to radiation. They now live with the knowledge that they and their coming generations are forever contaminated, that they could get cancer anytime and that their children and grandchildren and great grand children could be born with severe birth defects.
- Several independent studies have shown a many fold increase in leukaemia, brain tumours and other forms of cancer in the population of the affected regions in Belarus, Ukraine and Russia. Till 2005, there were at least 10,000 cases of thyroid cancer in Belarus alone, of which nearly a 1000 were in children—thyroid cancer in children is a very rare form of

cancer. Many of the genetic abnormalities and diseases caused by this accident are generations away and will not be seen by anyone alive today.

- Heavy radioactive fallout occurred over Austria, Bulgaria, Czechoslovakia, Finland, France, East and West Germany, Hungary, Italy, Norway, Poland, Romania, Sweden, Switzerland, Turkey, Britain, the Baltic States, and Yugoslavia. Evidence has already started surfacing of its health effects. There have been at least 10,000 additional cases of serious malformation in Europe due to Chernobyl. A recent study from Sweden showed an increase of 849 cases of cancers up to the year 1996 as a result of Chernobyl. There are now claims surfacing in France that people are suffering from thyroid cancer that may be related to Chernobyl fallout.
- Because cesium-137 and other isotopes such as strontium-90 and plutonium-239 have such long half-lives, food in contaminated parts of Europe will be radioactive for hundreds of years. In Britain, 1,500 miles from the crippled reactor, 382 farms containing 226,500 sheep are severely restricted because the levels of cesium-137 in the meat are too high. In the south of Germany, very high levels of cesium persist in the soil; hunters are compensated for catching contaminated animals, and many mushrooms and wild berries are still too radioactive to eat. Cesium-137 in some parts of France is as high as some extremely contaminated areas in Belarus, the Ukraine and Russia.

Even though, as the data above shows, food in many parts of Europe is still relatively radioactive, this terrible problem is rarely mentioned in the media or in daily conversation. In a form of psychic numbing, people continue to live their lives as if all were well, and the nuclear power industry continues to broadcast the myth that its product is clean and green.

A New Study on Chernobyl

Just as we were finishing this book, we came across a new study *Chernobyl: Consequences of the Catastrophe for People and the Environment*, published by the New York Academy of Sciences in

2009.[104] The book was originally published in Russian in 2007, and is authored by Dr Alexey Yablokov of the Center for Russian Environmental Policy in Moscow and a former environmental advisor to the Russian president, late Prof. Vassily B. Nesterenko, who was the director of the Institute of Nuclear Energy of the National Academy of Sciences of Belarus at the time of the Chernobyl accident, and Dr Alexey Nesterenko, a biologist and ecologist with the Institute of Radiation Safety, Belarus. The authors examined over 1,000 published scientific articles and over 5,000 Internet and printed publications, mainly in Slavic languages, and never before available in English.

The book is edited by Dr Janette D. Sherman-Nevinger, a toxicologist expert in the health impacts of radioactivity and presently working with the Environmental Institute, Western Michigan University, Kalamazoo, Michigan. Reflecting on her experiences while editing the book, Dr Sherman says: 'Every single system that was studied—whether human or wolves or livestock or fish or trees or mushrooms or bacteria—all were changed, some of them irreversibly. The scope of the damage is stunning.' The 327 page volume has a foreword by Dr Dimitro Grodzinsky, chairman of the Ukranian National Commission on Radiation Protection. He writes about how 'apologists of nuclear power' sought to hide the real impacts of the Chernobyl disaster from the time when the accident occurred. According to him, the book 'provides the largest and most complete collection of data concerning the negative consequences of Chernobyl on the health of people and the environment ...'

The book explodes the claim of the IAEA that the expected death toll from the Chernobyl accident will be at the most 4000. Some of the important findings of this book are:

- Radioactive emissions from Chernobyl accident, once believed to be 50 million curies, may have been as great as 10 billion curies, or 200 times greater than the initial estimate, and hundreds of times larger than the fallout from the atomic bombs dropped on Hiroshima and Nagasaki.
- One nuclear reactor can pollute half the globe. Chernobyl fallout covered the entire Northern Hemisphere. Apart from the Soviet Union, nations which received high doses of radioactive fallout include Norway, Sweden, Finland,

Yugoslavia, Bulgaria, Austria, Romania, Greece, and parts of the United Kingdom and Germany. About 550 million Europeans, and 150 to 230 million others in the Northern Hemisphere received notable contamination. Fallout reached the United States and Canada nine days after the disaster.

- Of the approximately 830,000 people who were in charge of extinguishing the fire at the Chernobyl reactor and deactivation and cleanup of the site (the so-called 'liquidators'), by 2005, between 112,000 and 125,000 had died.
- Nearly ten lakh people (985,000 to be more precise) have died worldwide due to Chernobyl fallout from 1986 through 2004, a number that has since increased.

These are absolutely numbing statistics. Just one reactor accident is enough to contaminate half the globe, for tens of thousands of years! And yet, the world wants to build new reactors!!

Post-Chernobyl

Today (that is, till 2010, before Fukushima), 24 years after Chernobyl, nearly 9,000 reactor-years experience has accumulated worldwide.[105] This post-Chernobyl period has passed without a major accident, large-scale contamination and severe radiological consequences. The question is, is this an achievement or just simply luck? Has the risk from nuclear power plants been mastered and safety been improved to 'acceptable standards'?

The nuclear industry claims that the Chernobyl catastrophe was the result of old technology and mismanagement within the old Soviet bloc; and that safety issues have been adequately addressed after the Chernobyl accident. Which is why no major accident has taken place since the Chernobyl disaster. This is one of the important arguments it has been making as a part of the counteroffensive launched by it over the last decade to resuscitate nuclear power after decades of stagnation.

The reality is quite the opposite. In the 24 years after Chernobyl, there have been several near-misses at nuclear power plants in the United States and other countries. It is only sheer fortune that another Chernobyl has not happened. This has very powerfully been brought out in a study which was initiated after one such near-miss, the

Forsmark incident in Sweden in 2006, when it was possibly only a matter of minutes by which an accident on the scale of Chernobyl was prevented from happening in this reactor. This accident led researchers at the Union of Concerned Scientists[106] and institutes in Germany and Austria to jointly carry out a study of the safety records of nuclear power plants in several countries, to identify whether any abnormal 'events'[107] took place in them after Chernobyl, and analyse these events.

The report,[108] released by the Greens in the European Parliament in 2007, concluded:

> Many nuclear safety-related events occur year after year, all over the world, in all types of nuclear plants and in all reactor designs. Some of these events and incidents could have evolved into serious accidents, had the defects, malfunctions, et cetera, not been discovered in time (near-misses)! Many of these have remained significantly under-evaluated when it comes to their potential risk. Therefore, the widespread belief that lessons learnt from the past have enhanced nuclear safety turns out ill-conceived.

The report mentions 16 'events' in 9 countries, all of which could have snowballed into a major accident: June 18, 1988, Tihange-1 (Belgium); February 9, 1991, Mihama-2 (Japan); April 3, 1991, Shearon Harris (USA); July 28, 1992, Barseback-2 (Sweden); February 7, 1993, Three Mile Island (USA); May 12, 1998, Civaux-1 (France); December 27, 1999, Blayais-2 (France); July 2000, Farley (USA); March 18, 2001, Maanshan (Taiwan); August 12, 2001, Philippsburg (Germany); December 14, 2001, Brunsbüttel (Germany); March 6, 2002, Davis Besse (USA); August 29, 2002, 17 TEPCO reactors (Japan); April 2003, Paks (Hungary); March 1, 2005, Kozloduy-5 (Bulgaria); July 25, 2006, Forsmark (Sweden).

Let us briefly discuss two of these 'incidents'.

David-Besse, 2002

In recent years, the most serious episode involving a US nuclear reactor took place at the Davis-Besse plant in Ohio in 2002. The reactor came within days or weeks of a major catastrophe.

To save money, the owner of the reactor, First Energy, had persuaded the NRC to delay inspection of a vital safety component. When the plant was finally shut down, safety inspectors found that corrosion had eaten away the outer 6 inch thick carbon steel cover of the reactor vessel. The inner lining of the reactor vessel was of 3/16 inch stainless steel, and high pressure inside the reactor vessel had caused the stainless steel lining to bulge outwards into the cavity caused by the corrosion. Fortunately, the stainless steel bent, but did not rupture. The emergency core cooling system was also not working. Had the stainless steel ruptured, the core cooling water would have leaked, and with the emergency cooling system inoperable, it would most probably have led to a cascade of events culminating in a reactor meltdown.[109]

First Energy knew about the corrosion but, in order to continue production, had delayed informing about it to the NRC. It was ordered to pay a fine of $28 million in 2004, which was barely one per cent of its profits in that year.[110]

Forsmark, 2006

In an even more serious accident on July 25, 2006, the Forsmark nuclear power station in Sweden came within just two hours of a meltdown.

The Forsmark accident was caused by the failure of back-up generators, following a problem with the main power supply. Without power supply, the reactor cooling system stops functioning, which can lead to sharp spike in temperature and a meltdown within just two hours. According to a former director of the plant, 'it was pure luck there wasn't a meltdown'.[111]

Conclusion: Nuclear Reactors are Inherently Accident-prone

The conclusion is inescapable: nuclear reactors are no more safer today than they were during the 1980s, when TMI and Chernobyl occurred. In the words of M.V. Ramana,[112] one of India's leading nuclear safety experts:

> The basic features of all nuclear reactors remain the same. It is a complex technology, involving large quantities of radioactive

materials, and relatively high temperatures and pressures. In such technologies, even minor failures or human errors can lead to a cascading chain of events culminating in a major accident. Sociologists and organisation theorists have come to the pessimistic conclusion that with such high-technology systems involving extremely hazardous materials, it is in the very nature of such systems that serious accidents are inevitable. In other words, that accidents are a 'normal' part of the operation of nuclear reactors, and no amount of safety devices can prevent them.[113]

More Vulnerability: Relicensing Old Nukes

As if this danger wasn't enough, nuclear authorities worldwide, under industry pressure, are granting permission to extend the operating lifetimes of the existing nuclear plants by 10-20 years.[114] More than half of America's nuclear plants have received new twenty-year operating licenses. Many of these plants have also received 'power up-rates' that allow them to run at up to 120 per cent of their originally intended capacity.[115]

In the words of Robert Alvarez, senior policy advisor at the US Department of Energy from 1993 to 1999, and presently executive director of the Standing for Truth About Radiation (STAR) Foundation, this is an 'invitation to disaster'.[116] That is because the risk of accidents increases as plants get older. The reason is obvious: as it is, there is deterioration in any machine as it gets old; but for nuclear plants, the aging is much more because nuclear reactors are subjected to unprecedented amounts of heat, pressure, stress, corrosion and radiation. All these make the various parts of the reactor brittle, increasing the possibility of mechanical failures of one or the other parts of the reactor, which could lead to massive releases of radioactivity into the atmosphere. This is why, as the reactor fleet in the US ages, the vulnerability of the reactors to failures has been increasing. According to David Lochbaum, one of the foremost nuclear safety engineers in the USA and Director of the Nuclear Safety Project for the Union of Concerned Scientists, there have been 47 instances since 1979 wherein nuclear reactors in the U.S. have had to be shut down for more than a year for safety reasons![117]

What makes the situation even more fraught with danger is that the NRC, the US government agency overseeing the nuclear industry, has gone about giving licenses for lifetime extensions and power upratings in a very lackadaisical manner. Instead of doing its own research on whether the plant is in a good enough condition to be given permission to operate beyond 40 years, it has simply accepted corporate assertions about safety, and even used industry language verbatim in its reports.[118] With such a casual approach towards relicensing, it shouldn't surprise anyone that the NRC has not rejected a single license-renewal application in the last many years.[119]

In an article published in *The Nation* aptly titled 'Zombie Nuke Plants', Christian Parenti, the well-known American investigative journalist, author and fellow at the Nation Institute, gives the example of the lifetime extension given to the Oyster Creek reactor located in New Jersey to illustrate this lax approach of the NRC. The Oyster Creek reactor is one of the oldest plants in the USA and was scheduled to be shut down in 2009. However, before that could happen, the NRC relicensed it on April 8, 2009, extending its life span by twenty years. Just seven days after that, workers at the plant found an ongoing radioactive leak of tritium-polluted water. Four months later, in August, workers found another tritium leak coming from a pipe buried in a concrete wall. The second leak was spilling about 25,000 litres a day and contained 500 times the acceptable level of radiation for drinking water. Obviously, radiation had made the pipe brittle, and so it had leaked. Which means that the pipe was old. But the licensing paperwork claimed that the pipe had been replaced! How many other mislabelled, brittle, old components remain in the plant's guts is difficult to say. Parenti writes: 'Unfortunately, stories like this are all too common: crumbling, leaky, accident-prone old nuclear plants, shrouded in secrecy and subject to lax maintenance, are getting relicensed all over the country.'[120]

Note: The situation in India is even worse. The Indian nuclear safety regulator, the AERB, keeps extending the life of the Tarapur 1 and 2 reactors, though they should have been shut down long ago. These reactors are of an even earlier make than the Mark-1 Fukushima reactors which exploded in March 2011. (See Chapter 9 for more on this.)

Are Generation-III Reactors Safer?

As we have discussed in Chapter 1, the bottom fell out of the nuclear reactor manufacturing industry in the USA and Europe after the Chernobyl accident. Not only did new reactor construction ground to a halt, plants ordered were also cancelled. Over the last decade, in a desperate bid to resuscitate itself, the Western nuclear reactor manufacturing industry has launched a huge propaganda offensive to usher in a 'nuclear renaissance'. One of the important arguments it is making is that it has drawn lessons from the Chernobyl accident and developed a new generation of nuclear power plant designs which are much safer than the older designs.

The nuclear industry describes its evolution in terms of 'Generations'. Generation-I reactors were developed in the 1950s-60s, and are primitive by today's standards. The majority of the reactors currently operating around the world are Generation-II reactors.

The latest generation of reactors, the Generation-III reactors or 'advanced reactors', were developed in the 1990s, after the Chernobyl accident. Within Generation-III, there is now also a Generation III+ design, but the distinction between them is unclear.[121] The World Nuclear Association claims that these reactors are safer, with reduced possibility of core melt accidents. The two European Pressurised Reactors (EPRs) under construction in Western Europe—the first reactors to be constructed anywhere in the USA, Canada and Western Europe (excluding France) in the last three decades—belong to this latter category. The EPR is supposed to be one of the most 'advanced' designs, having an improved safety level (in particular, it is claimed that the probability of a severe accident is reduced by a factor of ten), and it also has features to mitigate effects of severe accidents.[122]

Throughout the world there are around 20 different designs under development for Generation-III and III+ reactors. However, a 2005 Greenpeace study on nuclear reactor hazards by four eminent nuclear experts noted that most of these are 'evolutionary' designs that have been developed from Generation-II (that is, current) reactor types with some modifications, but without introducing drastic changes. A typical example is the EPR design: it is simply a slightly modified version of the French N4 and German Konvoi reactors (the two latest Generation-II PWRs currently in operation in France and

Germany respectively), with some improvements. The study noted that it is doubtful if the modifications which are hailed to be safety improvements will work as claimed. On the contrary, they have several other modifications which actually reduce safety.

The study concludes: 'All in all, "Generation III" appears as a heterogeneous collection of different reactor concepts. Some are barely evolved from the current Generation II.' The modifications are *primarily aimed at cost-cutting and better economics*, although the nuclear industry fallaciously claims that these new designs are safer as compared to currently operating reactors '*in the hope of improving public acceptance*' of nuclear power (our emphasis).[123]

More recently, US and UK nuclear safety regulators have raised serious concerns about the designs of some of these 'advanced' reactors (discussed in detail in Chapter 6).

Thus, these latest series of nuclear reactors are no more safer than the present Generation-II reactors. On the contrary, they are inherently more dangerous! That is because many of the Generation-III reactors are of large capacities, of 1000 MW and above, and so, they have much more radioactivity in their core. For instance, the EPRs being constructed in Finland and France are of 1600 MW, and so, in the event of a serious accident, the impact would be more devastating than Chernobyl—the Chernobyl reactor was of 1000 MW capacity! Therefore, the EPR needs more stringent quality control just to match the safety level of present day reactors; however, as will be discussed in detail in Chapter 6, it is doubtful if even present-day safety standards are being observed at Olkiluoto and Flamanville (the sites in Finland and France, respectively, where the EPRs are being constructed). In an attempt to reduce costs and complete the project on schedule, the nuclear companies constructing these reactors have selected cheap, incompetent subcontractors, and have overlooked safety-related problems!

Shut Down Every Single Reactor!

In her classic work, *Nuclear Power is not the Answer to Global Warming or anything else*, Dr Helen Caldicott, the pioneering Australian anti-nuclear crusader who has done such fantastic work to spread awareness about the hazards of nuclear energy all over the world, writes:

'Statistically speaking, an accidental meltdown is almost a certainty sooner or later in one of the 438 nuclear power plants located in thirty-three countries around the world.'[124] In its greed for profits, the world's nuclear industry is pushing to making her grim foreboding come true sooner than later.

Even assuming that there will not be another Chernobyl, the damage caused by radioactive waste from nuclear plants for our future generations will be no less worse than Chernobyl. In the words of Helen Caldicott again:

> Nuclear power produces massive quantities, hundreds of thousands of tons of radioactive waste, which will get into the water, concentrate into the fish, the milk, the food, human breast milk, foetuses, babies, children. Radioactive iodine causes thyroid cancer. Twelve thousand people in Belarus had thyroid cancer. Radioactive strontium-90 causes bone cancer and leukaemia, [it] lasts for 600 years. Cesium-137—all over Europe now—in the reindeer, in the lands, in the food, lasts for six hundred years, causes brain cancer. Plutonium, the most dangerous substance on Earth, 1 millionth of a gram causes cancer, lasts for 250,000 years. Causes lung cancer, liver cancer, testicular cancer, damages foetuses so they are born deformed.
>
> Therefore, nuclear waste for all future generations will cause cancer in young children because they are very sensitive, [will cause] genetic disease, congenital deformities. Nuclear power is about disease, and it's about death. It will produce the greatest public health hazard the world has ever seen for the rest of time. *We must close down every single nuclear reactor in Europe and throughout the world* ... (emphasis ours).[125]

4

IS NUCLEAR ENERGY CHEAP?

Components of Nuclear Power Cost

The cost of generating electricity from nuclear plants can be separated into the following components:

- cost of construction of the plant;
- operating cost of the plant;
- cost of disposal of the waste generated by the plant; and
- cost of decommissioning the reactor after its operating life is over.

Construction costs: These dominate the cost of nuclear power, constituting as much as 65-75 per cent of the total cost.

Operating costs: These include fuel, operating and maintenance costs. Because of the low fuel cost of nuclear plants, as compared to coal- and gas-fired plants, the total operating cost of nuclear power plants is generally a much smaller component of the total cost of nuclear power as compared to coal and gas power.

Waste disposal and decommissioning costs: These costs are unique to nuclear power. While these costs are going to be incurred in the future, they must be included in the cost of nuclear electricity generated by the plant. The problem is that these costs are difficult to estimate as the waste is intensely radioactive and will have to be managed for thousands of years, while decommissioning a nuclear plant is a very long term and complicated operation. In the case of decommissioning costs, in the US and Western Europe, the projected cost of decommissioning the plant is calculated and spread across the expected life of the plant. The waste disposal cost is much more difficult to estimate as there is no known method of safely disposing it, and so various countries have adopted different ways of including the waste disposal costs in the cost of nuclear electricity.

Part I: New Claims to Resuscitate Nuclear Power

Fuel costs of nuclear energy are remarkably low because a million times more energy is released per unit weight by fission than by combustion. And so, till the early 1970s, nuclear industry and governmental authorities the world over were claiming that nuclear energy would soon be 'too cheap to meter.'

That claim went kaput by the late 1970s. In the US, final construction costs of nuclear power plants were coming out to be several times their initial estimates. A 1986 DOE study of the

75 nuclear power plants commissioned between 1966 and 1977 found that while the predicted construction costs of these plants had been $45 billion, the actual costs turned out to be $145 billion—a total increase of $100 billion, and, that too, not inclusive of finance or interest charges.[1]

The same was true of operating costs—the actual operating costs were turning out to be much more than the estimated costs. A 1995 EIA (Energy Information Administration, DOE) study found that real (inflation-adjusted) average annual non-fuel operating costs for all plants in operation in the US in 1993 escalated from about $37 per kW of plant capacity (in 1993 dollars) to $126 per kW between 1974 and 1993.[2]

The construction cost overruns and sky-high operating budgets were important reasons, apart from safety concerns, that brought nuclear ordering to a halt in the United States and led to cancellations of more than 100 plants at various stages of construction.[3]

In Britain, the poor economics of nuclear energy became evident when the government attempted to privatise its nuclear power plants in 1987-90. Private investors rigorously calculated the costs and found that the operating costs alone were about double the expected market price for electricity, and refused to buy the nuclear assets.[4] The government finally managed to sell off the more modern eight nuclear power stations to British Energy in 1996. The deal was an absolute scandal. The British government sold the eight plants (each of around 1200 MW capacity) for just about £1.7bn, an amount which was about half the cost of building Sizewell B, the newest of the plants.[5]

Evidence from other countries, too, pointed in the same direction—that construction costs and operating costs of nuclear power plants were very high.

Apart from the accidents at Three Mile Island and Chernobyl, and the huge problem of disposal of nuclear waste, this unfavourable nuclear economics was another factor that contributed to the drying up of nuclear plant orders across the world in the 1980s-90s.

New Claims to Usher in a 'Nuclear Renaissance'

However, over the past decade, the nuclear industry has once again launched a huge propaganda offensive to convince the public about the supposed advantages of nuclear power over fossil fuel energy. It has now come up with a new argument—that it has developed a new generation of nuclear power plant designs, so-called Generation-III and III+ designs, which are safer and less expensive.

For instance, a report by the World Nuclear Association, an international organisation of private sector nuclear power equipment suppliers, aptly titled *The New Economics of Nuclear Power*, says that, taking into consideration the higher construction and lower operating costs of nuclear power as compared to fossil energy, the 'key development in the "new economics" of nuclear power is that, both costs considered, nuclear power has now become less expensive than fossil and any other form of electricity generation.'[6]

We have discussed the safety issues associated with nuclear reactors, including these new designs, in Chapter 3.[7] We discuss the economics of nuclear reactors, including the claims being made about Generation III and III+ designs, in this Chapter.

Part II: Reviewing the New Claims

Difficulties in Making Cost Estimates

The civilian nuclear power industry has been in operation for over fifty years. During such a long period, technological improvements and experience should have led to learning, and subsequently to enhancements in economic efficiency and falling costs. However, the nuclear industry has not followed this pattern. Like we saw above for the US, in country after country, the construction costs have gone considerably over budget. Similarly, construction time, instead of declining, has doubled—the average construction time for nuclear plants has increased from 66 months for completions in the mid-1970s, to 116 months (nearly 10 years) for completions between 1995 and 2000.[8] Therefore, it is difficult to draw lessons from past experience in order to make estimates of electricity costs for the new designs being promoted by the nuclear industry.

The other problem with making these estimates is that there is very little or no construction experience for these new designs. The only Generation-III reactors currently in operation are the Advanced Boiling Water Reactors (ABWR) developed in Japan.[9] By the end of 2009, four ABWRs were in service in Japan and two under construction in Taiwan. Total construction costs for the first two Japanese units were reported to be $3,236/kW for the first unit in 1997 dollars and about $2,800/kW for the second. These costs were well above the forecast range. These units have also suffered design problems in the turbine, implying that a new turbine design will be required, which might take several years.[10] So far as Generation III+ reactors are concerned, none has ever been built; only one Generation III+ reactor design, Areva's EPR, is under construction, one at Olkiluoto in Finland, one at Flamanville in France, and two at Taishan in China. Obviously, nothing is known about what will be the operating cost of this design; while all that we know about its construction cost is that it has sharply escalated as work has progressed (discussed later in this chapter).

Estimates by Independent Institutions

Over the past decade, many independent institutions have conducted systematic and detailed studies of nuclear electricity costs, and their assessments contradict the claims of the nuclear industry. We give below the results of some of these studies, done by well-respected bodies.

A June 2005 report, titled *Mirage and Oasis: Energy Choices in an Age of Global Warming,* from the New Economics Foundation, an independent British think tank founded in 1986 by the leaders of 'The Other Economic Summit',[11] concluded that the cost of nuclear power has been underestimated by almost a factor of three.[12] Another study by the US DOE's Energy Information Administration concluded that nuclear power is more costly than natural gas and coal plants.[13] A University of Chicago study in 2004 also came to the same conclusion.[14]

In May 2006, in response to the so-called 'Nuclear Renaissance', the Canada-based Centre for International Governance Innovation (an independent think tank led by a group of distinguished academics and supported by the government of Canada) initiated the 'Nuclear Energy Futures Project' to investigate the likely size, shape and nature of the purported nuclear energy revival to 2030. Its report in February 2010 concluded that one of the important factors constraining the expansion of nuclear energy was its economics: 'The economics are profoundly unfavourable and are getting worse. This will persist unless governments provide greater incentives ...'[15]

MIT Updated Study 2009

Probably the most sophisticated and widely cited study on the economics of nuclear power is a 2003 study by the Massachusetts Institute of Technology titled *Future of Nuclear Power*, which was updated in 2009 to take into consideration the sharp increase in construction costs of nuclear as well as coal and gas fired power plants. Even this study, by an institution that is known to be pro-nuclear, concluded that the 'levelised' cost of electricity[16] generated by a new nuclear power plant is about 30-35 per cent higher than the cost of electricity from a coal fired or combined cycle gas turbine plant (Table 4.1).[17]

Table 4.1: Cost of Electricity Generation Alternatives

MIT study update 2009 (in cents/kW)		
Nuclear	*Coal*	*Gas*
8.4	6.2	6.5

Even this update underestimates the real costs of nuclear energy as it is based on an estimated construction cost of $4000/kilowatt, or about $4 billion for a 1000 MW reactor,[18] whereas no new US reactor proposal is anywhere near this cost. The real world estimates are ranging from $6000-9000/kW, or 50-100 per cent higher than the construction cost estimates used by MIT to make its nuclear power cost estimates. A few examples:[19]

- Calvert Cliffs-3 reactor (proposed to be built by Constellation Energy and EDF in Maryland) is estimated to cost about $10 billion, or $6000/kW for this 1600 MW reactor.
- The proposed cost of the two new Turkey Point reactors (at the Turkey Point NPP in Florida) is $8200/kW.

- The Southern Company's proposal to build two new reactors at its Vogtle NPP in Georgia is currently estimated to cost about $6200/kW.

Opinion of Credit Rating Agencies and Financial Institutions

Credit rating agencies are out-and-out pro-corporate bodies; they will not make recommendations based on environmental considerations. Even they conclude that the economics of nuclear power is ruinous. Standard and Poor's stresses: 'no utility will commit to a project as large and risky as a new nuclear plant without assurance of cost recovery'. Similarly, Moody's 'believes that many of the current expectations regarding new nuclear generation are overly ambitious.'[20]

The World Bank has been willing to finance the most environmentally destructive projects, so long as corporations can make handsome profits. But even it is not willing to give loans for nuclear plants! The economics of nuclear power is so deleterious that with the exception of a 1959 loan to Italy, the World Bank has never financed a nuclear power plant and there are no signs that it has changed its financial risk analysis.[21] In fact, its lending advice explicitly states: 'Nuclear plants are thus uneconomic because at present and projected costs they are unlikely to be the least-cost alternative. There is also evidence that the cost figures usually cited by suppliers are substantially underestimated and often fail to take adequately into account waste disposal, decommissioning, and other environmental costs.'[22]

A statement signed by six of Wall Street's largest investment banks is even more revealing. In 2007, Citigroup, Credit Suisse, Goldman Sachs, Lehman Brothers, Merrill Lynch and Morgan Stanley informed the US DOE that they were unwilling to extend loans for new nuclear power plants unless taxpayers shouldered 100 per cent of the risks! In justifying this demand, the banks stated: 'We believe these risks, combined with the higher capital costs and longer construction schedules of nuclear plants as compared to other generation facilities, will make lenders unwilling at present to extend long-term credit ...'[23]

Real Life Scenario: Nuclear Reactor Costs Escalating

That the nuclear industry is deliberately understating cost estimates in order to promote a nuclear revival is obvious from the fact that real life plant construction costs have sharply escalated over the past few years.

In Canada, the Ontario government suspended a competitive process for purchase of new reactors in June 2009, because the price quoted in the only valid bid it received was three times the price it was hoping to pay. That bid was from Atomic Energy of Canada Ltd., and it quoted a price of $26 billion for the construction of two 1,200-megawatt Advanced CANDU Reactors, which worked out to $10,800 per kilowatt of power capacity![24]

Likewise, cost estimates for the two new 1350 MW ABWRs proposed to be built by NRG Energy at the South Texas Project (USA) have zoomed to $13.9bn, plus an additional $4.3bn as financial charges (for a total of $18.2bn), from a preliminary estimate of $5.4 billion. That works out to an overnight charge (that is, excluding financing charges) of roughly $5150/kW.[25] If at all the plant construction begins, what the final cost will be is anybody's guess!

These examples are not exceptions. *The Economics of Nuclear Reactors*, a report released on June 18, 2009 by Dr Mark Cooper, senior fellow for economic analysis at the Institute for Energy and the Environment at Vermont Law School, USA, analysed over three dozen cost estimates for proposed new nuclear reactors. It found that the projected price tags for these plants have quadrupled since the start of the so-called 'nuclear renaissance' at the beginning of this decade—a striking parallel to the eventually seven-fold increase in reactor costs estimates that doomed the 'Great Nuclear Boom' of the 1960s and 1970s, when half of the planned nuclear reactors had to be abandoned or cancelled due to massive cost overruns.[26]

Olkiluoto-3 and Flamanville-3

The order for the Olkiluoto-3 (OL3) reactor being built in Finland was placed in December 2003. It was a turnkey contract and Areva offered to build the 1600 MW plant for €3.2bn ($3.84bn), that is, for roughly $2400/kW (€1=$1.2).[27] Construction began in August 2005.[28] It was supposed to be completed by the summer of 2009.[29]

By early 2009, the project was acknowledged to be at least three years late and €1.7bn over budget, and was now expected to cost close to €5 billion.[30] In June 2010, Areva acknowledged that the cost had further escalated to €5.9 billion[31] (or more than $7.26bn, or $4500/kW, taking €1=$1.23 which was the rate in June 2010)—nearly double the contract price! This newly announced cost is based on the assumption that OL3 will be operational by end of 2012. That seems to be very doubtful, as the reactor's construction is only just halfway to completion, and the most challenging phases of construction are still to come.[32]

The fate of the other EPR ordered in France has been no better. In 2007, EDF (Électricité de France), France's national utility, after much convincing by Areva NP, ordered an EPR reactor, to be located at their Flamanville site. Construction commenced in December 2007. In May 2006, EDF had assessed that the plant would cost €3.3bn.[33] As construction progressed, its cost estimate too went through the roof. In end-2008, it acknowledged that the expected construction cost had increased to €4 billion.[34] Last year (2010), it admitted that the project is two years behind schedule—it has only been under construction for three![35] And that the cost estimate had escalated to €5 billion ($6.5 billion).[36]

For both these cases, the only Generation III+ nuclear reactors under construction (in Western Europe and North America), with cost estimates escalating to nearly double the contract price even before the reactor construction has reached halfway, making a guess about the final construction cost has become hazardous!

Let us compare these costs with the construction cost of setting up a coal fired plant in India. The construction cost of the OL3 1600 MW EPR was estimated at €5.9 billion in June 2010. Taking the Euro-Rupee conversion rate as it existed in May-June 2010 of €1 = Rs.57, that works out to Rs.21 crores/MW—nearly four and a half times the average cost of setting up a new coal power plant in India (Rs.4.5 crores/MW)!

With such astronomical construction costs, it is obvious that the cost of nuclear electricity from these new reactors is going to be huge, much more than the cost of electricity from fossil fuel plants.

Part III: Nuclear Subsidies

The real cost of nuclear electricity is actually much more than the above estimates. That is because the above calculations do not take into account the enormous government subsidies to nuclear energy.

All governments throughout the world which have a nuclear energy program subsidise nuclear energy. Thus, France claims that its nuclear power costs are 'the lowest in the world';[37] the reality is its entire domestic nuclear energy program has for decades profited from numerous government subsidies.[38] The French government has subsidised the cost of construction of France's nuclear plants,[39] which dominate the cost of nuclear electricity. It has nationalised the decommissioning and waste management costs: the waste management costs are estimated at between $21 billion and $90 billion;[40] the decommissioning cost estimates keep rising, and were estimated to be 65 billion euros in 2004.[41] It has also effectively taken over the accident risks—if Electricité de France (EDF), France's nuclear utility, had to insure for the full cost of a meltdown, the price of nuclear electricity would increase by about 300 per cent.[42]

The Campaign for Nuclear Phaseout, an alliance of anti-nuclear organisations of Canada, in a report prepared in 2003, estimated that the total subsidies given by the Canadian government to Atomic Energy of Canada Limited (AECL) over the 50-year period 1953-2002 totalled a whopping $17.5 billion! AECL is a Canadian government corporation that manages Canada's national nuclear energy research and development program, including designing and marketing of CANDU reactors. The calculations were based on figures given in AECL's own annual reports.[43]

In the UK, British Energy, which had already got a huge subsidy when it purchased eight nuclear plants from the government in 1996, got into financial difficulties and went bankrupt in 2002. The government, instead of allowing the company to close, intervened. The rescue is estimated by the government to have cost the taxpayers a mindboggling eleven billion pounds![44] On top of it, the government has taken over all the decommissioning and waste management costs, which amount to nearly a hundred billion euros and will keep on rising as more and more waste accumulates![45]

Most subsidies given by governments to the nuclear electricity industry are common throughout the world. We discuss below the most important of these subsidies, with examples mainly from the United States. The only reason for discussing USA in greater detail is not because it gives more subsidies, but because of greater availability of information.

Capital Subsidies

Investing in nuclear power is considered to be a high risk investment. In most countries around the world, the nuclear electricity sector is in the public sector, and therefore the high costs and huge risks associated with nuclear energy are guaranteed by the government.

In the US, even though the electricity industry has mostly been in private hands, till the 1990s, distribution costs were regulated by the states, and regulators allowed nuclear electric utilities to pass on their high costs to consumers—to the tune of half a trillion dollars, including:

i. more than $200 billion (in 2006 dollars) on account of cost overruns of completed nuclear power plants; [46]
ii. most of the $50 billion (in 2006 dollars) in construction costs of the abandoned nuclear plants;[47] and
iii. the high electricity generating costs from nuclear plants, which were on the average three cents per kWh more than electricity from fossil fuel plants, for the period 1968 to 1990—this totalled more than $225 billion (in today's dollars).[48]

A second wave of subsidies was given to nuclear reactors when the electricity sector was deregulated in the 1990s. Regulators allowed utilities to recover the difference between their remaining investments in nuclear plants and the market value of those plants—called 'stranded costs'—from consumers. These payments approached $100 billion in today's dollars.[49]

Without these subsidies, the present nuclear reactor fleet in the USA would never have been built![50]

Post-deregulation, that is, since the 1990s, Wall Street has been unwilling to provide capital to nuclear plant developers, except at very high interest rates, as these plants are going to find it very difficult to transfer their high construction costs to consumers. Therefore, over the past decade, the American nuclear industry has mounted pressure on the US government for a fresh round of subsidies—in the form of loan guarantees and other financial assistance—for building new nuclear plants. The introduction of government loan guarantees reduces the cost of financing a new nuclear power plant—and so the price of nuclear electricity—in two ways. First, now the lenders don't care about the risks and are willing to lend funds at low interest rates. Second, the guarantee enables plant owners to use much more of this inexpensive debt to finance the plant—up to 80 per cent in the case of the United States. The impact of loan guarantees on nuclear power generation costs can be dramatic: UniStar Nuclear Energy, which hopes to build a series of reactors across the USA, estimates loan guarantees will reduce its levelised costs[51] by nearly 40 per cent.[52] In fact, without loan guarantees, the nuclear industry will not even think of beginning construction of new plants, as was made very clear by Christopher Crane, President of Exelon Generation, one of the utilities that has expressed an intention to build new nuclear plants in the US: 'If the loan guarantee program is not in place … we will not go forward'.[53]

Obligingly, the US Congress in 2005 passed the Energy Policy Act (EPACT 2005), authorising loan guarantees of $18.5 billion for new nuclear plants of new designs over the next several years.[54] (The 2005 Energy Bill also gave several other financial handouts to the nuclear industry, including tax credits on electricity generation and additional support in case of delays in reactor construction.[55]) Even this huge sum was much below industry expectations, as that would have financed just three reactors. So industry has been pressurising the US Congress to increase this amount. Acceding to its demand, the Obama Administration, in its budget proposal for 2011, has proposed an additional $36 billion in new federal loan guarantees, for a total of $54.5 billion.[56]

The potential cost of this subsidy to the taxpayers is huge. In 2003, based on historical data, the US Congressional Budget Office estimated the risk of default on guaranteed loans for nuclear power plants 'to be very high—well above 50 per cent'.[57]

The US Senate is also considering two new bills—the American Power Act (APA) and the American Clean Energy Leadership Act (ACELA)—which propose to give another tens of billions of dollars in subsidies to the nuclear industry in the form of reduced accelerated depreciation periods, tax credits for investment and production, and so on.[58] These new subsidies are estimated to be worth between $1.3 billion and $3 billion on a net present value basis per new reactor.[59]

Apart from these federal government subsidies, the nuclear industry is pressurising the states to allow utilities to recover construction costs from customers even before the plant has come online. Thus, Georgia has approved a CWIP or 'Construction Work in Progress' law that will allow Southern Company to recover construction costs of the new nuclear plant that it is proposing to build in the state from ratepayers during the construction period itself, that is, even before the plant has generated a single unit of electricity! This effectively shifts the risk of building the plant onto the customers, because, in case the company abandons the plant for some reason, it will still be allowed to recoup 'prudent' costs from its customers. Florida also has such a law in place.[60]

Government Spending on Nuclear Related R&D

Another government subsidy to nuclear power is in the form of spending huge amounts of public money on research and development (R&D) related to the nuclear fuel chain. During the period 1961-2008, the US government invested 172 billion dollars in energy R&D; of this, the largest share, 36 per cent, or $61 billion, went into nuclear energy R&D. This was more than double the level of support to renewable energy and energy efficiency technologies ($26 billion).[61]

Capping Operator Liabilities in Case of Accidents

Accident risks have been the Achilles heel of the nuclear power industry since its birth. For most industries, even a large accident, while catastrophic to the immediately surrounding area, tends to be relatively well-contained geographically. However, a nuclear accident has the potential of rendering a much larger area uninhabitable for centuries, if not thousands of years!

And so, the insurance industry has not been willing to underwrite nuclear accident risks since the very inception of the nuclear industry. In the United States, Congress intervened early, in 1957, and passed the Price-Anderson Act to give the infant technology 'protection against potentially enormous liability claims in the event of a nuclear accident',[62] a benefit no other US industry has ever received. The Act has since been renewed (and modified) many times. This law sets a maximum cap on liabilities of nuclear power companies in case of claims arising from nuclear accidents.

Under this Act, power reactor licensees are required to obtain the maximum amount of insurance against nuclear related incidents available in the insurance market (as of 2010, $375 million per plant). Any monetary claims that fall within this maximum amount are paid by the insurer. In the event of an accident that exceeds $375 million in damages, the Price-Anderson fund is then used to make up the difference. This fund is financed by the reactor companies—each of the operating nuclear reactors in the US is obliged to contribute up to $112 million (as of 2008) in the event of an accident. As of 2008, the maximum amount in the fund was approximately $11.6 billion. Any claims beyond this, in the event of a major accident, would be covered by the federal government.[63]

The Price-Anderson Act thus provides a twofold subsidy to the nuclear industry. Firstly, it reduces the insurance nuclear power plants need to buy, thereby providing them a huge subsidy in terms of insurance premiums they don't have to pay. Secondly, it caps the total liabilities of nuclear plant operators in case of a major accident. According to an NRC study, damages from a severe nuclear accident could run as high as $560 billion (in 2000 dollars). The $12 billion provided by private insurance and nuclear reactor operators thus represents less than two per cent of the potential costs of a major nuclear accident. The remaining hundreds of billions of dollars would have to be paid by taxpayers.[64]

Without this liability shelter, nuclear reactors would never have split the first atom. Speaking before a committee of the Canadian House of Commons dealing with the Canadian Nuclear Liability Bill, Peter Mason, president and chief executive of nuclear supplier,

GE-Hitachi Nuclear Energy Canada, explained: 'If there was not a cap and if there was no suitable legislation insurance in place, then we wouldn't be in the nuclear industry.'[65] This has also been conceded by Dick Cheney (when he was the Vice-President of the USA): without the security of the Price-Anderson Act, 'nobody's going to invest in nuclear power plants.'[66]

While the US nuclear liability law limits the liability of the operator in case of a major accident, it does not at all mean that the victims of a nuclear accident automatically get their health costs reimbursed. As we have mentioned in Chapter 3, the US government refused to acknowledge that there was any significant radiation release from the Three Mile Island accident, told the victims that their health problems were pure imagination and denied them any compensation. The victims had to go to court and, as normally happens when ordinary people fight giant corporations, the case went on and on, and eventually they got tired and settled out of court.

A similar regime capping the maximum compensation to be paid by nuclear plant operators exists in every country in the world having nuclear reactors. For instance, in countries of the European Union which are signatories to the Vienna or Paris International Convention on Nuclear Liability, this is now at the most 700 million euros. Worse, no liability regime now in effect outside the USA provides more than $2 billion in aggregate cover, despite the large populations surrounding many of these plants.[67]

Nationalisation of Waste Management Costs

As it is, the cost of storing the highly radioactive waste generated by nuclear power plants is huge. On top of it, the very fact that this waste is intensely radioactive and is going to remain so for thousands of years leads to additional liabilities: one, the waste is bound to leak and contaminate the surrounding area (discussed in Chapter 3); and two, an accident or a terrorist attack at the waste storage site could have catastrophic consequences, much worse than a meltdown at a nuclear reactor.[68]

No private firm, howsoever big it may be, has the financial capacity to bear these risks. And so, national governments have stepped in and effectively nationalised both the financial costs and accident liability risks of waste management. Just like the insurance subsidy discussed above, without this subsidy too, it is doubtful if the nuclear power industry would have developed at all.

In the US, the government through the 1982 Nuclear Waste Policy Act has taken over the entire responsibility for permanently disposing of the nuclear waste generated by nuclear power plants, in return for which the utilities pay an artificially low flat fee of 0.1 cents per kilowatt-hour of nuclear generated electricity.[69] It is not known how this amount was arrived at, it is not based on actual experience as no long-term fuel disposal facilities exist anywhere in the world, but it is obvious that it is a huge underestimate. In other words, a huge subsidy is being given. To get a rough idea of this subsidy amount, here are some statistics: as of 2009, US nuclear power companies had paid a total of $16 billion for waste disposal services,[70] whereas the US Department of Energy (DOE) estimated that the Yucca Mountain nuclear waste repository alone was going to cost at least $96.2 billion;[71] the Obama government eventually abandoned the project in 2010,[72] but by then more than $13.5 billion had already been spent on it.[73]

Not only have governments nationalised nuclear waste management, they have also taken over the responsibility for managing contamination from the other parts of the nuclear fuel cycle, including the uranium mines and the enrichment and the reprocessing facilities, even when these facilities are in the private sector.

Nationalisation of Decommissioning Costs

Decommissioning a nuclear reactor once its operating life ends is a very long term and complicated operation as all the parts of the plant would have become radioactive; hence, it is also very costly. In countries like the US, where most nuclear utilities are in private hands, nuclear plant operators are required to set aside a certain part of their income during the working lifetime of the reactor, to meet future decommissioning expenses.

Decommissioning costs are difficult to estimate, because there is little experience with decommissioning commercial-scale plants. Estimates for decommissioning costs range from an average of $300 million in the US to £1 billion in the UK, per 1000 MW reactor. The French and Swedish nuclear industries expect decommissioning costs to be between 10 and 15 per cent of construction costs.[74]

Almost everywhere, private nuclear plant operators have underestimated decommissioning costs and have set aside insufficient funds to cover these expenses; they are confident that governments will step in and pay the deficit, in another subsidy to the industry. This has already happened in the UK, where the government has effectively taken over the future decommissioning costs of the nuclear reactors operated by the private sector company, British Energy, resulting in the transfer of liability of billions of euros onto future taxpayers; according to an estimate by Steve Thomas, Professor for Energy Policy at the University of Greenwich, this could be as much as €90bn.[75]

In the US, too, in June 2009, the Nuclear Regulatory Commission published concerns that the owners were not setting aside sufficient funds.[76] This shortfall is also expected to run into billions of dollars.[77] Obviously, considering the grip that the nuclear industry has over the US government, the owners are confident that once their plants shut down, any shortfall would be met by the government!

Blithely Ignoring Health Costs

So far, we have discussed the overt or covert subsidies given by governments to nuclear energy. Apart from these subsidies, probably the worst part of this nexus between governments and the nuclear industry is that governments have allowed the nuclear industry to simply ignore the health costs of nuclear energy. Nuclear electricity cost calculations do not take into account the health costs of the radiation leaked into the atmosphere at every stage of the nuclear fuel cycle, from mining to nuclear reactors to waste storage to reactor decommissioning. In fact, the nuclear industry does not even admit to these costs, blithely lying that it is clean and causes no health damage to its workers and the people living in the neighbourhood of its installations. And so, these costs are silently borne by the people. (This subsidy is given to coal power too. There too, the health costs are ignored.)

Note: As we see in Chapter 8, in India, besides all these subsidies, the government gives many additional subsidies to nuclear power, like heavy water subsidy, fuel subsidy, et cetera.

Part IV: Conclusion

A November 9, 2009 report, *New Nuclear—The Economics Say No* by Citi Investment Research & Analysis, a division of Citigroup GlobalMarkets Inc., says:

> Three of the risks faced by developers—Construction, Power Price, and Operational—are so large and variable that individually they could each bring even the largest utility company to its knees financially. This makes new nuclear a unique investment proposition for utility companies. Government policy remains that the private sector takes full exposure to the three main risks: construction, power price and operational. Nowhere in the world have nuclear power stations been built on this basis. We see little if any prospect that new nuclear stations will be built in the UK by the private sector unless developers can lay off substantial elements of the three major risks. Financing guarantees, minimum power prices, and/or government-backed power off-take agreements may all be needed if stations are to be built ...[78]

That's precisely the point we've been trying to make in this chapter—that nuclear energy is a very expensive way of generating electricity, is definitely much more costly as compared to electricity from fossil fuels and the only way it can be competitive with conventional electricity is if it is highly subsidised by the government. In fact, Stephen Thomas, the renowned independent energy policy researcher based in the UK, writes that studies by the British government in 1989, 1995 and 2002 all came to the same conclusion—that in competitive electricity markets, electric utilities would not build nuclear power plants without government subsidies.[79]

Apart from public opposition, this cost factor is one of the important reasons why nuclear electricity is on the decline the world over, especially in countries with competitive electricity markets.

Then how come Areva won an order for constructing the Olkiluoto-3 reactor in Finland? It is being claimed that this proves that the new Generation-III+ reactors are feasible in liberalised energy markets. However, a closer examination of the deal indicates that this is not a commercial order made in a free market without subsidies and guarantees, as the following facts about the order prove:

(i) Areva deliberately offered a low and fixed price for the project in order to get its first order in 15 years.[80] There were fears that if it did not get an order for its EPR reactor soon, it would lose key staff and the design would become obsolete.[81] (Areva is having to suffer heavy losses because of this. As mentioned above, the construction costs have zoomed to double the contract price.)

(ii) Areva is majority-owned by the French state.[82] So the French government went out of its way to organise low cost finance and loan guarantees for the project. [83]

(iii) The buyer, the Finnish electricity company TVO, is not a normal electric utility, but is an organisation unique to Finland. Its ownership structure is such that it will have a guaranteed market and will be able to pass on the high cost of nuclear electricity to its consumers; it will therefore not have to compete in the highly competitive Nordic electricity market.[84]

For all these reasons, Olkiluoto-3 is a special case. If anything, the experience with the construction of Olkiluoto-3 so far only serves to highlight that nuclear plants are extremely expensive to build. For this and other reasons, it is doubtful if other countries in Europe are going to follow suit and place orders for nuclear reactors very soon. (We discuss this in detail in Chapter 6.)

Similarly, while the nuclear industry is expressing optimism that it will soon renew construction of nuclear reactors in the USA, it is obvious from the above discussion that this is based on its hope that the US government will give it sufficient subsidies to make nuclear electricity competitive.

◆◆◆

5

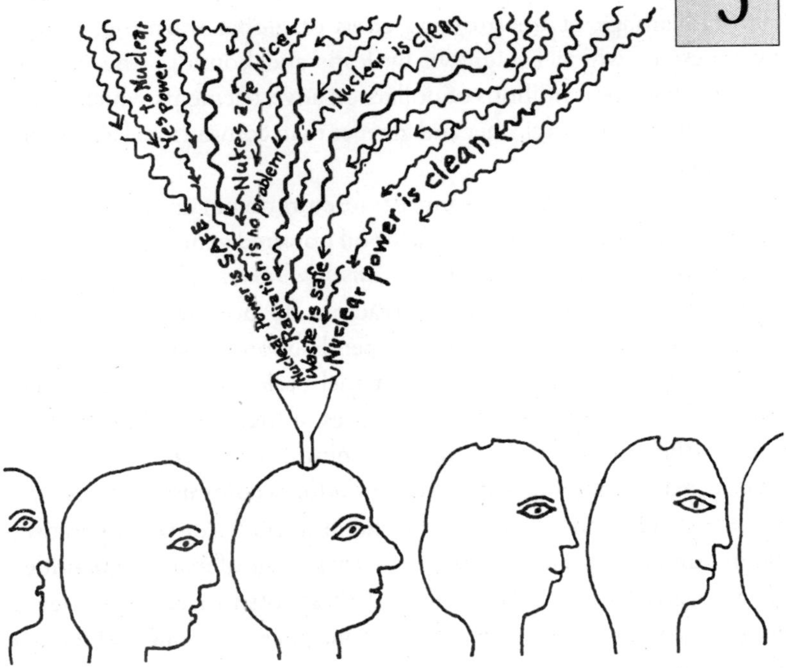

IS NUCLEAR ENERGY GREEN?

Speaking at the inauguration of the Pragati power project in West Delhi on May 24, 2008, Prime Minister Manmohan Singh said that 'the Government is committed to the development of nuclear energy as an environment friendly source of power'.[1]

The Principal Scientific Advisor to the Indian Prime Minister, R. Chidambaram, while delivering a lecture in Delhi on August 13, 2009, stated that nuclear energy was the only way India could achieve both energy security and combat climate change, and referred to a 2007 IAEA report which said that, for the world to keep global warming within two degrees Celsius, nuclear power would have to grow 80 per cent by 2030.[2]

One of the most important arguments being made by the nuclear industry during the past decade, in its attempt to revive nuclear energy, takes advantage of the growing crisis of global warming, public awareness of which has grown by leaps and bounds. The nuclear industry has spent millions of dollars on advertisements which claim that nuclear energy is cleaner and greener than conventional electricity from fossil fuels.

While it is true that the nuclear reactors do not emit greenhouse gases in the same quantum as coal or oil powered generating stations, to conclude that nuclear energy is *'an environment friendly source of power'* is a far stretch. Nuclear reactors do not stand alone; the production of nuclear electricity depends upon a vast and complex infrastructure known as the nuclear fuel cycle. And the fact is, the nuclear fuel cycle utilises large quantities of fossil fuels during all its stages—the mining and milling of uranium, the construction of the nuclear reactor and cooling towers, robotic decommissioning of the intensely radioactive reactor at the end of its 30 to 40-year operating lifetime and transportation and long-term storage of massive quantities of radioactive waste. The burning of these fossil fuels releases large quantities of carbon dioxide into the atmosphere, the source of today's global warming.

In fact, acting on a complaint by a group of law students from Queen's University, Advertising Standards Canada, the organisation which regulates the Canadian advertising industry, ruled in September 2010 that claims of nuclear power being 'emission free', made in adverts by the Power Workers' Union, 'were inaccurate, unsupported, and misleading'. The Union was told to remove its ads.[3]

We take a cursory look at the energy consumed during the various stages of the nuclear fuel cycle.

Carbon Emission and the 'Nuclear Fuel Cycle'

Uranium mining and milling are very energy intensive processes. The rock is excavated by bulldozers and shovels and then transported by truck to the milling plant, and all these machines use diesel oil. The ore is ground to powder in electrically powered mills and fuel is also consumed during conversion of the uranium powder to yellow cake.

In fact, mining and milling is so energy intensive that if the concentration of uranium falls to below 0.01 per cent, then the energy required to extract it from this ore becomes greater than the amount of electricity generated by the nuclear reactor; in other words, the nuclear fuel cycle becomes energetically non-productive. And most uranium ores are low grade; the high-grade ores are very limited—global high-grade reserves amount to 3.5 million tons—just enough to supply *three years of nuclear power* if all the world's energy needs were met by nuclear energy.[4]

The thousands and millions of tons of mine and mill waste should actually be chemically treated and buried deep in the ground where the uranium originally emanated. However, if this remediation process is scrupulously observed, as it should be, then extensive amounts of fossil fuel would be needed, making the energy costs of nuclear energy totally unviable.[5] And so the wastes are simply left dumped in the open, emitting radioactive elements into the air and water.

Similarly, the uranium enrichment process is also very energy intensive. For instance, in the US, the Paducah enrichment facility uses the electrical output of two 1,000 MW coal-fired plants for its operation, which emit large quantities of CO_2. The Paducah enrichment facility and another at Portsmouth, Ohio, also release from leaky pipes 93 per cent of the chlorofluorocarbon (CFC) gas emitted yearly in the US. This gas is the main culprit responsible for stratospheric ozone depletion. CFC is also a global warmer, 10,000 to 20,000 times more potent than carbon dioxide.[6]

The construction of a nuclear reactor is a very high-tech process, requiring an extensive industrial and economic infrastructure. Constructing the reactor also requires a huge amount of concrete and steel. Furthermore, construction has become ever more complex because of increased safety concerns following the meltdowns at Three Mile Island and Chernobyl. All this consumes huge quantities of fossil fuel. After the reactor's life is over, its decommissioning is also a very energy-consuming process.

Finally, constructing the highly specialised containers to store the intensely radioactive waste from the nuclear reactor also consumes huge amounts of energy. This waste has to be stored for a period of time which is beyond our comprehension—hundreds of thousands of years! Its energy costs are unknown.

Energy Balance

A superb study by senior scientists Jan Willem Storm van Leeuwen and Philip Smith, titled *Nuclear Power—the Energy Balance,* made in 2004 at the request of the Green parties of the European Parliament, attempts to make an estimate of the energy consumed during each stage of the nuclear fuel cycle. Excluding the energy costs of transportation and storage of radioactive waste, they estimate that under the most favourable conditions the nuclear fuel cycle emits one-third of the carbon dioxide emissions of modern natural gas power stations. This is assuming high-grade uranium ore is used to make the nuclear fuel. But these high-grade ores are finite. Use of the remaining poorer ores in nuclear reactors would produce more CO_2 emissions and nuclear energy's green choga will no longer remain green.[7]

Indian uranium ores are very poor. The official figures are that the average concentration of uranium in the Jaduguda uranium mines is a measly 0.06 to 0.07 per cent.[8] From the total uranium mined in Jaduguda over the last 40 years, Dr Surendra Gadekar has estimated, in an article published in the *Bulletin of Atomic Scientists,* that the ore quality at Jaduguda hasn't been better than 0.03 per cent for many years.[9] At such meagre concentrations, it is obvious that the total CO_2 emissions from the nuclear fuel cycle in India would be fairly high.

Potential of Nuclear Energy in Reducing Global Warming

However, this represents only half the argument. To estimate the full potential of nuclear energy in reducing global warming, we need to understand what are the various causes for the increase of greenhouse gas emissions. Increased burning of fossil fuels since the industrial revolution is only one of the causes, albeit a large one. It accounts for about 66 per cent of the total global greenhouse gas emissions (Table 5.1 below). The other causes are methane from agricultural operations and landfills, nitrogen oxides from the use of massive amounts of fertilisers in chemical agriculture, various chlorofluorocarbons used earlier as sprays and the large scale deforestation going on all over the world's jungles.[10] Thus, replacing burning of fossil fuels by nuclear energy can at best address only two-thirds of the problem (that too, assuming nuclear energy does not emit greenhouse gases, which is not true).

Table 5.1: Contribution of Various Sectors to Global Warming[11]

Fossil fuel burning	**66.5%**
consisting of	
Transportation	14.3%
Electricity and heat	24.9%
Other fuel combustion	8.6%
Industry	14.7%
Fugitive emissions	4%
Industrial processes	**4.3%**
Land use change	**12.2%**
Agriculture	**13.8%**
Waste	**3.2%**
Total	**100%**

Yet, there is a problem even with this estimate. Fossil fuels are burnt for various uses, of which generating electricity is only one (though it is the largest). Nuclear power can replace fossil fuels only in large scale electricity generation, and not in its other uses, like in the transportation sector. Worldwide, use of fossil fuels for electricity and heating contributes to only 25 per cent of the total greenhouse gas emissions (Table 5.1). Nuclear power can only marginally replace use of fossil fuels in heating supply systems. Therefore, replacing burning of fossil fuels with nuclear energy reduces only a very small percentage of the total greenhouse emissions and, that too, assuming that the nuclear energy is generated using high-grade uranium ore.

By how much? The International Energy Agency (IEA) has made an estimate.

IEA Estimates

The energy scenario produced by the IEA[12] estimates that even if existing world nuclear power capacity could be quadrupled by 2050, its share of world energy consumption would still be below 10 per cent. What is more significant for our present discussion, even such a massive expansion would help reduce CO_2 emissions by only 4 per cent![13] (Even so, we are not sure of this estimated reduction too, as we do not know whether, in calculating this figure, the IEA took into consideration the carbon emissions in the entire nuclear fuel cycle. But let us ignore this for the present.)

That is a very small reduction. The crisis of global warming is very acute, and to tackle it, what the world needs is not marginally reduced emissions, but deep cuts in them—40 per cent by 2020 and 95 per cent by 2050.[14] Obviously, nuclear power cannot significantly contribute to bringing about these reductions.

On the other hand, implementation of this scenario would require construction of 32 new 1000 MW nuclear reactors every year from 2008 until 2050. To put this into perspective, in the 1980s—the decade of nuclear industry's fastest growth—the industry built an equivalent of 17 large reactors a year. Investment costs for the 1,400 new reactors needed would exceed $10 trillion at current prices.[15] That is simply mind-boggling! Given the huge subsidies needed to build just one reactor, that would bankrupt even the richest countries!!

What About Renewable Sources of Energy?

The above discussion compared carbon dioxide emissions from the nuclear fuel cycle with that from gas and coal fired power plants. The entire nuclear lobby focuses on this comparison to make an argument for building nuclear power plants, and not only that, demanding huge subsidies for nuclear energy. However, there is another aspect of the whole issue, which the nuclear lobby very conveniently forgets: renewable energy sources emit even less greenhouse gases than nuclear plants! In comparison with renewable energy sources, power generated from nuclear reactors releases four to five times more CO_2 per unit of energy produced, when taking into account the entire nuclear fuel cycle.[16] Therefore, if the growing crisis of global warming is an argument in support of promoting nuclear energy as compared to electricity from burning fossil fuels, by this same logic, renewable energy should be promoted as compared to nuclear energy. Nevertheless, the advocates of nuclear energy conveniently overlook renewable energy in their passionate arguments about promoting green alternatives to electricity from fossil fuels. We discuss this alternative in detail in Chapter 10.

Not only does nuclear energy create a significant amount of greenhouse gases, and is on trajectory to produce at least as much greenhouse gases as conventional sources of energy within the next one or two decades, it additionally undermines the real solutions to climate change by diverting urgently needed investments away from clean, renewable sources of energy and adoption of energy efficiency measures. Olkiluoto-3 (OL3) has had a disastrous impact on Finland's renewable energy industry. Prior to the decision to build the new reactor, the Finnish renewable energy industry was thriving. Today, the renewable energy market has stagnated, as 85 per cent of planned investments in new power generation between 2006 and 2010 have been eaten up by OL3. Leading international business advisors Ernst & Young have ranked Finland as the third least-attractive among 25 countries for investments in renewable energy.[17]

6

GLOBAL NUCLEAR ENERGY SCENARIO: REVIEWING THE RENAISSANCE

Part I: Overview of the Global Scenario

Recent Changes in Global Nuclear Industry Scenario

Presently, thirty-one countries have nuclear power plants: nineteen in Europe (including Russia and the Ukraine), six in Asia (including

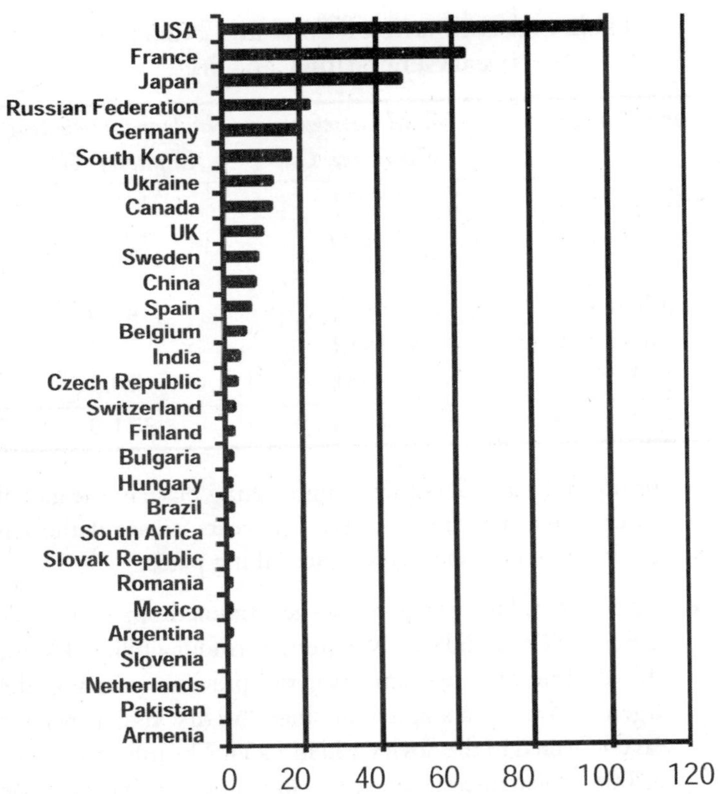

Graph 6.1: Nuclear Power Capacity by Country, end of 2009[1] (in GW)

China and Taiwan), five in the Americas and one in Africa (South Africa)—see Graph 6.1.

At the end of 2009, global nuclear generating capacity was roughly 370 GW. Of this, 33 per cent was in Western Europe, 30 per cent in North America, 21 per cent in the Far East, and 13 per cent in Eastern Europe and Russia. The rest of the world—Africa, Latin America, the Middle East and South Asia—accounted for only 3 per cent (Table 6.1).

Table 6.1: Estimates of Total and Nuclear Electrical Generating Capacity[2]

Country Group	*Total Generating Capacity, GW*	*Nuclear Generating Capacity, GW*
North America	1251	113.3
Western Europe	800	122.7
Eastern Europe and Russia	471	47.6
Far East (incl. China)	1412	77.9
Rest of World	981	10.3
Total	**4914**	**371.9**

Let us now take a look at the changes taking place in the global nuclear power scenario during the past few years, to see if there is indeed some kind of a nuclear renaissance taking place.

- In 2007, world nuclear electricity generation dropped by more than 50 TWh to 2608.2 TWh (terrawatthour = billion kWh). This decline of 2 per cent over the previous year was the largest decline in a single year since the first fission reactor was connected to the Soviet grid in 1954. The following year, in 2008, global nuclear generation lost another half percentage point (over the 2007 level).[3] The 2010 Edition of the IAEA report *Energy, Electricity and Nuclear Power Estimates for the Period up to 2030* records yet more decline: global nuclear energy generation in 2009 fell by another 1.5 per cent over the 2008 level to 2558.1 TWh.[4]
- Compared to total global electricity generation from all sources, world nuclear energy generation fell from 15.2 per cent in 2006 to 14.2 per cent in 2007, to 14 per cent in 2008 and to 13.8 per cent in 2009; that is, it had fallen for the third consecutive year in 2009.[5]
- Similarly, as we can see from the Graphs 6.2 and 6.3 below, from 1954, when the first nuclear reactor came online, to 1990, the total number of reactors worldwide and the total global generating capacity rapidly increased. However, after

that, the number of reactors has hovered around 430 and the increase in total capacity has also slowed down, and is presently hovering at around 360-370 GW.

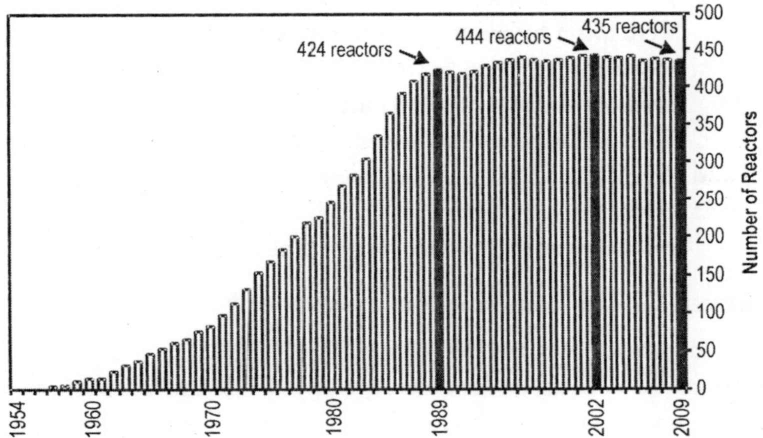

Graph 6.2: World Nuclear Reactor Fleet[6] from 1954 to August 1, 2009

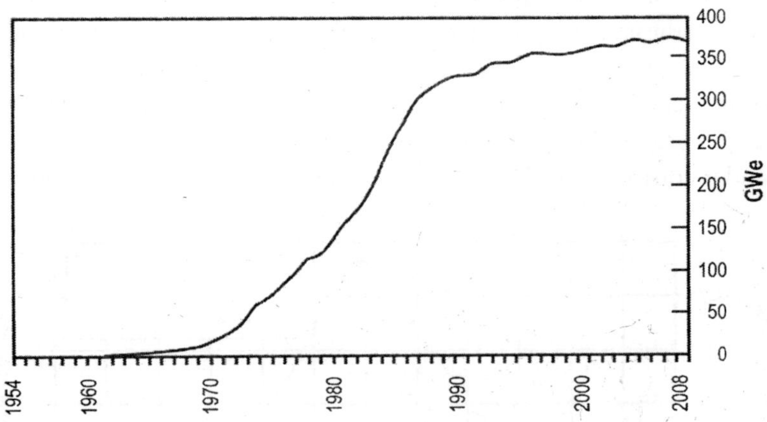

Graph 6.3: Growth of Global Nuclear Power Capacity, GW[7]

- At the end of 2009, there were 438 nuclear reactors operating in the world, six less than in 2002.[8] 2008 was the first year in the history of commercial nuclear power that no new nuclear plant came online (although two were connected to the grid in 2009).[9]

- Even though the total number of reactors has declined, the total installed capacity has increased slightly in recent years mainly because of technical alterations at existing plants, a process known as 'uprating'. However, in 2008, uprates too were offset by plant closures, resulting in a slight decline in world nuclear capacity by about 0.6 gigawatts over the 2007 level. At the end of 2009, the total installed capacity of the 438 operating nuclear reactors was 371.9 gigawatts.[10]
- As compared to the global electricity generation capacity, the global nuclear power capacity has consistently declined, from 8.7 per cent in 2006 to 7.6 per cent in 2009.[11]

Current Global Trend in Construction of New Reactors

As of August 1, 2009, the IAEA listed 52 reactors with a total capacity of about 46 GW as 'under construction'. While this represented a slight upswing over the previous five years (since 2004), on the whole, it was a huge decline from the peak reached in 1979 when there were 233 reactors of total capacity of over 200 GW being built concurrently. Even at the end of 1987, there were 120 reactors under construction.[13]

A comparison of the total capacity of the nuclear reactors under construction with the global power capacity under construction (from all sources) is also revealing. The total electricity generating capacity under construction (from all sources) in 2007 was estimated at over

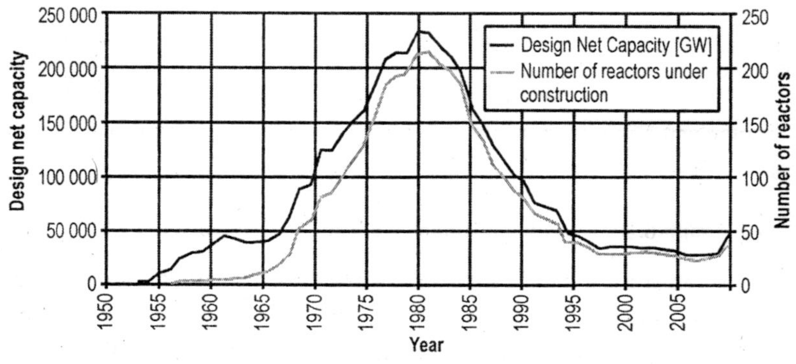

Graph 6.4: Number of Units and Total Nominal Capacity (in MW) Under Construction 1951-2008[12]

600 gigawatts. Of this, the vast majority was from coal, hydro and natural gas plants; the nuclear share was just around 4.4 per cent.[14]

Let us now take a closer look at the 52 nuclear projects under construction:[15]

- 13 reactors, one quarter of the total, have been listed as 'under construction' for over 20 years. Of these, 8 reactors do not have an official (IAEA) planned start-up date even today.
- In fact, of the 52 reactors under construction, 24 projects don't have an official (IAEA) planned start-up date.
- Over two-thirds (36) of the units under construction are confined to just four countries (China, India, Russia, South Korea), with China, alone, accounting for 16 of them. All of these nations have historically not been very transparent about the status at their construction sites.
- Past experience has shown that even a reactor in an advanced stage of construction is no guarantee for grid connection and power supply. The French Atomic Energy Commission (CEA) published statistics on 'cancelled orders' as of end–2002. The CEA listed 253 cancelled orders in 31 countries, many of them in advanced construction stage. After that, it stopped publishing statistics on cancellations.

The noted independent consultant, Mycle Schneider, along with professor for energy policy, Stephen Thomas, and consultants, Antony Froggatt and Doug Koplow, have, in their *The World Nuclear Industry Status Report 2009*, calculated the minimum number of plants that would have to come online over the next few decades in order to maintain the same number of operating plants as there were on August 1, 2009 (435). While many nuclear utilities envisage reactor lifetimes to be at least 40 years, and some have even applied for and obtained licenses for operating their reactors for more than 40 years, the report points out that these seem to be rather optimistic projections, considering the fact that the average age of all the 123 units that have closed down till date is about 22 years. Nevertheless, for their calculations, the authors assume that each of the presently operating and in-construction reactors will have a life of 40 years.

With these assumptions, the report finds that, in order to maintain at least 435 operating reactors worldwide (the same number as on August 1, 2009) in the coming decades, in addition to the 52 units currently under construction:

- 42 reactors (16,000 MW) would have to be planned, built and started up by 2015 (that is, one every month and a half);
- an additional 192 units (170,000 MW) would have to be constructed and brought online over the following 10-year period (which means, one reactor would have to come online every 19 days).[16]

Considering that it takes more than a decade of planning, regulatory processes, construction and testing before a nuclear reactor can produce electricity,[17] this means that it is going to be practically impossible to maintain, let alone increase, the number of operating nuclear power plants over the next 20 years.

Part II: Overview of the Nuclear Industry in North America and Western Europe

We now take a closer look at the prospects for a 'nuclear renaissance' in the United States, Canada and Western Europe, the region that was at the centre of the first boom in nuclear energy and where 63 per cent of the world's operating reactors are located (as on August 1, 2009). This is also the region where public opinion is most informed and the debate is most intense on nuclear energy.

We are deliberately refraining from examining the situation in China and Russia, the two countries where the maximum number of nuclear plants are under construction at present, for reasons discussed below.

Reasons for Ignoring China

Speaking at an International Ministerial Conference on Nuclear Energy in Beijing on April 19, 2009, Li Ganjie, the director of China's National Nuclear Safety Administration, warned that 'if we are not fully aware of the sector's over-rapid expansions, it will threaten construction quality and operation safety of nuclear power plants'.[18] He also stated that China's nuclear industry is challenged on all fronts: shortage of human resources; insufficient capability of nuclear power research, development, design and mastery of high-end technology; lack of capability in manufacturing and installing of facility; inadequate management; and weak nuclear safety supervision.[19]

Apart from this stray news, not much is known about safety at China's nuclear plants. The dictatorship prevents any information from coming out. The global nuclear industry is more than happy with this state of affairs, as it is only concerned with nuclear plant orders. Not only that, it can hold up China as an example for countries like India to duplicate. So it keenly spreads the belief that all is fine with China's nuclear plants. Energy consultant Mycle Schneider comments: 'Everything goes black when I consider that 16 nuclear plants are being built simultaneously in China, and all we hear is there are no problems there.'[20]

The reason for Schneider's harsh comment is not far to seek. China's attempt at constructing all kinds of giant projects at reckless speed has pushed the country to the edge of a monumental environmental crisis, perhaps the worst in world history. China's coal-fired plants and giant heavy industry complexes freely dump their toxic wastes into the environment, poisoning the land with a deadly brew of chemicals and metals. One major consequence is that 16 of the world's 20 dirtiest cities are located in the People's Republic. The inhabitants of every third metropolis are forced to breathe polluted air, causing the deaths of an estimated 750,000 Chinese each year. Half of China's 696 cities and counties suffer from acid rain. Two-thirds of its major rivers and lakes have become cesspools and more than 340 million people do not have access to clean drinking water. The Yangtze River, once China's proud artery of life, is biologically dead for long stretches. Cancer rates in many villages located near heavily polluting factories have shot up, earning them the name of 'cancer villages'. Cancer is now the nation's biggest killer, responsible for one in five deaths in 2007.[21]

It is therefore not at all surprising that such a dictatorship, that is so utterly unconcerned about environmental degradation and the health and well-being of its people, is making a huge push for setting up nuclear plants.

Another important reason as to why China is able to make a push for setting up so many nuclear plants is because it doesn't have to worry about costs—public money is being used to set up these projects, like in India.

Reasons for Ignoring Russia

Even though the IAEA says Russia was constructing 9 reactors as of end-2009,[22] we are ignoring Russia in our discussion on the 'global nuclear renaissance'. The reason is the criminal negligence shown by the Russian government towards disposal of radioactive waste from its nuclear facilities and the murderous apathy it has shown towards the victims of radiation leakages and nuclear accidents at its nuclear plants.

Thus, for example, Russia (and earlier the Soviet Union) has injected billions of gallons of liquid atomic waste deep into the earth —several hundred metres underground—at three widely dispersed sites: Dimitrovgrad near the Volga River, Tomsk near the Ob River and Krasnoyarsk on the Yenisei River. The total amount of radioactivity injected is more than 2 billion curies—to grasp the monstrosity of this dumping, this figure is several times the official estimate of the total radiation released from the Chernobyl accident, which was 50 million curies. While Russian scientists are claiming that the practice is safe, in reality nothing is known of its possible consequences, how far it will spread underground, what dangers it will cause—which is why no other country has ever tried this technique for waste disposal. Upon learning of these injections, Henry W. Kendall, a Nobel laureate at Massachusetts Institute of Technology, commented: 'Far and away, this is the largest and most careless nuclear practice that the human race has ever suffered.'[23]

An even more horrifying example is the way Russia (and earlier the Soviet Union) has handled radioactive waste leakages and the 1957 accident at a nuclear facility called the Mayak Chemical Combine, the biggest nuclear complex in the world. It makes the stomach churn. The Mayak nuclear facility, located near the city of Kyshtym in the province of Chelyabinsk in the Ural Mountains region, was built in the 1940s, and was a key facility for production of plutonium for the Soviet weapons program. From 1948 to the mid-1950s, high level radioactive waste from the plant was simply dumped into the Techa River, releasing nearly 3 million curies of radioactivity into the environment. The government restricted drinking of water and fishing in the river. However, because local residents were not told why the new restrictions were put in place, they continued using the river. While over 1,24,000 people were exposed to high levels of radiation, hardly anyone was evacuated.

The Techa River flows into the Ob, the great Siberian river which in turn flows into the Arctic Ocean. In the early 1950s, radioactivity from Mayak was detected in the Arctic more than a thousand kilometres away; after this, the Soviets stopped dumping high level waste into the Techa. However, dumping of low and medium level wastes into the river continued for many years!

Medium and low level wastes have also been dumped into many natural and artificial reservoirs, which are vulnerable to floods and droughts. Among them is probably the most radioactively contaminated body of water in the world, Lake Karachay. It is unclear exactly how much radiation was released into the lake. The US-based Natural Resources Defence Council, during a trip to the USSR in 1989, estimated that there is approximately 120 million curies of radiation present in the lake. That is more than twice the radioactivity released during the Chernobyl accident (the official estimate is 50 million curies). It further estimated that the lake holds more than 100 times the amount of strontium-90 and cesium-137 released at Chernobyl.

In 1967, a major accident occurred at Mayak. This region of the Urals is known for its wind storms. Following a partial drying of the lake due to two consecutive years of drought, in 1967, strong winds carried an estimated 5 million curies of radioactive dust from Lake Karachay across an area hundreds of square kilometres wide. Probably 4 lakh people were exposed to significant levels of radiation. However, as it happens with all nuclear accidents, only the most severely affected were evacuated. Since this disaster, the authorities have attempted to fill up the lake with rocks and concrete blocks. However, there are several other reservoirs in the region which also contain radioactive waste from the Mayak complex. Seepage from these reservoirs has contaminated the groundwater over an area of several square kilometres, and continues to spread.

The worst accident at Mayak occurred in 1957. After the dumping of high level liquid reprocessing wastes into the Techa stopped in 1951, they now started being stored in large stainless steel tanks. In 1957, the cooling system of one of these tanks failed, resulting in a massive explosion: more than 20 million curies of radioactivity was released into the atmosphere. The radioactive cloud travelled for miles. 217 towns and villages with a combined population of 270,000 were significantly contaminated. However, only 10,000 people were evacuated.

Until the Chernobyl disaster in 1986, Mayak was the worst radiation accident the world had ever seen. The Soviet Union kept news about the accident hidden from the world for nearly twenty years. In a strange conspiracy of silence, even though the CIA knew about the accident, it, too, hid the news, for fear that its release would fuel anti-nuclear sentiment in the US and jeopardise the US nuclear energy program! The first report about the accident came out only in 1976, when a Soviet scientist wrote about it in an article in the popular British science magazine *New Scientist.*

Despite these numerous tragedies, the last of Mayak's five plutonium production reactors was shut down only in 1991. Even after that, the spent fuel reprocessing plant at Mayak has continued to operate; worse, in 2001, the Russian Parliament overturned a ban on import of spent fuel from other countries for reprocessing at the Mayak complex! Mayak has reprocessed over 1,540 tons of spent nuclear fuel from several countries including Hungary, Bulgaria, Germany and Finland so far. Russia is also planning to sign reprocessing contracts with Switzerland, Spain, South Korea and other countries. As a result of this, over 3 million cubic meters of liquid low-level and middle-level radioactive waste has been generated, which continues to be dumped into nearby lakes like Karachay, Old Marsh and several artificial reservoirs, from where it leaks into the River Techa, and further to the River Ob and to the Arctic Ocean.

None of the countries shipping their dirty nuclear waste to Russia would allow Mayak to continue operating on their own land. They are exploiting Russia's weak environmental and health standards to dump their radioactive waste on people who have already suffered the devastating consequences of nuclear contamination for half a century. Mayak is a horrific example of the true face of the global nuclear industry.

Mayak is probably one of the most contaminated places on Earth. Despite the high radiation levels in the region, the Russian authorities are making no efforts to evacuate the people living here. More than fifteen lakh people living in this region (out of a total population of thirty two lakhs) have suffered radiation exposures equivalent to 20 times that suffered by the victims of the Chernobyl accident. Thousands must have died the most painful death, while tens of thousands of those alive must be living in great suffering: a recent study by Greenpeace found the rates of malignant cancers among local people to be significantly higher compared to the rest of Russia, while another study found that genetic abnormalities were 25 times higher than in other areas.[24]

1. Reviewing the Nuclear Renaissance in the USA

The United States has 104 operating nuclear power plants (as of 2009). While this number is more than any other country in the world, the number of cancelled projects is even larger. Of the 253 nuclear plants ordered in the US since 1953, 71 were cancelled before construction started, 50 were cancelled after construction began and another 28 were permanently shut down before their 40-year operating licenses expired.[25]

No new order for a nuclear reactor has been placed in the United States since 1978, and even that plant was later cancelled. In fact, all US reactor orders after October 1973 were eventually cancelled—that is, it is now 37 years since a new order (that was not subsequently cancelled) has been placed. The last reactor to be completed was Watts Bar-1, in 1996. Its completion took 23 years.[26]

Despite these dismal statistics, the nuclear industry has been claiming that a nuclear renaissance is underway in the USA. It is basing its claim on the following achievements:

- The most important success achieved by the US nuclear industry, without which actually no nuclear renaissance is possible, is that it has been able to win billions of dollars of loan guarantees and other financial handouts from first the Bush and now the Obama administrations (discussed in Chapter 4).
- Buoyed by these subsidies, in 2007, for the first time in three decades (since the Three Mile Island accident in 1979), utilities in the US applied for a license to build a nuclear plant. As of July 2009, the NRC had received 17 applications for a total of 26 units.[27]
- In February 2010, the Obama administration announced the authorisation of the first loan guarantee of $8.3 billion to the Southern Company to build two new 1,000 megawatt Westinghouse AP-1000 nuclear reactors at its Plant Vogtle site in Georgia.[28]
- Construction on the 1,200 MW Watts Bar-2 reactor has been restarted by Tennessee Valley Authority, a federally owned corporation in the USA. Its construction began in 1972 but was frozen in 1985. The reactor is now expected to be completed by 2012.[29]
- The US nuclear power industry has also been successful in getting plant life extensions. Originally, reactor life was envisaged to be 40 years. But now, utilities are seeking permission to operate reactors for up to 60 years. As of July 2009, 54 nuclear plants had been granted a life extension license by the Nuclear Regulatory Commission (NRC), 16 applications were under review and around 21 had submitted letters of intent.[30]

Despite these gains, the overall future prospects for the nuclear industry are not as rosy as it is claiming them to be. Despite its multi-billion dollar propaganda campaign to convince the people about the benefits of nuclear energy, public opposition to nuclear energy continues to remain strong and it has led to big setbacks for the nuclear industry. Let us take a look at the other side of the picture.

Setbacks

The most important of these defeats has been on the question of the quantum of loan guarantees. As discussed in Chapter 4, without loan guarantees, industry cannot even think of constructing a new nuclear reactor. The nuclear industry had lobbied hard during the Bush presidency to get the Congress to give loan guarantees for $50 billion. But campaigning by public interest and anti-nuclear groups got that amount knocked down to $18.5 billion.[31] This amount is barely enough to cover loan guarantees for 3 reactors. Utilities are now asking the DOE for $122 billion in loan guarantees for the 26 new reactors they propose to construct![32] Obviously, only if Congress overrides strong public opposition and sanctions a huge increase in loan guarantees will the nuclear renaissance ever take off!

Many states in the United States have laws which either explicitly or effectively ban the construction of new nuclear plants. The nuclear industry has done intense lobbying to get these states to lift their ban, but has so far completely failed. Thus, for instance, Minnesota has a moratorium in place on construction of new nuclear power plants; while California, West Virginia, Wisconsin and some more states have laws according to which no new nuclear plant can be constructed in these states until there is a national facility which safely disposes of high level nuclear waste. In 2009, the nuclear industry tried in six states to get these laws repealed, but its efforts came to naught everywhere. Similarly, the nuclear industry failed to get the Missouri legislature to pass a CWIP law that would have enabled costs to be imposed on the state's ratepayers to finance construction of a new nuclear plant, which was then promptly mothballed. Industry efforts to get nuclear declared 'renewable' by the states of Indiana and Arizona also failed to achieve results.[33]

Growing public opposition to the expansion plans of the nuclear industry is also putting at risk one of the important recent successes of nuclear utilities—lifetime extensions of their operating plants. In Vermont, because of a huge grassroots campaign, an overwhelming 26 members of the 30-member state Senate voted in February 2010 against giving a life extension to Entergy's Vermont Yankee Nuclear

Plant for another 20 years after its scheduled closing in 2012. Of course, the fight isn't over; Entergy is a powerful corporation and has declared that it has not thrown in the towel. The House still has to vote and it is to be seen whether it will vote the same way and retire Vermont Yankee. The Vermont Senate vote was the first time a state legislative body has voted to retire a nuclear plant.[34]

Apart from these setbacks at the policy level, even the ambitious plans made by the nuclear industry for construction of new reactors have suffered serious setbacks.

- President Bush's National Energy Policy had set a target of constructing two reactors by 2010. However, construction on even the first of these reactors has yet to begin even in the first half of 2011.
- Of the 26 reactors for which applications had been received by the NRC till the end of 2009, 19 have been cancelled or delayed and every project has suffered a downgrade by credit rating agencies.[35]
- The nuclear plant construction applications received by the NRC cover 5 designs. However, so far, only one of these designs—the ABWR—has been certified by the NRC. Even its certification runs out in 2012, and major modifications are likely to be needed for it to be re-certified.[36]
- 14 of the 26 reactors whose construction applications are pending before the NRC are of Westinghouse's AP-1000 design. In October 2009, the NRC raised serious concerns about this reactor design. The NRC stated that Westinghouse has failed to demonstrate whether the AP-1000 nuclear reactor structure can withstand hurricanes, earthquakes, tornadoes and other external impacts. It also stated that the unsuccessful efforts to secure information had gone on for a year.[37] Additional concerns about the AP-1000 design were voiced in a report released in April 2010 that was commissioned by the AP-1000 Oversight Group, which involves more than a dozen nuclear watchdog organisations. The report, prepared by Arnie Gundersen, a nuclear engineer

and a former senior executive in the nuclear power industry, stated that the design was particularly vulnerable to through-wall corrosion.[38] These concerns put in doubt the future of all projects involving the AP-1000 reactor design. And they constitute more than half the reactors proposed!

- One of these AP-1000 projects whose future has now become uncertain is the Vogtle nuclear plant in Georgia. This was the first project to win a loan guarantee from the Obama administration just a few months ago.
- The DOE has shortlisted three more projects for a second loan guarantee: South Carolina Electric and Gas' proposal for two AP-1000s at the Summer Nuclear Power Plant in South Carolina; EDF-Constellation Energy's proposal to build a EPR reactor at its Calvert Cliffs site in Maryland; and NRG Energy's proposal for two ABWRs at the South Texas Project Nuclear Plant in Texas.[39] However, all three projects have become crisis-ridden:
  (i) South Carolina Electric and Gas' proposal is for AP-1000 reactors, so obviously its future is in doubt.
  (ii) In October 2010, Constellation Energy announced that it was withdrawing from its joint venture with France's EDF to build the Calvert Cliffs-3 reactor in Maryland; it blamed the US government for insufficient subsidies for the decision.[40] The project is now virtually dead, with their being only an outside chance that EDF would go ahead with it.[41]
  (iii) Cost estimates for building two additional ABWR's at the South Texas nuclear plant have ballooned to $18.2 billion from a preliminary estimate of $5.4 billion, even before the first stone for the project has been laid![42] There is no knowing what the final costs are going to be! Indications are that this may well have stalled the project.[43]

Clearly, the nuclear renaissance in the United States is stuck in quicksand.

2. Reviewing the Renaissance in Canada

Canada was one of the first countries to invest in nuclear power. It developed the CANDU design, a heavy water reactor. Officially, there are 18 CANDU units in operation in Canada; another 4 units are in what the IAEA calls 'long-term shutdown'. The reactors in operation have been plagued by technical problems that have led to construction cost over-runs, shut downs for long periods, and reduced annual capacity factors. In the mid-1990s, one-third of Canada's nuclear plants were shut down for technological reasons, the largest shut down in the world.[44]

No nuclear plants have been ordered in Canada since 1978.[45] However, like in the USA, a number of its operating plants have been refurbished to extend their operating lifetimes. While refurbishing usually takes less time and is less costly than building a new plant, for several of Canada's reactors, the cost overruns have been so large that the refurbishing cost has become almost as much as expensive as new construction.[46]

Over the past few years, there have been several proposals to build new nuclear plants in Canada. These would have been Canada's first nuclear plants in 3 decades. However, all have come to naught, because of strong public opposition and high financial risks. The President of the Canadian Nuclear Safety Commission (CNSC) has stated that CNSC is 'facing many of same issues as the rest of the nuclear industry'.[47]

3. Reviewing the Renaissance in Western Europe

There is no fixed definition of which countries constitute Western Europe. We have for our purpose defined it to include the following 18 countries: Austria, Belgium, Denmark, Finland, France, Germany, Greece, Iceland, Ireland, Italy, Luxembourg, Netherlands, Norway, Portugal, Spain, Sweden, Switzerland and the UK.[48]

Nine of these 18 countries—Belgium, Germany, Finland, France, Netherlands, Spain, Sweden, Switzerland and the United Kingdom—operated 129 nuclear power reactors with a total installed capacity of 125 GW as of August 1, 2009. This was 33 units less than in 1988-9 when the number of operating units peaked.[49]

Two reactors are currently under construction in this region, one in Finland and one in France. Except for France which has continued to build nuclear reactors even after the Chernobyl accident (the latest being the Civaux-1 and 2 units which got underway in 1991 and 1993 respectively and were coupled to the grid in 1999, and the recent EPR reactor project in Flamanville), and the order for the EPR reactor by Finland in 2003, no new reactor order has been placed in Western Europe since 1980—that is one order outside France in 30 years. On the other hand, dozens of reactors will go offline in the coming years—at least one-third of Europe's nuclear plants would be decommissioned by 2025.[50]

Despite this, apologists for the global nuclear industry are claiming that a nuclear renaissance is underway in Western Europe. Let us take a closer look at this so-called revival. For our discussion, we divide the 18 countries of Western Europe into three categories: countries with no nuclear plants which are still anti-nuclear (8), countries with nuclear plants which had earlier planned to phase them out and are now considering reversing this phase-out (6), and countries with nuclear plants and without phase-out policies (4).

(i) The Anti-nuclear Countries

Austria, Denmark, Greece, Iceland, Ireland, Luxembourg, Portugal and Norway are declared non-nuclear countries. They do not operate any nuclear plants. While most of them have never had a nuclear power program,[51] Austria and Denmark did have one but decided to scrap it many years ago. Austria had in fact built a nuclear plant in the 1970s, and there were plans to build two more reactors, but in 1978 a referendum against nuclear power succeeded because of which the technically finished reactor was never started. Since then, a majority of the people and all major political parties are against nuclear power.[52] Denmark too was once in the forefront of nuclear research and had planned on building nuclear power plants. However, in 1985, the Danish parliament passed a resolution that nuclear power plants would not be built in the country.[53]

In recent times, there have been rumours that Greece was planning to go nuclear. However, in a statement in February 2009, the Greece Development Minister trashed these rumours and in fact ruled out investment in coal fired plants too.[54]

(ii) Countries where Nuclear Policy is in Flux

Rising public consciousness about the terrible environmental consequences of nuclear energy, especially after the Chernobyl accident, led the Netherlands, Sweden, Belgium and Italy to impose moratorium on construction of new nuclear plants and decide to phase out their operating nuclear plants. Italy shut down its last nuclear reactor in 1990;[55] the Dutch parliament voted in 1994 to phase out its only nuclear power plant by 2003;[56] while Sweden planned to complete the phase out of its 10 reactors by 2010.[57] Belgium decided to shut down its 7 reactors after 40 years of operation, which meant they would shut down between 2014 and 2025.[58]

A powerful anti-nuclear movement ultimately led to the German Parliament voting in 2002 to pass the Nuclear Exit Law, whereby all the 19 operating nuclear reactors would gradually shut down and all civilian uses of nuclear power would cease by 2020, meaning construction of new nuclear plants would be prohibited. In accordance with this law, two units have been shut down so far.[59]

Spain imposed a moratorium on construction of new nuclear plants in 1983. In 2008, the centre-left government of Jose Luis Zapatero came to power on an election manifesto which pledged to gradually replace nuclear energy with renewable energy and also phase out Spain's nuclear plants once they reached the age of 40 years.[60]

Nuclear Revival

Powerful lobbying by the nuclear industry has got five of these countries—Belgium, Sweden, Italy, Netherlands and Germany—to reconsider their decision to phase out nuclear energy. Spain has not reversed its decision but is going slow on its implementation. The policy reversals in these countries are at the centre of what the nuclear industry is proclaiming to be a 'nuclear renaissance' taking place in the world. Let us take a look at the extent of nuclear revival in these countries.

Reviewing the Nuclear Revival

The **Dutch** decision to phase out its sole 480 MW nuclear plant was later abandoned by a conservative government, and in 2006, the government granted permission to extend the life of the plant from 2013 to 2033.[61] While there is pressure on the government for allowing the construction of a second nuclear plant, current Dutch policy on nuclear new build remains uncertain with the government putting off a decision on its formal stance until at least 2011.[62]

In 2009, the **Belgium** government announced, without overturning the Nuclear Phase-Out Law, that it is postponing the phase-out by 10 years implying that the phase-out would not begin until 2025.[63] Before it could get its decision ratified by the Parliament, the government fell in April 2010. Despite fresh elections, as of October 2010, a stable government had yet to be formed. The Greens, who are very strong in Belgium, have announced that they will only join the new government if it agrees to keep the Nuclear Phase-Out Law in place.[64] Therefore, the situation is much in flux. But one thing is clear—even if the existing plants get a lifetime extension, no new nuclear plants are going to come up in Belgium in the near future.

In **Sweden**, in 2009, the centre-right coalition government announced a decision to scrap the Nuclear Phase-Out Law, and begin construction of new plants to replace Sweden's aging reactors from 2011.[65] The Parliament finally passed their proposal only in June 2010, and that too by a narrow margin of just 2 votes.[66] But soon after, the government lost its majority, in the elections held in September 2010.[67] With the opposition staunchly opposed to nuclear power, the future of the 'renaissance' looks uncertain. The government has also set ambitious targets for renewable energy and energy conservation, which actually leave very little space for nuclear energy. This makes it very unlikely that a nuclear plant is going to be ordered in Sweden in the near future.[68]

In **Italy**, a new right wing government came to power in April 2008, and announced plans to start rebuilding nuclear plants within five years. But implementing these plans is not going to be easy, as public opposition to nuclear plants continues to be strong.[69] In March 2010, a majority of the regional councils of Italy voted against any return to nuclear power. This greatly increases the central government's problems, as many of the governors belong to the ruling party. With the Berlusconi government continuously besieged by scandals, and having a very thin majority in Parliament, the strong opposition to nuclear power makes the future of the Italian renaissance very uncertain.[70]

In **Germany** too, the new center-right government of Angela Merkel has abandoned the commitment to phasing out nuclear power by 2021. In October 2010, it used its majority in the lower house (Bundestag) to pass a bill to extend the working lives of its reactors by an average of 12 years.

However, Merkel does not have a majority in the upper house (Bundesrat) where Germany's states are represented, so she has not submitted the legislation for approval there as it is sure to get defeated. The opposition parties and the state governments are claiming that this is unconstitutional and are planning to move the Federal Constitutional Court over this. The Social Democrats have declared that they will overturn the legislation if they come to power.[71] Considering the intense hostility to nuclear power in Germany—recent polls indicate that a majority of Germans are in favour of phasing out nuclear power as soon as possible[72]—it is doubtful if there is going to be any significant revival of nuclear power there.

The **Spanish** government is going slow on its election pledge to phase-out nuclear energy; in 2009, it extended the operating license of Spain's oldest plant by two years, allowing it to operate until 2013. However, at the same time, the government has also made energy conservation and promotion of renewable energy its top priority. The emphasis on renewable energy has made Spain the world's second largest producer of solar power and the third largest of wind power (in 2009).[73] With such a huge push towards renewable energy, there is very little scope for revival of nuclear power in Spain; at the most, the existing nuclear plants will be given a lifetime extension by a few years.

(iii) Countries without Nuclear Phase-out Policy

These are Switzerland, Finland, France and the United Kingdom.

The **United Kingdom** operates 19 reactors, all of which except one are scheduled to be shut down by 2025. The UK nuclear industry is trying hard to get the government to agree to construction of new plants. Nuclear utilities and fuel industries have faced huge troubles in the UK, moving between scandal and bankruptcy. Nevertheless, first Tony Blair's government and then Gordon Brown's government have attempted to keep the nuclear option open, and in 2009, the British government took the first steps towards building of new reactors.[74] But these plans have apparently got a setback with the coming to power of the Conservative-Liberal government in May 2010. While the Conservatives are all for nuclear energy, the Liberal Democrats have long campaigned against it. The new energy minister, who is from the Liberal Party, has said that the Liberals have compromised and will support Conservative proposals to build new reactors, but only on the condition that no subsidies are given to nuclear energy. If the government sticks to its promise, this condition makes it virtually certain that no new nuclear power plant will be built in the UK.[75]

Switzerland operates five reactors. Switzerland's nuclear operators have initiated a debate over building replacements for the country's aging nuclear reactors, but the short-term prospects look dim. Referenda over phasing out nuclear energy have never won a majority in the country, but because they were defeated only by a very thin margin, they have effectively acted as a moratorium on the building of new nuclear plants.[76]

Finland and France are the two clear cut exceptions as far as nuclear energy policy goes in Western Europe. **Finland** currently operates four units. In December 2003, Finland became the first country to order a new nuclear reactor in Western Europe after more than a decade (the last one being the Civaux Nuclear Plant in France). The 1600 MW EPR being built by Areva in Olkiluoto under a turnkey contract was supposed to be constructed in four years, but is already more than four years behind schedule, and its cost has escalated to at least double the contract price. Despite these troubles, in July 2010 the Finnish Parliament approved a government proposal to construct two new nuclear power plants in the country.[77]

France is probably the most pro-nuclear country in the world. In 2008, the 59 French reactors accounted for a little more than half of West Europe's nuclear capacity (63.2 GW). France also accounts for one of the two reactors presently under construction in Western Europe. French nuclear reactors produce over 75 per cent of the country's electricity, although only about 55 per cent of its installed electricity generating capacity is nuclear. In other words, France has a huge overcapacity that has led it to dumping electricity on neighbouring countries. It also means that France does not need to build any new nuclear plants for a long time; the only reason why the French government and EDF have decided to go ahead with the construction of a new unit, Flamanville-3, is because the nuclear industry desperately needs new orders to survive.[78]

Part III: Reviewing the Renaissance in Real Life: Olkiluoto-3 and Flamanville-3

The flagships of the 'Nuclear Renaissance' being proclaimed by the global nuclear industry are the two Generation-III+ EPR reactors being constructed in Finland and France, Olkiluoto-3 and Flamanville-3 (respectively). However, both of them have got holed below the water line ...

Olkiluoto-3

Areva, the largest nuclear builder in the world, in its marketing of EPR worldwide, has promoted it as a nuclear power plant that is safer, cheaper, more mature and more reliable than any other reactor in the world. Its promotional material states: 'The EPR is the direct descendant of the well proven N4 and KONVOI reactors, guaranteeing a fully mastered technology. As a result, risks linked to design, licensing, construction and operation of the EPR are minimised, providing a unique certainty to EPR customers.'[79] However, what is certain about its Olkiluoto-3 (OL3) project is that none of these promises are being delivered.

Till November 2009, the Finnish nuclear safety authority STUK had detected about 3000 safety and quality problems in the OL3 project![80] Alarmingly, these include problems with several key components:

- Control and instrumentation system: This is the nerve centre of the reactor and controls every aspect of reactor operation as well as emergency systems. In November 2009, Finnish, UK and French nuclear safety authorities raised questions about its design, saying it was at odds with basic principles of nuclear safety.[81]
- Primary circuit: This is probably the most crucial part of the reactor, as it contains the water coolant which is responsible for safety of the reactor. The primary circuit is subject to extreme heat, pressure and radiation for decades. Its components are hard to replace, some actually impossible to replace, once the reactor is in use. And so its manufacture needs the highest quality standards. Yet, there have been quality problems with almost all the components of the primary circuit: all eight primary coolant pipes had to be recast, and STUK found the refabricated pipes to be inferior too; most of the components of the reactor pressure vessel and the pressuriser had to be remanufactured as they did not meet safety standards; and repairs had to be made in the steam generator too![82]
- Containment steel liner: There were serious defects in the welding of the containment steel liner, which constitutes the last barrier against leaks of radioactive substances into the environment in case of an accident. While STUK has acted to get these defects rectified, it has been forced to lower quality standards while doing so.[83]
- Concrete base slab: STUK found that even the concrete base slab of the reactor was of inferior quality: the water content of the concrete was too high, because of which its compressive strength as well as chemical resistance were below requirements, and could cause the base to crack in the long run. Even this defect cannot be fully rectified![84]

TVO, the Finnish electric company which has ordered the EPR, and Areva failed to spot many of these failures, and only timely detection by STUK enabled them to be corrected. However, STUK itself has admitted that the number of problems is so large that it is possible that it will not be able to detect all of them. The problem of detection has become more difficult because contractors on several occasions are known to have attempted to hide their mistakes by fabricating measurements, covering up defective structures, failing to record shoddy repair work, et cetera.[85]

According to a study done for Greenpeace by renowned nuclear expert Dr Helmut Hirsch,[86] while STUK has acted to get these defects rectified, there are many issues on which it has not provided satisfactory answers, including: the extent of problems during manufacture of the reactor pressure vessel, the procedure for re-manufacturing the primary piping, the overall state of the containment liner after repairs and the effectiveness of the counter-measures planned because of the higher water content of the base slab concrete. He also points out that there are several instances where STUK has relaxed safety requirements and allowed installation of faulty components.[87]

These are scary facts! Because, the EPR being of 1600 MW capacity, is the largest reactor ever built, and so its core contains more radioactive elements than any other reactor. In addition, for reasons of economy, it is designed to burn fuel longer, leading to increased radioactivity and greater production of dangerous nuclear isotopes. This will obviously mean greater thinning of the fuel cladding and more cracks resulting in higher radioactive releases from the reactor. A high burn-up will also lead to much higher toxicity of the radioactive waste; according to an EDF study, EPR waste will have about four times more radioactive bromine, iodine, cesium, et cetera as compared to ordinary Generation-II PWRs; other reports put this figure to be much higher.[88] All these facts make the EPR potentially more dangerous in case of an accident as compared to almost any operating nuclear reactor. In the event of a serious accident, the impact would be cataclysmic, many times more devastating than Chernobyl! As a result, more stringent construction and quality control is needed for the EPR to be able even to match the risk levels of operating reactors. However, the quality control problems at the Olkiluoto-3 site, as discussed above, indicate that it is highly questionable whether even present-day safety standards will be maintained at this plant.

That is one part of the Olkiluoto-3 fiasco. The other part is that the project has turned into a financial disaster: as of mid-2010, quality control problems and design defects have led to construction running four years behind schedule, resulting in estimated costs escalating to double the contract price ...![89]

Flamanville-3

The other European order for an EPR, Flamanville-3 in France, is doing no better, despite the fact that construction here started two and a half years after Olkiluoto-3. It is being built by Electricité de France (EDF), which is majority owned by the French government. It has far more nuclear construction experience than any other utility in the world. Work on this plant started in December 2007. Two and a half years later, in June 2010, EDF admitted that the project was running two years late and the cost overrun was more than 50 per cent.[90]

The reason for the delay and cost overruns is the same as that for Olkiluoto-3: quality control problems. The initial blasting to prepare the site had problems. The reinforcing of concrete was not done properly. Cracks were found in the reactor's foundations. In April 2008, the French nuclear watchdog, ASN, announced that it had found a quarter of the welding in the reactor's steel liner to be defective.[91] A year later, it asked for two out of three pressuriser forgings to be remanufactured.[92]

Because of the inherently dangerous nature of the EPR reactor, France has witnessed fierce protests against it, with tens of thousands coming out on the streets in the cities of Rennes, Lyon, Toulouse, Lille and Strasbourg, as well as in Flamanville.[93]

What is Going Wrong?

What are the reasons for the quality control problems encountered in construction of both the Olkiluoto-3 and Flamanville-3 reactors?

One reason is that both Areva and EDF have tried to cut corners in safety and quality standards in order to reduce costs.[94] The Finnish

Safety Authority (STUK), in a report on the reasons for the delay in construction schedule of the OL3 reactor, stated: 'The major problems involve project management ... The power plant vendor has selected subcontractors with no prior experience in nuclear power plant construction to implement the project. These subcontractors have not received sufficient guidance and supervision to ensure smooth progress of their work ...'[95] For instance, Areva got the containment steel liner manufactured in a Polish machine yard which had no earlier experience of nuclear construction![96]

The second and more important reason is design problems with the EPR reactor. In order to cut down lead time, the Finnish and French authorities allowed construction of the Olkiluoto-3 and Flamanville-3 reactors to begin before the design was finalised and fully approved by them. Whereas the correct procedure is that designs should be complete and full safety regulatory approval given before construction is allowed to begin so that, in case there are design changes, these do not disrupt construction.[97]

Over time, the Finnish and French regulators realised that there were serious design problems with the reactor. They found the design of the control and instrumentation system—the nerve center of the reactor—to be at odds with basic principles of nuclear safety. Its back-up system was not sufficiently independent of the main system for it to be able to provide reliable back-up if the main system fails. In other words, if the main control system fails, there is a risk that the back-up system will fail for the same reason.[98] And so they asked for design modifications. However, because construction had already begun on the basis of the old design, making these modifications was difficult.[99] This is probably the most important reason for the serious quality control problems, construction delays and cost overruns of the two reactors.

The Nuclear Installations Inspectorate (NII) of the UK, which is conducting a detailed review of the EPR design, has also expressed similar concerns about the design of the control and instrumentation (C&I) system of the EPR in a letter to Areva. In its letter, the NII has stated that the EPR technology was significantly compromised because of the interconnectivity of what were meant to be independent systems designed to operate the plant and ensure its safety. The letter also highlighted concerns about the absence of safety display systems or manual controls that would allow the reactor to be shut down, either in the station's control room or at an emergency remote shutdown station.[100]

The US Nuclear Regulatory Commission, which too is carrying out a review of the EPR design, has, in a communiqué issued as recently as July 23, 2010, also expressed reservations on the control systems and other issues.[101]

The Roussely Report

As the problems with construction of the EPRs at Olkiluoto and Flamanville mounted, the French government ultimately acknowledged that all was not well with the French nuclear industry and, in October 2009, commissioned a former CEO of EDF, Francois Roussely, to review what was going wrong with the EPR. The report, *The Future of the French Civilian Nuclear Sector*, was published in July 2010. Roussely stated in his report that experience with Olkiluoto and Flamanville had 'seriously shaken ... the credibility of the EPR model and of the capacity of the French nuclear industry to succeed in new nucleaı plant construction.'

He attributed the problems to the complexity of the EPR model, 'including ... the redundancy of safety systems.' The report suggested that this complexity 'is certainly a handicap for its implementation and therefore its cost', and partly explains the difficulties encountered in Olkiluoto and Flamanville.[102]

Future of EPR in Doubt

This is a damning diagnosis. One of the selling points of the new generation plants is that they claim that their designs have incorporated the lessons of Chernobyl and Three Mile Island accidents. Now, one of the lessons from Three Mile Island accident is that if a safety system fails, there should be an independent—redundant—back-up system available. Further, the design must rationalise the various layers of safety systems, so as to reduce the complexity of the design. The Roussely report, however, says that the complexity of the EPR is because of its extra back-up safety system. This criticism therefore raises questions on one of the most important advancements in design that is supposed to be incorporated in the EPR—that even while having an independent back-up safety system, the complexity of the design should be reduced.[103] (Additionally, the UK and US safety regulators are raising questions about the independence of this back-up safety system too.)

Stephen Thomas, professor of Energy Studies at the University of Greenwich and a researcher in the area of energy policy, especially nuclear policy, for over 30 years, says that reducing this complexity in design is not going to be easy; it would require major modifications in design, which means that Areva would have to seek authorisation of its new design from nuclear regulators all over again. This whole process would probably take a decade![104]

To add to the EPR's woes, at both Olkiluoto and Flamanville, the cost of construction has sharply escalated. With all the design and other problems, there is no knowing what the final cost is going to be. Even if they ignore the design problems, no European country, nor the USA, is going to order another EPR at such an astronomical cost.

Clearly, the EPR is in trouble ...

Part IV: Conclusion

From the detailed discussion above, it is obvious that, despite intense lobbying and propaganda campaign by the nuclear industry, the 'nuclear renaissance' is turning out to be a damp squib. Even though the US administration has expressed its willingness to dole out billions of dollars in new subsidies for new reactors, and the first loan guarantees for construction of two reactors in Georgia have been announced, all the proposals for construction of new reactors have run into trouble. Eventually, after all the huffing and puffing by the nuclear industry and their spokespersons in the administration, at the most one or two reactors might start getting constructed. Clearly, there is no nuclear renaissance in the USA. This is also true of Western Europe, where the construction of two new reactors after nearly two decades has become such a fiasco that it is doubtful if any more reactors are going to be built there in the near future. All proposals for constructing new reactors in Canada, another country with a large nuclear power program, have been cancelled. Russia has announced plans to build a few nuclear reactors but, given its huge gas and oil resources, it is unlikely that it will invest huge amounts in nuclear power. China and India are likely to build a few reactors; Korea, Japan and Eastern Europe might also add a reactor or two but, considering that dozens of nuclear plants are scheduled to shut down in the next two decades, it is obvious that the overall worldwide trend for nuclear power is going to be downwards. In all likelihood, the sun is setting for nuclear power globally.

The reasons for this dismal future are the colossal problems with nuclear energy. Apart from skyrocketing costs, difficulty in raising loans due to high financial risks, construction delays and design problems, mankind has yet to find answers to the terrible safety issues with nuclear energy—the deathly radioactive pollution of the environment caused by leakage of radiation from the nuclear reactor, the as yet intractable problem of safe storage of high level wastes, and the potential for catastrophic accidents. Because of these problems, public opposition to construction of nuclear plants in their neighbourhoods is intense and so, even if governments have been willing to support the construction of new nuclear plants, they have been forced to scuttle these plans due to the powerful protests of people.

Paying the huge environmental and health costs associated with nuclear energy, especially because mankind will continue to pay these costs for thousands of years, for power from plants whose life will at the most be 60 years, becomes even more meaningless when we contrast this with the potential of energy from renewable sources like sun, wind and biomass to meet all our future energy needs. As we see in Chapter 10, the cost of energy from renewable sources is rapidly falling, some technologies have already become competitive with electricity from conventional sources, and their economies will further improve as they develop technically. Realising this, many countries, especially in Europe, are making a huge push to replace polluting coal and gas fired plants and dangerous nuclear power plants with renewable sources of energy.

Other Independent Assessments of 'Nuclear Renaissance'

Prognos Institute Report[105]

The Swiss 'Prognos' Institute, based in Basel, was commissioned by Germany's Federal Agency for Radiation Protection in Salzgitter to carry out a realistic estimate of the future development of nuclear energy worldwide till the year 2030. In its report, submitted in November 2009, it has come to the same conclusion as that drawn by us above:

> The world-wide renaissance of nuclear power that has so often been predicted will not take place in the next few decades. Nuclear energy will be on the decline till the year 2030, and will continue to decline in importance globally.

The study finds that although the number of announcements of new nuclear power stations is on the increase, and everything seems to have been prepared for the big renaissance of nuclear power, it is only so in theory. Many nuclear projects worldwide are already at a standstill. In view of the growing financing problems and political instability, at best only a third of the planned new projects will be realised worldwide. Even if construction begins, there are going to be many problems. The study concludes:

- Shutdowns of aged plants will lead to a decrease in the total number of reactors and there will be a significant decline in installed capacity and electricity generation from nuclear power plants.
- Compared to the reference level of March 2009, the number of nuclear power stations in operation worldwide is likely to decrease by 22 per cent by the year 2020 and by about 29 per cent by the year 2030.
- Even though worldwide electricity consumption is forecast to grow rapidly, nuclear energy will decline significantly in importance by the year 2030. The percentage of worldwide electricity generation accounted for by nuclear energy will decline from 14.8 per cent in the year 2006 to an estimated

9.1 per cent in the year 2020, and to 7.1 per cent in the year 2030.

CIGI Report[106]

In February 2010, the Centre for International Governance Innovation (CIGI), an independent non-partisan thank tank based in Canada and supported by the government of Canada, released the main report of its Nuclear Energy Futures (NEF) project: *The Future of Nuclear Energy to 2030 and Its Implications for Safety, Security and Nonproliferation.* The report was the culmination of three-and-a-half years of research into the purported nuclear energy revival and its implications for global governance.

The report concludes that there are significant barriers to the revival of nuclear energy in the near future, till at least 2030. The key barriers identified by it are the same as those that we have discussed above:

- unfavourable economics compared to other sources of energy;
- nuclear energy is too slow to address climate change and to compete with cheaper alternative means of tackling it;
- demands for energy efficiency are leading to fundamental rethinking of how electricity is generated and distributed;
- the nuclear waste issue remains unresolved, with no country currently implementing a sustainable solution;
- growing fears about safety, security and nuclear weapons remain in the public consciousness; and
- developing countries face additional constraints, including inadequate infrastructure, poor governance, deficient regulatory systems and finance.

7

INDIA'S NUCLEAR ENERGY PROGRAM

Part I: History

Pre-Independence Period

Much before Independence, during the period 1930-48, a series of outstanding Indian scientists, most notably Drs C.V. Raman, S.N. Bose and Meghnad Saha, had done pioneering research work in the field of fundamental physics, including nuclear physics, in various Indian universities. Prof. Satyendranath Bose's work on quantum mechanics paved the way for the formulation of Bose-Einstein statistics and the theory of Bose-Einstein Condensate. Some of his papers on nuclear fundamental researches had been translated by Einstein himself into German for publication in Europe. The particle boson is named after him. C.V. Raman had won the Nobel Prize in Physics for his work on the scattering of light in 1930, the first Asian to win the Nobel Prize in the sciences. By the mid-1930s, Professor Meghnad Saha had already become internationally renowned for his work in astrophysics. His theory of thermal ionisation is ranked as one of the most important landmarks in the history of astronomy. Saha first introduced the teaching of nuclear physics in the curriculum of higher studies of science in the country. It was due to his pioneering efforts that the first cyclotron was brought to India in 1941 by his brilliant student Dr B.D. Nag Chaudhuri, who had received his doctorate from the University of California at Berkeley in the cyclotronic sciences. Saha was a great institution builder and established numerous institutions including the renowned Institute of Nuclear Physics in Calcutta, which later was named after him as Saha Institute of Nuclear Physics.

Post-Independence Debate: Centralism vs Democracy

India's first Prime Minister, Jawaharlal Nehru, was very keen to advance research in atomic energy, because, in his words, if India had "to remain abreast of the world, [it] must develop this atomic energy".[1] And so he initiated India's nuclear program just a few months after independence with the passage of the Atomic Energy Act of 1948. However, ignoring the whole galaxy of brilliant scientists who had done such wonderful work in nuclear physics and even established institutions of research during the trying years of the freedom struggle, Prime Minister Nehru handed over the reins of India's nuclear energy program to Dr Homi Bhabha. Dr Bhabha was also a gifted physicist who had made his mark while a student at Cambridge University in the 1930s, but he was much younger to Saha and others.[2]

One probable reason for this was that Saha was critical of the Atomic Energy Act of 1948, which imposed secrecy on India's entire nuclear energy program. He emphasised that France had been able to make progress in atomic energy research in large measure, despite American obstructions, due to open disavowal of secrecy. The program there was "a great national effort in which the knowledge and skills of all available scientists of the country ... were utilised for the objective." Even in the USA, he pointed out, only procurement of minerals, production of fissile materials and the weapons development were entirely under the AEC. The program of generating power from nuclear reactors was pursued in collaboration with industrial firms, whereas programs related to peaceful use of atomic energy were carried out in collaboration with associations of universities and research organisations. During the debate in the Lok Sabha on peaceful uses of atomic energy on May 10, 1954, he stated: "First of all there should be no secrecy. If you read out the Atomic Energy Act, you find that it does not tell us what to do but it simply tells us what is not to be done ... I would ask our honourable friends on the Treasury Bench to read the Atomic Energy Acts of England and America and see how broad-based they are ..."[3]

Nehru not only sidelined Saha and other leading scientists, he virtually made Bhabha the dictator of India's nuclear energy program. Bhabha was Chairman of the Atomic Energy Commission (AEC), Secretary of the Department of Atomic Energy (DAE), Director of the Atomic Energy Establishment, Trombay (now named after him as Bhabha Atomic Research Centre), and also the Founder-Director of the Tata Institute of Fundamental Research (TIFR). Bhabha's enormous powers and his centralistic style of functioning gradually led to taking away of all research initiatives from the various Indian universities, and their centralisation under Bhabha in Bombay.

Saha was critical of this centralisation of powers, even more so of research, under a single person. He wanted universities to do research, including in nuclear physics and engineering, and be supported in their efforts.[4] Speaking in the Lok Sabha in 1954, he stated:

> ... if you analyse the work done in other countries, you find that atomic energy cannot be developed unless you enlist the services of thousands of scientists in your own country... In this particular case, for five years, the scientists of India have been precluded from taking any part in the development of atomic energy. I

> throw it as a challenge to the party in power, let them justify why they did not take the scientists of this country into confidence in this great work.

He went on to make a fervent appeal for democratisation of India's atomic energy establishment and research: "If our young scientists are entrusted with this great task of atomic energy, they can deliver the goods. I would, therefore, request the government to make our atomic energy establishment more broad-based than it has been so far."[5] As is obvious from the subsequent history of India's atomic energy program, Saha's fervent appeal was ignored.

Dr Bhabha as Dictator

The Atomic Energy Act of 1948 made atomic energy the exclusive responsibility of the state and allowed for a thick layer of secrecy. It authorised the creation of the Atomic Energy Commission (AEC), which was set up in August 1948. The AEC was to be the apex body in charge of nuclear policy in India. Dr Bhabha became the Chairman of the AEC.

There was some criticism of the secrecy provisions in the Parliament when Nehru introduced the bill. One member, Krishnamurthy Rao, compared the bill with the British and American acts and pointed out that the bill did not have mechanisms for oversight, checks and balances as the US Atomic Energy Act. He also pointed out that in the UK, secrecy is restricted only to defence matters, and questioned the rationale of extending secrecy to research for peaceful purposes also.[6]

In 1954, the government through a Presidential Order set up the Department of Atomic Energy (DAE) as the overall body responsible for research, technology development and commercial reactor operation. The DAE was to be under the direct charge of the Prime Minister. Policy making was to be done by the AEC. In the same year, the government set up the Atomic Energy Establishment (AEE), as India's primary centre for nuclear research (later renamed Bhabha Atomic Research Centre or BARC after Bhabha's death in 1966). Bhabha was made its first Director.

On March 1, 1958, the government passed a new official resolution, establishing the AEC in the DAE. The then Prime Minister (late Pandit Jawaharlal Nehru) laid a copy of this resolution on the table of the Lok Sabha on March 24, 1958. The resolution also made the Secretary to the Government of India in the Department of Atomic Energy ex-officio Chairman of the Commission. While the other members of the AEC were to be appointed on the AEC Chairman's recommendation and after the Prime Minister's approval, the Chairman was empowered to overrule all other members of the AEC (except in financial matters). Thus, a single person, as the Chairman of the AEC and Secretary of the DAE and reporting directly to the Prime Minister, became all powerful in nuclear matters in the country. The resolution also further strengthened the AEC with full executive and financial powers.[7]

Bhabha thus became the moghul of the nuclear establishment of India. He had total powers to initiate and regulate plans, formulate and execute his own procedure and rules, and had an open ended budget.

In 1962, the government granted yet more powers to the AEC by passing the totally undemocratic Atomic Energy Act of 1962, which replaced the weaker Act of 1948. No democratic country has given such authoritarian powers to its atomic energy establishment. The Act of 1962 grants absolute powers to initiate, execute, propagate and control exploration, planning and manufacture of atomic material and its related hardware and all nuclear research and developmental activities to the sole authority of the Chairman of the AEC. Despite having such immense powers, the AEC does not report to the Cabinet, but directly to the Prime Minister.[8]

The Act also empowers the AEC to restrict disclosure of any information related to nuclear issues. Under Section 18 (1) of this Act, the government is empowered to restrict the disclosure of information, whether contained in a document, drawing, photograph or in any other form whatsoever, which relates to or illustrates: (a) an existing or proposed plant used or proposed to be used for the purpose of producing, developing or using atomic energy, or (b) the purpose or method of operation of any such existing or proposed plant, or (c) any process operated or proposed to be operated in any such existing or proposed plant. Section 20 denies any person or organisation not authorised by the AEC to invent or patent anything which the AEC believes as relating to atomic energy. Section 21 (5) gives the AEC absolute authority over any legal or formal arbitration.[9]

Eminent jurists like Justice Krishna Iyer have termed these powers given to the AEC as unconstitutional and undemocratic.[10]

Dr Bhabha's Ambitious Plans

From the very beginning, plans for the Indian nuclear program were ambitious and envisaged covering the entire nuclear fuel cycle. Bhabha initiated the development of India's first research reactors at BARC (Trombay, near Mumbai): Apsara, a swimming pool research reactor, was set up in 1956, and CIRUS, a 40 MW heavy water moderated, light water cooled, natural uranium fuelled reactor, was set up in 1960. India also developed facilities for mining uranium, fabricating fuel, manufacturing heavy water, reprocessing spent fuel to extract plutonium and, on a somewhat limited scale, enriching uranium. For executing these plans, the DAE set up a number of subsidiary organisations: the Nuclear Power Corporation of India Limited (NPCIL), which is responsible for designing, constructing, and operating nuclear power plants; the Uranium Corporation of India Limited (UCIL) which is in charge of mining and milling of uranium; the Heavy Water Board, which is in charge of the many plants that produce heavy water; and the Nuclear Fuel Complex, which manufactures fuel for the nuclear reactors.[11]

Three Stage Program

Simultaneously, Bhabha in 1954 also announced a grand three stage program for the development of nuclear energy in the country. The logic behind this was that India has very little uranium, and the little it has is of poor quality. What India does have is plenty of the element thorium, about 32 per cent of the world's deposits. While thorium cannot fuel a nuclear reactor by itself, thorium-232 (Th-232) can be converted into fissile uranium-233 in a plutonium fuelled breeder reactor—the neutrons released during the fission of Pu-239 are captured by Th-232, converting it into U-233. To make use of India's thorium reserves to create fissionable uranium-233 and generate electricity from this, Bhabha announced a three phase strategy for the development of this technology.

The first stage of this involved natural uranium fuelled Pressurised Heavy Water Reactors, followed by reprocessing the spent fuel to extract plutonium. In the second stage, this plutonium is used in the cores of fast breeder reactors, with the nuclear cores surrounded by a 'blanket' of U-238 or natural uranium to produce more plutonium. Subsequently, the blanket would be of thorium, which would produce fissionable uranium-233. But before introducing thorium in the blanket, a sufficiently large fleet of breeder reactors with uranium blankets would have to be commissioned to ensure that there is adequate plutonium to fuel the follow-on second stage thorium blanketed breeder reactors.

Once there is enough uranium-233, then the third stage can be launched, which involves breeder reactors using uranium-233 in their cores and thorium in their blankets.[12]

The First Atomic Power Plants

Bhabha also initiated discussions with US, Britain, Canada and the Soviet Union for assistance for setting up atomic power plants in the country. The AEC selected the CANDU type heavy water reactors which use natural uranium as fuel as best suited for India's atomic power program. It was decided to go in for natural uranium fuelled reactors as the enrichment process was very costly and also because these reactors made the most efficient use of uranium, whose reserves were limited in the country. These reactors were under development in Canada, and it was willing to offer generous technological and financial assistance for setting up such reactors, as a part of its Colombo Plan. Those were the days of the Cold War, and the aim of this Plan was to prevent newly independent third world countries from going over to the Soviet bloc.[13]

While all of India's initial reactors were to be of this type, Bhabha negotiated an agreement with the United States for setting up a Boiling Water Reactor also. Since this runs on enriched uranium, the US agreed to supply this fuel for the entire life of the reactor.[14]

In addition to water moderated reactors, the AEC was also keenly interested in fast breeder reactors as they were a central part of the three-stage program. And so it initiated efforts in that direction very early, in the mid-1960s. We discuss India's fast breeder reactor program in Chapter 9.

Targets and Achievements

On the basis of these plans and assuming optimistic development times, Bhabha announced in 1954 that there would be 8000 MW of nuclear power in the country by 1980.[15] As the years progressed, these predictions were to increase. By 1962, the prediction was that nuclear energy would generate 20-25,000 MW by 1987; and in 1969 the AEC predicted that by 2000 there would be 43,500 MW of nuclear generating capacity. All of this was before a single unit of nuclear electricity was produced in the country—India's first reactor, at Tarapur, was only commissioned in 1969![16]

The achievements have been quite different. While total electricity generation capacity in the country (from all sources) has seen a huge increase, from a meagre 1800 MW in 1950 to 90,000 MW in 2000[17] and 1,69,749 MW as on December 31, 2010[18], installed capacity of nuclear power generation has grown much more slowly: it was about 600 MW in 1979-80, 950 MW in 1987 and 2720 MW in 2000.[19] By December 2010, it had grown to 4560 MW,[20] which was less than 3 per cent of the total electricity generation capacity in the country! The AEC had set a target of achieving 43,500 MW of total nuclear power capacity by the year 2000; even by 2010, it had been able to achieve just 10.5 per cent of this target!

This utter failure has not been because of paucity of resources. Practically all governments have favoured nuclear energy and the DAE's budgets have always been high. The high allocations for the DAE have come at the cost of promoting other, more sustainable, sources of power. In 2002-03, for example, the DAE was allocated Rs.3350 crores, dwarfing in comparison the Rs.470 crores allocated to the Ministry of Nonconventional Energy Sources (MNES), which is in charge of developing solar, wind, small hydro and biomass-based power.[21] In 2009-10, DAE's budget had ballooned to Rs.6030 crores, while the MNES (now Ministry of New and Renewable Energy, MNRE) had been allocated just Rs.600 crores.[22] Despite the much smaller allocations for the latter, installed capacity of renewable energy was 16,787 MW in December 2010,[23] nearly four times that of nuclear energy (4560 MW)!

Part II: India's Present Nuclear Facilities

Uranium Resources and Mining

The four most promising uranium mining areas in India are: the East Singhbhum district (Jharkhand), West Khasi hills (Meghalaya), the Bhima Basin area (Gulbarga district of Karnataka) and Nalgonda district (Andhra Pradesh). India's uranium resources are modest, with 54,000 tons as reasonably assured resources and 23,500 tons as estimated additional resources in situ.[24]

Mining and processing of uranium is carried out by the Uranium Corporation of India Ltd. (UCIL), a subsidiary of the DAE. Presently, it operates five underground mines, all in Jharkhand, at Jaduguda and Bhatin (since 1967), Narwapahar (since 1995), Turamdih (since 2002) and Bagjata (commissioned in Dec 2008). The last three are modern mechanised mines. In December 2009, it also commissioned India's first open cast mine, at Banduhurang, also in Jharkhand. UCIL has also begun construction of a seventh mine in the area, the Mohuldih underground uranium mine, located in Saraikela-Kharswan district. Processing of the ore is carried out at two mills, one located near Jaduguda, which processes 2090 tonnes of ore per day, and another at Turamdih, with 3000 t/day capacity.[25]

UCIL has also begun work on a new underground mine at Tummalapalle near Pulivendula in Kadapa district of Andhra Pradesh. This is expected to start producing uranium this year (2011). This would be the first mine outside Jharkhand. A second mining project in the state is planned in the Lambapur-Peddagattu area in Nalgonda district.[26] UCIL is also planning a small mine and uranium processing unit at Gogi in Gulbarga area of Karnataka.[27]

Outside of the Singhbhum area, Meghalaya has the largest reserves of uranium in India. The state is estimated to have 9.22 million

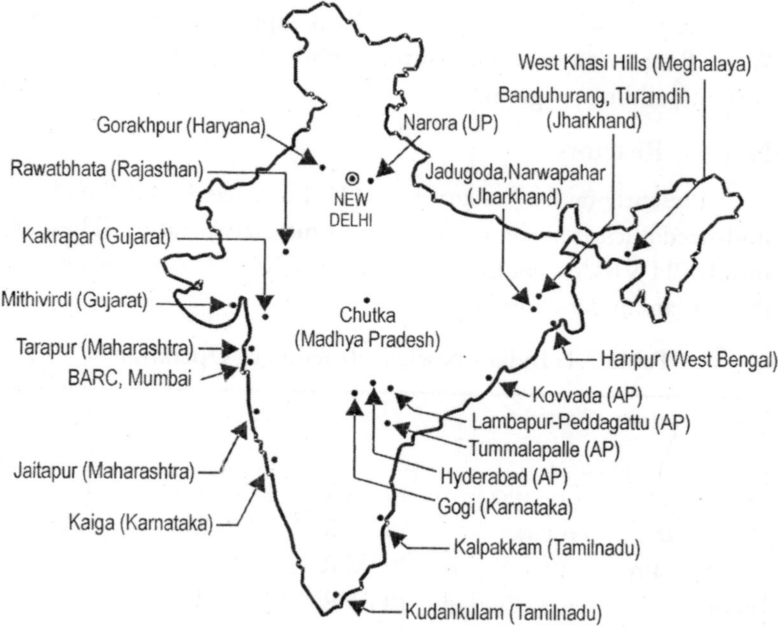

India: Uranium Mines, Nuclear Fuel Complex, Nuclear Reactors

tons of uranium ore deposits, which are also supposed to be of high quality.[28] Despite having the necessary clearances to begin mining in the West Khasi Hills district of the state, UCIL has been unable to begin mining in the area, due to strong people's opposition.

Fuel Fabrication[29]

The yellow cake from UCIL's milling plants in Jharkhand is sent to DAE's Nuclear Fuel Complex at Hyderabad for refining and conversion into nuclear fuel. The main 400 t/yr plant fabricates PHWR fuel (which is unenriched). A small (25 t/yr) fabrication plant makes fuel for the Tarapur BWRs from imported enriched (2.66 per cent U-235) uranium. Mixed carbide fuel for the Fast Breeder Test Reactor (FBTR) was first fabricated by BARC in 1979.

A very small enrichment plant, insufficient even for the Tarapur reactors, is operated by DAE's Rare Materials Plant at Ratnahalli near Mysore.

Heavy water for India's PHWRs is supplied by DAE's Heavy Water Board, and its seven plants are working at capacity due to the current building program.

Nuclear Reactors

Presently (as on January 1, 2011), India has 17 small and two mid-sized nuclear power reactors in commercial operation. These are mostly PHWRs, except for two units of BWRs in Tarapur. Another 7 reactors are under construction.

Table 7.1: India's Nuclear Reactors in Operation[30]

Power station	*State*	*Type*	*Units*	*Total capacity (MW)*
Kaiga	Karnataka	PHWR	220 x 3	660
Kakrapar	Gujarat	PHWR	220 x 2	440
Kalpakkam	Tamil Nadu	PHWR	220 x 2	440
Narora	Uttar Pradesh	PHWR	220 x 2	440
Rawatbhata	Rajasthan	PHWR	100 x 1, 200 x 1, 220 x 4	1180
Tarapur	Maharashtra	BWR, PHWR	160 x 2, 540 x 2	1400
		Total	**19**	**4560**

Of the 19 in operation, the newest are Rawatbhata-5&6, both 220 MW PHWRs, which attained criticality in December 2009 and January 2010 respectively.[31] Tarapur-3&4 are also new, both are 540 MW (490 MW net) PHWR nuclear reactors, and they started commercial operation in 2005-06. Kaiga-3 started up in February and went into commercial operation in May 2007.

Many of these reactors are facing problems and have been downrated (discussed in Chapter 9). Due to these problems and shortage of uranium fuel, in mid-2008 India's nuclear power plants were running at about half their rated capacity.

Reactors under Construction

Table 7.2: Nuclear Reactors Under Construction in India[32]

Power station	*State*	*Type*	*Units*	*Expected date of commercial operation*	*Total capacity (MW)*
Kaiga-4	Karnataka	PHWR	220 x 1	2010	220
Rawatbhata-7&8	Rajasthan	PHWR	700 x 2	2016	1400
Kakrapar-3&4	Gujarat	PHWR	700 x 2	2015	1400
Kudankulam	Tamil Nadu	VVER-1000	1000 x 2	2011-12	2000
		Total	7		**5020**

Apart from these seven, a 500 MW prototype Fast Breeder Reactor (FBR) is under construction at Kalpakkam by BHAVINI, a government enterprise set up to focus on FBRs. It was expected to start up in 2010 and produce power in 2011, but as we see in Chapter 9, this timeline is most probably way off the mark.

Reprocessing

Unlike most other countries, the DAE pursues reprocessing as a way of dealing with spent fuel—to extract plutonium for use in Fast Breeder Reactors and for nuclear weapons. India has three full-scale reprocessing plants, at Trombay, Tarapur and Kalpakkam, to extract reactor-grade plutonium for use in fast breeder reactors. The Trombay plant was commissioned in 1965, the Tarapur plant was commissioned in 1977 but has been functioning much below capacity, while the Kalpakkam Atomic Reprocessing Plant (KARP), with a capacity of 100 tons per annum, was commissioned in 1998.[33]

Radioactive Waste Management

The DAE does not have enough reprocessing capacity to reprocess all the waste from its reactors. So, most of the remaining waste is stored in spent fuel pools near the reactors.

As discussed in Chapter 3, reprocessing results in large quantities of waste. The DAE classifies the waste from its reprocessing plants into Low Level Waste (LLW), Intermediate Level Waste (ILW) and High Level Waste (HLW).

Gaseous wastes produced during routine operations at nuclear reactors and reprocessing plants are released through stacks (75-100 metres tall) into the environment after filtration. Likewise, low-level liquid wastes—consisting mostly of tritium but also small quantities of cesium-137 and strontium-90—are released into nearby water bodies, such as the sea in the case of coastal reactors. Data on such releases are scarce—and often conflicting—but suggest that releases at Indian reactors are much higher as compared to similar reactors elsewhere. Intermediate-level liquid wastes generated in reprocessing plants are concentrated and fixed in cement.[34]

Geological Disposal of HLW Waste

Because it contains the bulk of the radioactivity in spent fuel, the greatest concern is HLW. There is no agreed solution to the problem of disposal of HLW. DAE presently deals with this waste by immobilising or vitrifying it—the waste is mixed with glass at a high temperature and allowed to cool, which slows down the diffusion of radionuclides from HLW. These blocks are stored at the Solid Storage & Surveillance Facility at Tarapur.

The DAE has proposed disposing of vitrified HLW in geological repositories about 500-600 metres below the ground in some appropriate host rock such as granite or basalt. Initially, deep geological formations in the southern Indian peninsula were explored as likely burial sites. A number of bore holes 0.6 miles deep were dug in an abandoned chamber of the Kolar gold mines to test the formation's behaviour under simulated radioactive decay heat. Those tests evidently did not yield the desired results. Then, in 1999, it was reported that an area in Rajasthan had been identified as suitable for burying wastes. This led to public protests from local communities. Shortly afterwards, the government announced in Parliament that it had not taken any decisions on the disposal of nuclear waste, and that such a decision might 'take another two decades of research and development'. So far no geological disposal site seems to have been finalised.[35]

Part III: US-India Agreement and Nuclear Suppliers Group

Background

One of the biggest symbols of the unjust world order we live in is the Nuclear Non-Proliferation Treaty (NPT), which came into force on March 5, 1970. Currently, 189 countries have signed it.[36] The ostensible purpose of this treaty is to limit the spread of nuclear weapons, but it allows five countries to have nuclear weapons—the United States, Russia, the United Kingdom, France and the People's Republic of China; they are officially recognised as 'nuclear weapon states.' Coincidentally, these five nations also serve as permanent members of the United Nations Security Council. The NPT recognises the right of its signatory countries to develop programs for peaceful uses of nuclear energy, and allows transfer of nuclear technology and materials to these countries for this purpose, as long as they can demonstrate that their nuclear programs are not being used for the development of nuclear weapons.

India did not sign the NPT arguing that the treaty did not advance the goal of universal disarmament and instead divided the world into a club of 'nuclear haves' who alone are free to possess and multiply their nuclear stockpiles, and a larger group of 'nuclear have-nots'.

Despite not signing the NPT, India managed to access nuclear technology from Western countries for its nuclear energy program in the 1960s. However, following India's nuclear tests in 1974, the Western countries decided to further tighten rules governing international nuclear trade. Many of the signatories of the NPT now additionally formed an informal group, the Nuclear Suppliers Group (NSG), to further limit export of nuclear materials, equipment and technology. They formulated a set of guidelines which condition such exports on comprehensive safeguards by the IAEA, which are designed to verify that nuclear energy is not diverted from peaceful use to weapons programs. Consequently, whatever little collaboration that was taking place between India and the Western countries was terminated by them, and uranium imports also ceased.

By the turn of the twenty-first century, momentous changes had taken place in the world. In this changed world scenario, India decided to abandon the Nehruvian model of development and globalise the Indian economy. Simultaneously, India also decided to abandon its non-aligned foreign policy and independent defence policy, and align with the United States. As a reward, the US offered India an agreement on nuclear cooperation, which was greedily accepted by India's rulers.[37]

Indo-US Nuclear Deal

On July 18, 2005, US President Bush and Indian Prime Minister Manmohan Singh issued a joint statement, wherein among other things, they announced their intention to enter into a nuclear deal. According to this 'Joint Statement', India agreed to separate its civil and military nuclear facilities and place all its civil nuclear facilities under IAEA safeguards, continue its unilateral moratorium on nuclear testing, and refrain from transfer of enrichment and reprocessing technologies to countries that do not have them and support international efforts to limit their spread; in exchange, President Bush promised to get the US Congress to adjust US laws and policies and also work with other countries to modify international regimes to enable full civil nuclear energy cooperation and trade with India.[38]

The deal took more than three years to come to fruition as it had to go through several complex stages. In its final shape, the deal places under permanent safeguards those nuclear facilities that India has identified as 'civil' and permits broad civil nuclear cooperation, while excluding the transfer of 'sensitive' equipment and technologies even under IAEA safeguards. On August 18, 2008 the IAEA Board of Governors approved the safeguards agreement with India, and on February 2, 2009 India signed an India-specific safeguards agreement with the IAEA. Once India brings this agreement into force, inspections will begin in a phased manner on the 35 civilian nuclear installations identified by India in its Separation Plan.

Following the approval by the IAEA board, the United States approached the Nuclear Suppliers Group (NSG) to grant a waiver to India to enable it to commence civilian nuclear trade. The 45-nation

NSG granted the waiver to India on September 6, 2008 allowing it to access civilian nuclear technology and fuel from other countries, without becoming a party to the NPT.

The US House of Representatives passed legislation allowing civil nuclear trade with India on September 28, 2008. On October 1, 2008 the US Senate also approved the agreement, which was signed into law by President George W. Bush on October 8, 2008. Two days later, the Indian External Affairs Minister Pranab Mukherjee and his counterpart the US Secretary of State Condoleezza Rice formally inked the agreement.

While the US Congress discussed the deal threadbare before approving it, Prime Minister Singh blocked the Indian Parliament from scrutinising the deal. Not only that, the Bush administration's replies to questions raised by US Congressmen on the nuclear deal (which were kept under wraps and only made public in September 2008) reveal that the Indian Prime Minister has blithely lied to the Indian Parliament while defending the nuclear deal.[39]

Part IV: New Nuclear Plans after the Deal

Uranium Imports

Following the clearance given by the Nuclear Suppliers Group, India has signed agreements with a number of countries for uranium supplies, including France, Russia, Kazakhstan, Namibia and Mongolia. On September 1, 2009, unit 2 of the Rajasthan Atomic Power Station (Rawatbhata-2), which had been shut down for some repairs, became the first reactor to supply power to the grid using uranium imported from France and Russia.[40] During the period January-July 2010, India imported 868 tonnes of uranium from France, Russia and Kazakhstan; as of August 2010, according to the DAE, seven reactors were using imported uranium.[41]

New Reactors

Even before the Indo-US nuclear deal opened up the prospects of reactor and uranium imports, the DAE had entered into an agreement with Russia to supply India's first large nuclear power plant, comprising two VVER-1000 (V-392) reactors, under a Russian-financed $3 billion contract. The nuclear plant is coming up in Kudankulam in Tamil Nadu; construction began in 2001. Russia will supply all the enriched fuel, and allow India to reprocess it and keep the plutonium. The first unit was due to start supplying power in March 2008 and go into commercial operation in late 2008, but this schedule has slipped.[42] The NPCIL website now says that the first unit will be commissioned in June 2011, and the second unit is about 9 months behind it.[43]

In 2005, the government granted approval to set up two more imported VVER-1000 reactors at Kudankulam alongside the two already being built there by Russia, and site clearance was also given to set up two imported 1000 MW LWR units at Jaitapur in Ratnagiri district of Maharashtra.

After getting approval from the Nuclear Suppliers Group to begin civil nuclear trade in September 2008, the Indian government has moved quickly to sign civil nuclear cooperation agreements with a large number of countries, including France, Russia, UK and Kazakhstan. The agreement with France was signed as early as September 30, 2008, that is, even before India and the US formally signed the nuclear deal on October 10, 2008.[44]

With India now able to import uranium as well as reactors from other countries, the Indian government has announced plans to set up a string of nuclear reactors all over the country. These include some of the biggest nuclear power plants in the world. Many of these so-called 'Nuclear Parks' would be having a number of big size imported reactors each, with a total capacity of around 8000–10,000 MW at a single location. As of January 2010, the government had given 'in principle' approval to over 38,000 MW new reactor capacity, indicating that it is looking to scale up India's nuclear capacity nearly ten-fold over the next decade.

So far, in principle approval has been given for the following 'Nuclear Parks':[45]

- Kudankulam in Tamil Nadu: Two more pairs of Russian VVER-1000 units, making it a total of 6 reactors of total capacity 6000 MW.

- Jaitapur in Maharashtra: A total of six EPR reactors from Areva, 1650 MW each, for a total capacity of 9900 MW.
- Mithivirdi in Gujarat: Six LWR reactors, 1000 MW each, to be set up by US-based corporations, either GE-Hitachi or Westinghouse.
- Kovvada in Andhra Pradesh: Six LWR reactors of 1000 MW each, also to be set up by US-based corporations, either GE-Hitachi or Westinghouse.
- Haripur in West Bengal: Six Russian VVER-1000 reactors to be set up here.

Table 7.3: Nuclear Reactors Planned or Firmly Proposed

Reactor	*State*	*Type*	*MW each*
Kudankulam 3-6	Tamil Nadu	PWR - VVER	1000 x 4
Jaitapur	Maharashtra	PWR - EPR	1650 x 6
Kaiga 5&6	Karnataka	PWR	1000 x 2
Fatehabad	Haryana	PHWR	700 x 4
Chutka	MP	PHWR	700 x 2
Mithivirdi	Gujarat	PWR (US)	1000 x 6
Kovvada	Andhra Pradesh	PWR (US)	1000 x 6
Haripur	West Bengal	PWR (VVER)	1000 x 6

Note: For many of the above projects, the reactor size mentioned is only tentative, the final capacity would depend upon the rating of the reactor selected.

The NPCIL has obtained all the statutory clearances for beginning construction of Kudankulam-3 and 4. It has also completed grading and levelling of the site for these two units, and was hoping to finalise the agreement during Russian President Dmitry Medvedev's visit to India in December 2010, so as to begin construction of the reactors in 2011,[46] but this has got stalled over Russian objections to India's nuclear liability law.

The NPCIL has also obtained all the necessary clearances for the Jaitapur nuclear power project. Land acquisition for the project was completed forcibly, despite massive opposition by the people,[47] even before the plant received the mandatory environmental clearance, indicative of the fact that the clearance was always going to be a mere formality. The government is in such a tearing hurry to begin construction that land acquisition has been done even though the final project agreement has yet to be signed with France! According to newsreports, during French President Sarkozy's visit to India in December 2010, only a framework agreement for the supply of the first two reactors by Areva was signed, and negotiations on issues like pricing are still underway.[48] That is some deal. We've agreed to buy the reactor, and begun preparations for its construction, without finalising the price! Another indicator of the powerful vested interests behind the project.

According to the DAE, the process of obtaining the necessary clearances for the Haripur, Kovvada and Mithivirdi projects has also begun; once this is done, the land acquisition process would begin.[49] There are also plans to set up another 6000 MW Nuclear Park at Markandi (Pati Sonapur) in Orissa.[50]

In addition, the NPCIL has got in-principle approval to set up 4 indigenous PHWR reactors of 700 MW each at Gorakhpur village in Fatehabad district of Haryana, and another 2 similar reactors at Chutka in Madhya Pradesh.[51]

8

INDIA: THE TRUE ECONOMIC COSTS OF NUCLEAR ELECTRICITY

In a statement before the Indian Parliament on July 29, 2005, Prime Minister Manmohan Singh stated that the proposed nuclear cooperation agreement with the United States would allow India to import nuclear fuel and nuclear reactors, enabling us to produce 'cheap and affordable' nuclear energy ...[1]

While dedicating Tarapur-3 and 4 reactors to the nation on August 31, 2007, the Prime Minister again emphasised that one of the reasons why India is placing 'so much importance on nuclear energy' is because it is financially 'affordable' ...[2]

The Prime Minister is lying. Nuclear energy is anything but 'cheap and affordable'.

We have discussed the worldwide costs of nuclear power in Chapter 4, where we had concluded that nuclear energy is one of the most expensive ways of generating electricity, and is definitely much more costly as compared to electricity from fossil fuels. The only way it can be competitive with conventional electricity is if it is highly subsidised by the government.

The situation is the same with India too. It is the enormous hidden subsidies that nuclear energy gets that allows the Prime Minister to claim that nuclear energy is and will continue to be 'cheap' and 'affordable'.

Even officially, nuclear electricity in India is costlier than electricity from conventional sources. The government claims that nuclear electricity in India costs between Rs. 2.70 and 2.90 a kilowatt-hour (that is, per unit) from its reactors built since the 1990s, a price which is far higher than the cost of electricity from coal-fired plants.[3]

The actual cost of electricity from indigenous nuclear reactors is much more than the above figures. This is because the subsidies given to nuclear power are huge—we don't even know the extent. And the cost of electricity from the proposed new imported reactors—especially the EPR reactors proposed to be installed at the Jaitapur Nuclear Park—is going to be simply mind-boggling! Read on ...

Part I: Subsidies for India's Indigenous Reactors

The subsidies given by the government of India to nuclear energy are even more than the massive subsidies given by the US government discussed in Chapter 4.

There is no need to give loan guarantees for nuclear reactor construction in India, as the nuclear industry is in the public sector. With the government standing guarantee to pay irrespective of the escalation in construction cost, this greatly reduces the capital cost of the reactor. Apart from this implicit subsidy, the government of India, through the DAE, also gives numerous explicit subsidies to the NPCIL, the public sector corporation that runs all of India's nuclear reactors.

The DAE subsidises the NPCIL in nuclear fuel price, by supplying it fuel bundles from its Nuclear Fuel Complex at much less than the cost of production.[4] It also supplies heavy water (HW) from its heavy water plants for use in NPCIL's CANDU reactors at subsidised rates. Most of India's reactors are CANDU reactors, and heavy water is a major cost component of producing electricity from these reactors. M.V. Ramana, an eminent nuclear physicist who is presently a research fellow with Princeton University, has made some estimates of the subsidy involved. His calculations show that even conservatively, as per standard and required accounting practices, a subsidy of over Rs.12,000 per kg is being offered.[5] Let us make a cursory estimate of the total subsidy given. Heavy Water Reactors need HW initially to attain criticality; once they start operating, they need HW periodically to make up for losses. The initial coolant and moderator inventory requirement for each 220 MW reactor is 70 tons and 140 tons of HW respectively (for a total of 210,000 kgs). Once the reactor begins operation, the reactor loses about seven tons of HW per year, which must be therefore replenished.[6] This means that the total subsidy given by the DAE in the capital cost of a 220 MW heavy water nuclear reactor would be of the order of 210,000 kgs x Rs.12,000 per kg = Rs.252 crores. Let us compare this with the capital cost of a 220 MW reactor. The estimated capital cost of the Kaiga-1 and 2 reactors[7] (220 MW each), when they attained criticality in 1999, was Rs.2,896 crores,[8] or Rs.1450 crores per reactor. This means that the DAE has been subsidising the capital cost of NPCIL's CANDU reactors to the tune of around 17 per cent of the capital cost—and the capital cost of the reactor is the dominant component of the cost of nuclear electricity. Similarly, the total subsidy given by the DAE in the annual operating costs of a 220 MW Heavy Water

Reactor would be of the order of 7,000 kgs x Rs.12,000 per kg = Rs.8.4 crores.

Another important subsidy given to nuclear electricity in India is in the cost of dealing with the radioactive waste from the nuclear power plants. As discussed in Chapter 4, this cost is huge, and most countries subsidise it; without this subsidy, the nuclear industry wouldn't exist. For instance, in the USA, the government has taken over the entire responsibility of dealing with the nuclear waste, for which it charges the nuclear utilities a highly subsidised fee of $277,000 per ton of spent fuel waste, which amounts to roughly 0.1 cents per kWh of nuclear electricity generated by them.[9] The US government charges the nuclear utilities at least something for taking care of their nuclear waste; in India, the DAE bears all the waste management expenses, and does not charge the NPCIL a single paisa for managing the waste generated by its reactors![10]

On top of it, the DAE reprocesses this spent fuel, which as we have seen in Chapter 3 is an even costlier way of dealing with spent fuel waste.

Additionally, the DAE is going to bear the entire decommissioning expenses of India's reactors, as and when they are closed down.[11] To get a rough idea of the amounts involved, let us use the estimates made by US Nuclear Regulatory Commission. It estimates that typical decommissioning costs of nuclear reactors would be around 20-30 per cent of the capital costs.[12] For the 2 × 220 MW Rawatbhata-3 and 4 reactors,[13] whose capital cost was Rs.2,511 crores,[14] this means that decommissioning them would cost at least Rs.500 crores.

So far as making a provision for insurance liability against accidents is concerned, there is no need for the NPCIL to do so. For, the NPCIL does not even acknowledge minor accidents! And in the case of a major accident, which by 'God's grace' has not happened so far, it assumes that the government will bear the costs.

Finally, like all countries with nuclear power programs, in India, too, the health costs of nuclear energy are simply brushed under the carpet. (This subsidy is given to coal power too. There, too, the government simply ignores the terrible health costs of burning coal.)

The DAE refuses to admit that radiation from any of its nuclear energy related installations is affecting the health of people living around them. Even the courts have refused to hear petitions on these issues. So the question of making a provision for health care of radiation affected victims and paying them compensation does not arise!

Part II: Imported Reactors: Even More Subsidies

Following the signing of the Indo-US nuclear deal as a part of its nuclear energy expansion plans, the government is going in for imported nuclear reactors in a big way (see Chapter 7 for details). As discussed in Chapter 6, the foreign equipment suppliers are desperately short of orders, and the Indian government could have done some hard bargaining with them. Instead, it is bending over backwards to give them multiple subsidies.

There has been no competitive bidding for any of these reactors. The government has one-sidedly announced that it is reserving one 'Nuclear Park' for each of its favoured foreign vendors: Jaitapur for Areva, Mithivirdi and Kovvada for Westinghouse/GE-Hitachi, Kudankulam and Haripur for Atomstroyexport. It is an unparalleled giveaway: the government has announced these reservations even before the terms of the reactor contracts have been negotiated![15] The foreign suppliers have been assured that they will be given the contract irrespective of the price they quote!!

To add to the pampering, the foreign firms don't have to acquire land for these projects, the government of India is doing so, under British era undemocratic laws, wherein land can be compulsorily acquired from the people at a cost determined arbitrarily by the government.

Irrespective of the cost of electricity that would be produced by these imported reactors, the government will be buying it—because the plants are going to be run by the government-owned NPCIL. Let us take a look at the estimated cost of electricity from the Jaitapur Nuclear Plant which is going to be built by the French nuclear corporation Areva. (We are not discussing the cost of electricity from the reactors being built at Kudankulam because we do not have any studies of electricity cost estimates from this plant.)

Jaitapur Nuclear Plant

On December 6, 2010, during French President Sarkozy's visit to India, India signed a framework agreement with France's state-run nuclear group Areva for the purchase of two reactors for the Jaitapur Nuclear Plant to be set up at village Madban in Ratnagiri district of Maharashtra (Jaitapur is a village near Madban). So eager has been the government to sign this deal, so powerful are the vested interests behind this deal, that the government is not willing to even discuss the economics of producing electricity from these reactors. While announcing the agreement at a press conference, the Prime Minister stated that pricing issues are still 'subject matters of negotiations',[16] meaning, that the government has agreed to buy the reactors, without finalising the price! Clearly, the government has something big to hide. Let us take a closer look at the Areva deal.

Newsreports have estimated the cost of the two reactors to be about 7.0 billion euros (9.3 billion dollars).[17] However, this seems to be a big underestimate. The contract price of the 1600 MW EPR being constructed in Finland was €3.2bn when the agreement was signed in December 2003; by June 2010, its cost had escalated to around 5.9 billion euros, and the reactor was only halfway to completion.[18] Obviously, the final cost is going to be much more. Let us calculate the cost in rupees. Even assuming that each Jaitapur reactor is going to cost 5.9 billion euros, then, taking the current Euro-Rupee exchange rate (in January 2011) of Rs.59 to €1, this means each reactor should cost at the minimum Rs.34,800 crores! That works out to Rs.21 crores per MW, as compared to Rs.5 crores per MW for coal-fired plants!!

The total installed capacity of the Jaitapur plant after all six reactors are constructed is going to be 9900 MW. At Rs.21 cr/MW, this means the plant is going to cost a mind-boggling Rs.2 lakh crores! The cost of an equivalent coal-fired plant would be just Rs.50,000 crores, implying a saving of Rs.1.5 lakh crores!!

Given this huge capital cost, what will be the unit cost of electricity from the plant? No one, right from the Prime Minister to the Chairman of NPCIL, is willing to discuss it. Areva's CEO, Anne

Lauvergeon, in an interview to *The Hindu* on November 25, 2010, asserted that it would definitely be below Rs.4 a unit![19] That is obviously ludicrous. More realistic estimates put the cost of electricity from the Jaitapur Nuclear Plant at least Rs.7-9 per unit: Prabir Purkayastha, the well-known power sector analyst, estimates the cost of electricity assuming the plant is going to cost Rs.20 cr/MW to be Rs.7-8 per unit[20]; while Dr Vivek Monteiro, a well known physicist who holds a doctorate from Harvard University, estimates that the cost would be at least Rs.9 a unit.[21]

Of course, these cost estimates do not take into account the subsidies to nuclear power discussed above, like decommissioning costs, waste management costs, et cetera.

Clearly, the cost of electricity from the imported reactors is going to be even more than that from Enron's Dabhol Power Plant in Maharashtra, which virtually bankrupted the Maharashtra State Electricity Board (MSEB). If the government succeeds in its plans of setting up a string of Nuclear Parks along the coast, India is going to be saddled with dozens of nuclear Enrons.

Nuclear Liability Bill: Indemnifying Foreign Suppliers

The costs of nuclear electricity are so huge that the foreign vendors are still not satisfied with these subsidies. American reactor suppliers like General Electric and Westinghouse grunted that they would not invest until India gave them a sovereign guarantee that the *entire liability of any potential catastrophe would be borne solely by Indians*; in other words, that they will not be held liable for any accident at the plants supplied by them—they are aware that it could bankrupt them. And so, their concubine, the US government, ratcheted up the pressure on the Indian government to pass such a law. On June 25, 2009, the US Assistant Secretary of State for South and Central Asian Affairs Robert Blake told a committee of the US House of Representatives: '... we are hoping to see action on nuclear liability legislation that would reduce liability for American companies and allow them to invest in India ...'[22] Obligingly, the government has now got the 'Civil Liability for Nuclear Damage Bill 2010' passed by a pliant Parliament. The provisions of the bill are absolutely outrageous.

The nuclear liability bill has two key clauses. Firstly, in the event of an accident, it indemnifies the supplier from all liabilities. Further, while the US liability law (Price Anderson Act) permits 'economic channelling', but not 'legal channelling', of liability, thereby allowing criminal proceedings and other lawsuits against any party in courts, the Indian liability law channels all financial and legal liability to the operator.[23] The law does allow the operator—and only the operator, not the victims of the accident who have been completely denied all rights—to sue the foreign vendor in courts in case the nuclear incident has taken place because of 'supply of equipment or material with patent or latent defects or sub-standard services.'[24] The problem is, the operator is going to be the government-owned NPCIL. Considering the extent to which the Indian government is selling out the interests of the people of India to foreign corporations and governments—an example being the complete betrayal of the victims of the Bhopal gas tragedy in the case against Warren Anderson, Union Carbide and its successor company Dow Chemicals—it is obvious that foreign reactor suppliers can sleep easy.

Secondly, the Indian law caps the liability of the nuclear plant operator in the event of a nuclear accident at a laughable Rs.1500 crores. Beyond this cap, if necessary, the government assumes responsibility for paying the damages, but subject to a maximum cap of 300 million Special Drawing Rights,[25] equivalent to roughly Rs.2100 crores.[26] (As on October 29, 2010: 1 SDR = $1.57; while $1 = Rs.44.40; therefore 1 SDR = Rs.69.71; 300 million SDRs = Rs.2091 crores.) Since the operator of the reactors is going to be the NPCIL, a public sector corporation, it essentially means that all the damages for an accident in the foreign reactors are entirely going to be paid by you and me who are not even remotely responsible for the accident; the foreign supplier is completely let off the hook!

That is precisely the reason why the Indian government pushed the Nuclear Liability Bill through the Indian Parliament, overriding the objections raised by a wide range of pro-people intellectuals. This has in fact been admitted to by an Indian minister, 'The Nuclear Liability Bill ... will ... indemnify American companies so that they don't have to go through another Union Carbide in Bhopal.'[27] For

Delhi's moghuls, the perpetrator of the world's worst chemical accident is unfortunately being victimised, and so they are pledging legal protection for a possible nuclear Bhopal.

A secondary beneficiary of the liability law is Indian big business, which is expecting subcontracts to the tunes of hundreds of crores of rupees from the foreign plant suppliers (See Chapter 11). The Federation of Indian Chambers of Commerce and Industry (FICCI), the oldest and biggest lobbying arm of Indian big industry, had extended its strong support to the passage of the bill through the Indian Parliament, saying that it was essential for participation of both domestic and foreign suppliers in India's nuclear program. And when the debate between the government and the opposition intensified over the inclusion of Clause 17 (b), which allows the operator to sue suppliers in case of defective equipment, FICCI came out strongly for deletion of the clause, arguing that the clause 'is neither implementable' nor 'justified' and 'not desirable', and that it will 'completely undo the government's efforts to accelerate nuclear power generation in our country.'[28]

Supreme Court Decisions Violated

By freeing the foreign suppliers of all liabilities in case of an accident at a reactor supplied by them, the Nuclear Liability Law violates the principle of absolute and strict liability laid down by the Supreme Court wherein the court ruled: 'Once the activity carried on is ... potentially hazardous, the person carrying on such activity is liable to make good the loss ... irrespective [of] whether he took reasonable care ...'[29] Since a nuclear reactor is inherently hazardous, by an extension of this principle, at the very least, the foreign supplier of the reactor should be held equally responsible for an accident along with the operator, irrespective of whether there was a design fault or not.

The Law also violates the right to full compensation which has been interpreted by the Supreme Court to be a part of Right to Life guaranteed under Article 21 of the Constitution. Known as the Polluter's Pay Principle, according to this tenet, a polluting industry has to not only fully compensate the victims for the accident, it must also fully bear the costs of restoring the environmental degradation.

The Nuclear Liability Law violates this principle by artificially capping the total compensation that would be paid out in case of an accident at 300 million SDRs. This translates into a measly $460 million, which is even lower than the compensation of $470 million approved by the Supreme Court of India for the victims of the Bhopal gas disaster way back in 1989, and which is universally considered shamefully inadequate. If exchange-rate changes and inflation are taken into account, the sum works out to about one-third of what the Bhopal victims got,[30] whereas a nuclear accident can be many hundreds of times bigger than the Bhopal gas tragedy! An accident like the Chernobyl reactor core meltdown of 1986 can wreak damage running into hundreds of billions to several trillions of dollars, and make huge swathes of land uninhabitable for centuries. Among the countries having a liability law (which limits the liability of the nuclear operators), US, Germany, Finland, Japan, South Korea and Switzerland have not placed any cap on maximum liability (the excess amount will of course be paid by the government).[31]

Liability Legislation: No International Obligation

Even assuming that the country needs nuclear energy and needs to import nuclear reactors, there was no need to enact a liability legislation. The government of India and the media have created the impression that India needed to pass a liability legislation in order to become compliant with international nuclear liability instruments (like the Convention on Supplementary Compensation for Nuclear Damage, which India signed on October 27, 2010, two months after the Nuclear Liability Bill cleared Parliament), and that this was a necessary condition for India to able to engage in nuclear trade with the world. The reality is, India is under no obligation to enact a liability law or become a signatory to an international liability convention to become eligible for engaging in nuclear trade with other countries. For instance, Russia has refused to pass legislation to waive or cap accident liability for its foreign suppliers. While many countries have liability laws, the powerful World Nuclear Association—the lobbying arm of 180 nuclear firms, including GE, Westinghouse and Areva—admits, 'States with a majority of the world's 440 nuclear power reactors are not yet party to any international nuclear liability convention, relying on their own arrangements.'[32]

US-Russia Still not Happy

According to newsreports, both the USA and Russia are not happy with the Nuclear Liability Law passed by the Indian Parliament. They want India's nuclear liability regime to channel 'absolute and exclusive liability to nuclear power plant operators', even if an accident occurs due to negligence on the part of the foreign supplier![33] As mentioned above, India's liability law allows for the Indian operator to sue the foreign supplier in case the accident occurs due to defective equipment supplied by the latter.

Both the US nuclear industry and the Obama administration have mounted pressure on the Indian government to either make amendments to the law, or find some way to circumvent it. According to *The Wall Street Journal*, the US government, recognising the difficulty of the Indian government in sending the freshly passed law back to the Parliament for an amendment, is trying for a government-to-government agreement with India that could take precedence over the Law.[34] Considering the extent to which the Indian government is surrendering to imperialist interests, it is certain that very soon, as soon as the political atmosphere in Delhi gets more congenial, the Indian government will take steps to address US concerns.

The bloodthirsty imperialists are not satisfied if you just bow before them. They want you to show your absolute submission by doing a 'Sashtanga-dandavat-pranam', that is, pay obeisance by lying down flat on the ground.

To Conclude ...

As we have argued elsewhere, India's ruling classes have totally divorced themselves from the people of the country. The country is now being run solely to maximise the profits of foreign multinationals and their Indian collaborators, India's big business houses.[35] The nuclear policy of the government of India is just another instance of this betrayal ...

9

INDIA'S NUCLEAR INSTALLATIONS: SAFETY AND OTHER ISSUES

Part I: India: Nuclear Dictatorship

The nuclear industry is notorious all over the world for suppressing information. Even then, in the US and West European countries, at least some information is officially available on the release of radioactivity into the atmosphere from uranium mines and nuclear power plants (some of which we have given in Chapter 3). In India, however, no such information is available. The nuclear authorities in India take refuge behind the draconian Atomic Energy Act of 1962 to deny all information about the state of India's nuclear installations and the various accidents taking place in them, under the plea that all such information is 'classified', and cannot be disclosed in the interests of national security![1] Obviously, the DAE has mixed up national security with nuclear safety to cover up its safety lapses, and the Indian courts including the Supreme Court have gone along with its interpretation (we discuss this in greater detail in the next section). The Indian Atomic Energy Act of 1962 is so authoritarian that the DAE is not accountable even to the Parliament![2] In fact, the DAE has used this law to prevent nuclear plant workers from accessing their own health records too!![3]

India's nuclear establishment has become a dictatorial entity lording over the people of the country. Even when major accidents have occurred and news about them have leaked out through the media to become public knowledge, the AEC / DAE / NPCIL / UCIL have tried their best to suppress information about the accidents, have blithely lied about the extent of the accidents, have denied that any radiation releases occurred from their installations due to these accidents, and have tried to play down the impact of these accidents on people as well as on workers. A few examples:

- A major mishap in Tarapur in 1980 resulted in thousands of litres of irradiated water gushing out from the reactor. But the Chairman of the AEC reluctantly acknowledged only a 'pinhole' leak, even though the water had gushed out from a 15-cm tube![4]
- On March 26, 1999, six tons of highly radioactive heavy water leaked out from the Madras Atomic Power Plant (at Kalpakkam). The accident was serious enough for the plant management to declare an emergency, which means the plant was just one step away from being evacuated. The plant authorities initially tried to suppress the news. But the workers'

union revealed it to the press and it created a furore. The plant authorities then played down the accident, claiming that the leak was 'insignificant' and 'anticipated', and made the ludicrous declaration that the coolant water which spilled was not radioactive![5]

- On December 25, 2006, the pipeline carrying radioactive waste from the uranium mill to the tailing pond in Jaduguda (Jharkhand) burst and continued to spew toxic sludge into a creek for nine hours before the flow of the radioactive waste was shut off. Consequently, a thick layer of toxic sludge on the surface of the creek killed scores of fish, frogs and other riparian life. The waste from the leak also reached a creek that feeds into the Subarnarekha River, seriously contaminating the water resources of communities living hundreds of kilometers along the way.[6] However, all that the UCIL website admits is: 'The pipe burst spilling tailing slurry in December 2006 .... was attended in the shortest possible time and corrective measures were also taken.'[7] That's all, not a word more!

Kaiga 'Incident': Lies Galore

Since science is all about pursuing truth, most people find it difficult to believe that India's top scientists are lying. Let us therefore discuss a recent accident in greater detail. On November 24, 2009, urine tests revealed that more than 90 workers at Unit 1 of the Kaiga nuclear plant had high levels of radioactive tritium in their bodies[8] (according to Kaiga township residents, 250 workers were affected[9]). The incident only came to light on November 28, after the media got hold of the story (the NPCIL failed to suppress the news probably because too many workers needed hospitalisation). We don't know for how long these workers were exposed to this radiation, and what was their exact medical condition. Under public pressure, the authorities released the radiation count levels, and this showed radiation levels of as high as 59-74 megabecquerels in some of the workers. The maximum permissible amount of tritium in a litre of urine is 3.7 megabecquerels. However, nothing beyond that was revealed.[10]

After the media broke the story, India's top nuclear authorities immediately went into denial mode, claiming that all was well with the reactor, no accident and no leakage of radioactive tritium had taken place at the plant. Even Prime Minister Manmohan Singh grandly declared that there was 'nothing to worry about the small matter of contamination'.[11] India's leading scientists made public statements that there was nothing to fear about the safety of workers as tritium is not poisonous and its presence in the human body would come down on its own.[12] They were lying; they are no longer scientists in search of the truth, they have become prizefighters. Tritium is a very dangerous material. Tritiated water, being chemically similar to water, is easily absorbed and then quickly distributed throughout the body via the blood. While half of it is out of the body within 10-12 days, some of it gets bound to organic molecules and spends much longer time in the body. Tritium is a beta emitter and very mutagenic: it can cause cancer, genetic defects and developmental abnormalities. (We've discussed its effects in greater detail in Chapter 3, Part III.) This risk increases in direct proportion to the radiation exposure, and there is no threshold below which the risk is zero.[13]

If there was no accident, then how did tritium enter the bodies of workers? The official story is: the workers got poisoned after they drank tritiated water (that is, water contaminated by tritium) from a water cooler. And, pray, how did tritiated water enter the cooler? The story goes that it was due to internal 'sabotage' or 'mischief-making' by unidentified employees: these employees, 'it appears', added tritium-contaminated heavy water to a drinking-water cooler through its overflow pipe.[14] But that would have required a pump! How can an employee do that without being discovered?

The story becomes even more farcical if one considers that tritium is an extremely costly material. The estimated costs of producing tritium vary from about $30,000 (about Rs.14 lakh) per gram in Canada to $100,000 (about Rs.46 lakh) per gram in the US. Tritium is also a strategically important material, being used in nuclear weapons as a booster.[15] Clearly, it's a cock-and-bull story.

India's atomic scientists not only deny any accidents in Indian nuclear establishments, they also deny that any major accidents have occurred at nuclear plants around the world, including at Three Mile Island and Chernobyl! Dr P.K. Iyengar, one of India's leading nuclear scientists, and a former Chairman of the AEC, while speaking at the Bangalore Science Forum in September 1988 (he was then the Director of the BARC), expressed the view that the Three Mile Island accident was not an accident at all as 'no one died there'! And according to Dr Raja Ramanna, another former Chairman of the AEC, Chernobyl was not a nuclear accident, but a 'curious fire accident'!![16]

Challenging India's Nuclear Dictatorship in the Courts

It is common sense that safety issues at India's nuclear installations have nothing to do with the security of the country, and information about these issues should be shared with the people. If India's atomic energy establishment is denying us information about these issues, it is clearly violative of the democratic fabric of our country. If it is using the Atomic Energy Act of 1962 to deny us this information, then such an interpretation of this Act is violative of our fundamental rights, and is unconstitutional. Because an accident at a nuclear reactor can kill lakhs of people and render huge areas uninhabitable for centuries and because it is affecting our very Right to Life, we have a fundamental right to know whether all is well at India's nuclear installations, it cannot be left to the whims of our bureaucrats and politicians. Why don't we invoke the Right to Life guaranteed to us by the Constitution and ask the Supreme Court of India to end the nuclear dictatorship prevailing in our country? It has been tried by some of India's most eminent lawyers but, unfortunately, our courts have upheld the dictatorial powers conferred on the AEC under this Act.

Following newsreports in mid-1996 that the Atomic Energy Regulatory Board (AERB), India's nuclear safety regulator, under the chairmanship of Dr Gopalakrishnan had compiled more than 130 issues affecting the safety of our nuclear establishments, the well-respected human rights organisation People's Union for Civil Liberties (PUCL) filed a Public Interest Litigation (PIL) in August 1996 in the Bombay High Court. Citing the grounds of Right to Life and Right to Know, the PUCL petition sought, amongst others: (i) disclosure of the adverse report of the AERB; (ii) for a direction to the Union

Government to make the AERB an independent body so that it might act as an effective watchdog for nuclear safety in the country (this issue is discussed later in this Chapter); and (iii) for declaring Section 18 of the Atomic Energy Act unconstitutional as it confers on the central government untrammelled powers for withholding from the public information about the working of India's nuclear power plants. The Sarvodaya Mandal of Mumbai represented by Dr Usha Mehta, the noted Gandhian and freedom fighter, also filed a petition in support of the PUCL petition.

The two PILs were heard at length at the stage of admission by a Bench presided over by the then Chief Justice M.B. Shah. Senior officials of the country's nuclear establishment, including Dr R. Chidambaram, then chairman of the AEC and secretary to the DAE, filed bulky affidavits, opposing the petitions. In his affidavit, Dr R. Chidambaram invoked the provisions of Section 18 of the Atomic Energy Act and stated that the AERB document was a secret document as it related to nuclear installations, and went on to say that publication of the document 'will cause irreparable injury to the interests of the State and will be prejudicial to national security.'

It is important to note that the petitioners did not ask for any information about India's nuclear arsenal or its storage site or anything related to India's national security. They only expressed a genuine concern that there were not enough safety precautions in nuclear power stations in the country and any accident could have a disastrous effect on human beings, animals, environment and ecology. However, accepting the blatantly false arguments made in the statements and affidavits of officials of the DAE, the Bombay High Court dismissed the writ petition at the admission stage itself.

Both the Bombay PUCL and the Bombay Sarvodaya Mandal filed appeals before the Supreme Court against the decision. The atomic energy department again repeated its argument that disclosure of the documents would harm national security. On January 6, 2004, a Bench of the Chief Justice of India and Justice S.B. Sinha dismissed both the appeals.[17]

Some months later, another PIL petition was filed in the Supreme Court regarding the impact of uranium mining being done by UCIL in Jaduguda, Jharkhand. The petition made a prayer to the court to direct the UCIL, the AEC and other concerned authorities to take all possible steps under a time bound program to ensure that the radioactive effluents generated by the mining and allied activities of the Jaduguda uranium mines are controlled and treated properly so that the same do not cause serious hazards to the health and lives of those working or living in or around the mines. In response, the Chairman of the AEC filed an affidavit stating under oath that adequate steps had been taken to check and control radiation from uranium waste. Accepting this bare-faced lie, on April 15, 2004, the Supreme Court dismissed the petition stating that it did not see any merit in it.[18]

Let us take a look at the state of India's nuclear installations. It should give all of us sleepless nights!

Part II: Uranium Mining

The Uranium Corporation of India Ltd. (UCIL), a subsidiary of the DAE, has been mining uranium in Jharkhand for over four decades now. It presently operates five underground mines and an open cast mine in the region.

Untruths Unlimited

The corporation's website claims that 'UCIL has a track record of adopting absolutely safe and environment friendly working practices in Uranium Mining and Processing activities.' It claims that it regularly monitors external gamma radiation, radon concentration and concentration of radionuclides in surface and ground water; while refusing to make radiation data public, citing the Atomic Energy Act of 1962, it nevertheless asserts that there is no radioactive contamination of the area due to uranium mining.[19] Its website claims that 'the diseases prevalent in the villages around UCIL workings are not due to radiation but attributed to malnutrition, malaria and unhygienic living conditions, et cetera.'[20]

Shocking Carelessness

The reality is the exact opposite. There are three tailing ponds in the Jaduguda (also spelled as Jadugoda, from the word 'Jar agora' which means a grove of the castor oil tree) region, spread over an area of 100 acres; they are estimated to contain crores of tons of radioactive waste.[21] Seven villages stand within one and a half kilometers of these tailing ponds; one of them, Dungardihi, begins just 40 meters away. More than 30,000 people live within a 5 km radius from the tailing ponds.[22] The corporation has not taken the slightest of precautions to protect the health of these people from radiation releases from the mines and tailing ponds; its mining practices completely disregard the fact that uranium is radioactive, that the waste from the mines continuously emits the highly carcinogenic radon-222 gas and that the mill tailings contain uranium decay products like the highly radioactive thorium-230 with a half-life of 80,000 years. (See Chapter 3 for a more detailed discussion.)

This is eloquently brought out in numerous surveys of these villages and the area around the mines by independent experts, including: a survey by the well-known physicist Dr Surendra Gadekar and medic Dr Sanghamitra Gadekar in 2000;[23] field trips in 2001 and 2002 by Professor Hiroaki Koide from the Research Reactor Institute, Kyoto University, Japan;[24] and the more recent survey by a team of doctors from Indian Doctors for Peace and Development (IDPD), the Indian chapter of 1985 Nobel Peace Prize recipient International Physicians for Prevention of Nuclear War (IPPNW) in 2008.[25] They found that:

(i) The uranium mines are located on adivasi land. The adivasis were forcibly removed from their lands, but no attempt was made to resettle them at a suitable distance from the mines. Consequently, they continue living on the edge of the mines. Without explaining the risks, they were offered employment as uranium miners, which they willingly accepted as they had been deprived of their lands. They were thus doubly exposed to radiation: as miners and, along with their families, to the radioactive dust blowing from the tailing ponds.

(ii) No safety measures have been taken by the company. The waste is carelessly dumped in the open; the ore is transported to the mills in uncovered dumpers; the tailing ponds are not fenced off properly, and people freely walk across them, not knowing that they are thus getting exposed to gamma radiation.

(iii) The company is so utterly callous that it has supplied waste rock from the mines to the local people for construction of roads and houses!

(iv) It is suspected that radioactive wastes from India's nuclear reactors have been abandoned in these tailing ponds. The tailing ponds in Jadugoda have a high concentration of cesium-137, one of the fission products of uranium which is found in the spent fuel. There is no other logical way in which this radionuclide could have found its way into the tailing ponds!

The state of UCIL's newest mine, the Banduhurang open cast mine, which was commissioned in 2009, is no better. A *Tehelka*[26] reporter visited the mine in September 2010. He found no prohibitory signs, no warnings about radiation, no barbed wire and no demarcation of territory. Mounds of radioactive waste from the mines lay scattered everywhere, sometimes inside the villages surrounding the mine. The radioactive waste water released from the mine simply joined a stream flowing through the villages where children were found bathing and women washing clothes. Trucks carrying uranium ore were loosely covered with plastic sheets, radioactive dust flying in the wind. This, when just a few years ago, the Chairman of the DAE had filed an affidavit in the Supreme Court claiming that all precautions were being taken to check and control radiation from the waste from UCIL's uranium mines. UCIL officials tried their best to prevent the reporter from visiting the mine and filing his report, slapping all kinds of charges on him, putting him behind bars for 12 hours, seizing his equipment ...[27]

Accidents Galore

As if this was not enough, there have been numerous accidents at the mines due to UCIL's faulty technical and management practices. Tailing pipelines, carrying uranium mill tailings from the Jaduguda uranium mill to the tailing ponds, have repeatedly burst, causing spillage of the radioactive sludge into nearby homes and water bodies. The latest such pipeline burst took place on August 16, 2008; before this, bursts had taken place on February 21, 2008 and April 10, 2007.[28]

One of the worst such accidents took place on December 25, 2006—the toxic spillage from the burst pipeline continued for nine hours before it was finally shut off. (See Part I of this Chapter for more details.)

Terrible Health Costs

The impact of these radiation releases on the health of the people of the nearby villages has been colossal. In 1993, the Bindrai Institute for Research Study and Action (BIRSA)[29] conducted a survey in seven villages within a kilometre of the tailing dams. The field health workers who conducted the survey were trained by Dr Imrana Qadeer, Professor at the Centre of Social Medicine and Community Health, Jawaharlal Nehru University, New Delhi. It took two years to complete the survey. The report revealed that a shocking 47 per cent of the women in the area suffered disruptions in their menstrual cycle, 18 per cent said they had suffered miscarriages or given birth to stillborn babies in the last 5 years and 30 per cent suffered from fertility problems. Nearly all women complained of fatigue, weakness and depression. Further, the survey found a high incidence of chronic skin diseases, cancer, tuberculosis, bone, brain and kidney damages, nervous system disorders, congenital deformities, nausea, blood disorders and other chronic diseases. Children were the most affected. Many were born with skeletal distortions, partially formed skulls, blood disorders and a broad variety of physical deformities, most common being missing eyes or toes, fused fingers or limbs incapable of supporting them. Brain damage often compounded these physical disabilities.[30]

The Gadekars, in their medical survey, found similar health

impacts. They found a high incidence of congenital deformities and mental retardation among infants in the vicinity of Jaduguda. They also found extremely high levels of chronic lung diseases, which were most likely to be silicosis or lung cancer, in the company's mine and mill workers—the company termed these cases as tuberculosis so as to avoid compensation payments.[31] The more recent health survey by a team of doctors from the IDPD also found clear evidence of increased incidence of sterility, birth defects and cancer deaths among people living in the nearby villages.[32]

While the UCIL management doesn't accept the fact that radioactivity has in any way been harmful to either people, animals, trees or plants, senior officials of the corporation have made arrangements for their own food to come from a government farm about 44 km away![33]

Expanding the Toxic Trail

After destroying the Jaduguda region, UCIL is now proposing to start uranium mining in Andhra Pradesh, Karnataka and Meghalaya (see Chapter 7 for more details).

UCIL and DAE's utter disregard for the impact of uranium mining and processing on the environment is evident from its proposal to start mining in the Lambapur-Peddagattu area in Nalgonda district of Andhra Pradesh. This mining site is right above the Nagarjunasagar Reservoir and is in the vicinity of the Akkampally Reservoir. The Nagarjunasagar Reservoir is an important source of irrigation for the districts of Nalgonda, Guntur, Krishna and Prakasham and is also the drinking water source for many towns, while the Akkampally Reservoir is the pumping station of Krishna River water to the twin cities of Hyderabad and Secunderabad.[34] If at all uranium mining begins in this area, it is absolutely certain that the Nagarjunasagar and Akkampally Reservoirs are going to get radioactively polluted, ultimately polluting the food chain of the people of Andhra Pradesh! And as we have discussed in detail in Chapter 3, this is not like industrial pollution which can be remedied. Radioactive pollution will contaminate the state's water sources for thousands of years!!

Fortunately, protests by local people have so far prevented the UCIL from making any headway in its diabolical plan.

UCIL is particularly desperate to begin mining in Meghalaya, which is supposed to have the largest reserves of uranium in the country after Jharkhand. Exploratory mining work done in this state in 1991 had led to the emergence of strange diseases among people, fish began to die in nearby rivers, and so the people organised and forced UCIL to close down its operations.[35] UCIL again revived the project some years ago but, once again, a powerful movement has put a spanner in its plans.

Part III: Nuclear Fuel Complex, Hyderabad

UCIL processes the uranium ore in its mills in Jharkhand and sends the yellow cake to the Nuclear Fuel Complex (NFC) in Hyderabad. Here, the uranium fuel rods are fabricated from the yellow cake, and supplied to all the nuclear plants in India.

The NFC churns out 50,000 tons of contaminated waste water, containing radioactive materials and chemical wastes, every day. This is discharged into a waste storage pond located in the complex. Seepage from this pond has contaminated the underground water, and with the NFC/DAE sublimely unconcerned, this radioactive contamination is going to increase with time.

As a result, the situation in and around Hyderabad is becoming grave. Mysterious and painful diseases have already visited residents in the vicinity of NFC. The DAE has prohibited residents of Ashok Nagar, a locality near NFC, from drinking water from underground wells in the area. Eleven villages near NFC also face the same problem. As the contamination spreads, it will affect the underground water supply to the entire city.

Hyderabad has an acute shortage of drinking water, and so many residential complexes in the city install their own borewells. A day may come when it will be highly dangerous to use the underground water and people may have to desert Hyderabad as has happened in the area near Hanford Works in the USA.[36]

PART IV: INDIA'S NUCLEAR POWER REACTORS

As discussed in Chapter 3, release of small or large quantities of radioactivity from nuclear power plants (NPPs) occurs quite often, at every nuclear reactor around the world. These releases can be planned, that is, the nuclear plant authorities purposefully decide to vent radioactive gases into the air or release radioactive water into nearby seas and rivers. Or they can be because of human or mechanical error, which the nuclear industry euphemistically refers to as 'incidents', in order to downplay the severity of the accident and mollify public concerns. Several of these 'incidents' have snowballed and have had catastrophic ramifications, the biggest of course being the Three Mile Island and Chernobyl disasters. (And now of course, an even bigger disaster has taken place, at Fukushima!) As discussed in Chapter 3, the technology of nuclear reactors is complex and events can spin out of control in a very short time, all possible accident modes cannot be predicted, all of which means that there is no way to ensure that reactors will not have major accidents.

Thus, even though the nuclear industry claims it is emission-free, nuclear power plants collectively release lakhs of curies of radiation into the atmosphere every year, with deathly consequences for life on planet Earth, consequences which will be with us till the end of time, as many of these radioactive materials released into the atmosphere have half-lives of up to half a million years!

India: World's Most Unsafe Reactors

India's nuclear reactors are even more unsafe. Some years ago, a survey in *Nuclear Engineering International*[37] listed India's reactors in the lowest bracket in terms of efficiency and performance.[38] Helen Caldicott, one of world's best known anti-nuclear-energy activists, writes that India's nuclear plants are amongst the most contaminated in the world, exposing hundreds of workers to excessive doses of radiation.[39] The US-based watchdog group—the Safe Energy Communication Council (SECC)[40]—has also described India's nuclear energy program, especially its reactors, to be the 'least efficient' and the 'most dangerous in the world'.[41] Molly Moore's report in the

Washington Post published in 1995 is even more damning:

> Four decades after India launched a full-scale nuclear power program ... it operates some of the world's most accident-prone and inefficient nuclear facilities. During 1992 and 1993, its most recent two-year monitoring period, the Indian government reported 271 dangerous or life-threatening incidents, including fires, radioactive leaks, major systems failures and accidents at nuclear power and research facilities. Eight workers died in that period.[42]

In what may appear to be astonishing, the same opinion was expressed by Dr A. Gopalakrishnan, Chairman of the AERB, the body responsible for overseeing safety at India's nuclear installations, in an interview to the media while remitting office in 1996. He stated: 'Many of our nuclear installations have aged with time and have serious problems', and that the current safety status of the nuclear installations under the DAE 'is a matter of great concern'![43]

But why didn't Dr Gopalakrishnan do anything about it when he was in office? The shocking answer is, he had very little authority to do so! The AERB is a toothless body!!

India's Safety Watchdog: A Lapdog

India's rulers are so unconcerned about nuclear safety, they are so indifferent to the possibility of a Chernobyl in India, that they have the most ineffective nuclear safety regulator in the world!

Like other countries having a nuclear power program, India also has a nuclear safety regulator, the Atomic Energy Regulatory Board (AERB). It was set up in 1983 by the DAE to lay down safety standards, frame rules and regulations in regard to public and worker safety under the provisions of the Atomic Energy Act of 1962, and enforce their compliance in all DAE and non-DAE installations. However, unlike nuclear safety regulators of other countries, India's nuclear regulator is not independent of the bodies it is supposed to oversee, but is subservient to them! The AERB reports to the AEC. The Chairman of the AEC is also the head of the DAE. The Chairman of the NPCIL and the director of BARC are also members of AEC. However, the chairman of the AERB is not a member of the AEC! Thus, the regulatory authority is subordinate to the NPCIL and BARC, bodies it is supposed to regulate! This makes the regulatory process a complete sham. Among nuclear and threshold nations of the world, India is the only country where such a situation prevails.[44]

This lack of independence of the regulatory authority in India contravenes the international Convention on Nuclear Safety (CNS), of which India is a signatory. According to this convention, the regulatory body should be provided with adequate authority, competence and financial and human resources to fulfill its assigned responsibilities. There should be an effective separation between functions of the regulatory body and those of any other organisation concerned with the promotion or utilisation of nuclear energy. Additionally, the regulatory body must communicate its regulatory decisions and their bases to the public.[45] The functioning of the AERB violates the CNS on all these three counts. Not only is the AERB subordinate to the body it is supposed to regulate, it is completely dependent on the DAE: the AERB depends, to a major extent, on the DAE for funds, manpower, technical expertise and material resources. And, as we see below, the AERB and DAE do not share any information regarding safety at India's nuclear installations with the people, and are empowered by law to maintain this secrecy.

Probably the only time the AERB has attempted to function as an independent safety regulator was in the period 1993-96, when Dr Gopalakrishnan was the Chairman of the AERB. During his tenure, the AERB undertook to prepare a comprehensive document on DAE's safety status. However, all his efforts to improve the safety situation of India's nuclear installations were stonewalled by the DAE. In Gopalakrishnan's own words (in an article published in the Chennai fortnightly *Frontline* in 1999):

> After four months of serious effort by the AERB staff and after referring to more than 700 of the DAE's own documents, the AERB prepared a report titled *Safety Issues in DAE Installations.* It covered about 130 safety issues, of which 95 are of top priority. This document was discussed and approved by the AERB at its 46th meeting on November 7, 1995 and then submitted to the

> AEC ... To date, however, it is not known whether any concrete action has been taken on this report, even though the present Chairman of the AERB asserts to the press that 'every issue is being seriously looked into'.[46]

Failing to make any reforms in the system, all that Gopalakrishnan could do was to voice his concerns before the people of the country upon his retirement. In an interview to *The Times of India*, Mumbai (June 18, 1996), he stated:

> During my six-year-old association with the AERB (three years as a member and the remaining period as chairman), I was able to study the nuclear regulatory process thoroughly. I discovered that it was a total farce. I was of the opinion that the government and the public should know this because ultimately they finance the nuclear establishment. My straightforward attitude was not liked by the top bosses of the establishment. The DAE wants the government and the people to believe that all is well with our nuclear installations. I have documentary evidence to prove that this is not so.[47]

Defending his employer, G.R. Srinivasan, Director, Health and Safety, NPCIL, stated: 'Even if a highly unlikely accident takes place, our nuclear power plants are so designed that the public domain would suffer no harmful exposure.'[48] Probably the NPCIL has designed its reactors in such a fantastic way that in case a Chernobyl-type accident occurs in India, the radiation released would go like a rocket straight into the stratosphere.

Accidents at Nuclear Reactors

Let us now take a look at the performance of India's reactors so far, based on the little information that has come out through unofficial and occasionally official sources.

There have been hundreds of accidents, of varying degrees of severity, at the nuclear reactors and other facilities operated by the DAE. In 1989, Prof. Dhirendra Sharma, author of *India's Nuclear Estate* and Director of Centre for Science Policy, Dehradun, estimated that the Indian nuclear industry has suffered from at least '300 incidents of a serious nature ... causing radiation leaks and physical damage to workers.' He added, 'These have so far remained official secrets.'[49] DAE's hair-raising efficiency is also brought out in another study made by researchers at the American University. They calculated at least 124 'hazardous incidents' at nuclear plants in India between 1993 and 1995.[50]

That none of these led to catastrophic radioactive releases to the environment is not by itself a source of comfort. According to safety theorists, this absence of evidence of 'accidents should never be taken as evidence of the absence of risk' and 'just because an operation has not failed catastrophically in the past does not mean it is immune to such failure in the future'.[51] In fact, quite a many of these caused significant radioactivity releases into the atmosphere and, on at least one occasion, at the Narora NPP in the state of Uttar Pradesh, the accident very nearly led to a Chernobyl-like meltdown. Had the catastrophe occurred, it is impossible to imagine its consequences in a densely populated state like Uttar Pradesh.

We give below a brief summary of the little information that is available about the actual state of affairs with India's nuclear reactors and the accidents that have taken place.

(i) Tarapur-1 and 2

The Tarapur Atomic Power Station (TAPS), located about 100 miles north of Mumbai, was commissioned in 1969. The two Boiling Water Reactors at the Tarapur station are of vintage US design. All similar reactors around the world have been shut down long ago for safety reasons.

The Tarapur reactors suffer from so many problems that they have earned the distinction of being amongst the 'dirtiest reactors in the world'.[52] The two reactors share the same subsystems, including the same emergency core cooling system, in violation of all safety standards. Even more disturbing is that the use of nitrogen to make the containment inert has been discontinued. Therefore, if the coolant does not perform its function, an explosion is quite likely to occur, leading to reactor meltdown. Besides, many parts of TAPS are uninspectable, and the DAE lacks the equipment and/or technology to correct its problems. The secondary steam generators in each unit are totally disabled owing to extensive tube failures, and because of this both the reactors have long since been de-rated from 210 MW to 160 MW. The plant has suffered innumerable radioactive releases. Radiation contamination of the reactor building and its environs is hundreds of times higher than the design intent. According to Gopalakrishnan, the TAPS reactors 'should have been shut down in the interest of public safety long back.'[53]

(ii) Tarapur-3 and 4[54]

These PHWRs are amongst India's newest reactors, and they are also the biggest, of 540 MW each. On June 28, 2010, panic spread in the villages around Tarapur after people learnt that an accident had occurred at Tarapur-4, due to which the reactor had to be shut down. A snag in the fuelling machine led to a spent fuel bundle getting stuck in it, endangering the lives of thousands of people living around the plant. It took 8 days for experts to rectify the problem.[55]

(iii) Rawatbhata-1 and 2 and Kalpakkam-1 and 2

All of these are PHWRs of Canadian design, also known as CANDU reactors.

Rajasthan Atomic Power Station (RAPS) at Rawatbhata in Rajasthan has 6 units. All these 6 units were designed to be of 235 MW installed capacity (and not just these 6, but all of India's 15 operating CANDU reactors including the 2 Kalpakkam reactors, 3 Kaiga reactors, 2 Narora reactors and 2 Kakrapar reactors were originally designed to be of 235 MW each). Till Narora-2 went online, this was the official description of this design. Then, all of a sudden, at a press conference announcing the opening of the Narora-2 reactor in end-1991, Dr P.K. Iyengar, the Chairman of the DAE, announced that Narora-2 was of 220 MW. He also stated that Narora-1, which had started up one year ago, was also of 220 MW. No explanation was given for this sudden derating of both these reactors by 15 MW each. In fact, he did not even acknowledge that these reactors were being derated. They were of 220 MW. Period. Since then, that has been the official description of all these 15 CANDU reactors.[56]

We discuss the oldest two units here, Rawatbhata-1 and 2.[57] The first unit went critical in 1972, and the second in 1980. They have faced so many technical problems that neither unit has ever worked at its installed capacity. Rawatbhata-1 was derated to 100 MW very early in its life, shut down for many years in the 1980s for repairs, asked to cease operation again in 2002 for two years, and finally has been shut down since 2004 while the government mulls over its future. It is plagued by a number of serious defects, ranging from turbine blade failures, cracks in the end-shields, a leak in the calandria overpressure relief device, and leaks in many tubes of the moderator heat exchanger. Seven years have gone by, and the DAE continues to ponder over its future. Unit-2 has also had tube leakage and other technical problems and could never operate continuously at its rated capacity; it too has suffered shut down for many years for repairs.[58]

The RAPS reactors have suffered several dangerous accidents: in 1976, the reactors were flooded due to construction errors, because of which the emergency core cooling system got obstructed and this could have led to a meltdown; the reactors were once again flooded in 1982; in 1985, a fire disabled four out of eight pumps of Rawatbhata-2;[59] on February 12, 1994, Rawatbhata-1 was shutdown for the repair of its calandria overpressure relief device which leaked radioactive heavy water.[60]

Madras Atomic Power Station (MAPS) at Kalpakkam, around 70 kms from Chennai, also has two units of 220 MW (originally 235 MW) each. These two reactors have created a world record of sorts by being in the gestation phase for more than 15 years. Within a couple of years of commissioning, the reactor inlets of both reactors cracked because the DAE had not heeded Canadian advice on how to fabricate them. Both the reactors have been de-rated to 175 MW because of this. Their continued operation even in this mode is not considered safe. The various safety issues at Kalpakkam-1 and 2 puts this Tamil Nadu station in a risk category unacceptable anywhere else in the world.[61]

The MAPS reactors have suffered several heavy water leaks (like the one on March 26, 1999 discussed in Part I of this Chapter).[62] In

1990, the turbine blade of Unit-1 cracked, and industrial robots were required to solve the problem.[63] On December 26, 2004, a tsunami hit the plant, to what effect nobody knows. All that is known is that the plant was profoundly affected, and many employees lost their lives.[64]

The biggest deficiency in the RAPS and MAPS reactors is the absence of a high-pressure emergency core cooling system (ECCS) for avoiding core meltdown in the case of a loss-of-coolant accident. No pressurised heavy water reactor anywhere in the world currently operates with such an obsolete and unsafe ECCS, according to Gopalakrishnan.[65]

(iv) Narora-1 and 2

The two 220 MW (235?) CANDU reactors of Narora Atomic Power Station (NAPS) in Uttar Pradesh went critical in 1989 and 1991 respectively.

The most serious accident that has occurred at an Indian nuclear reactor took place at this plant on March 31, 1993. We discuss it in greater detail to again illustrate the point made in Chapter 3 that because of the inherently complex nature of nuclear reactor technology, even minor failures or human errors can lead to a cascading chain of events culminating in a major accident.

Early that morning, two blades of the turbine at the first unit broke off due to fatigue. These sliced through other blades, destabilising the turbine and making it vibrate excessively. The vibrations caused pipes carrying hydrogen gas that cooled the turbine to break, releasing the hydrogen which soon caught fire. Around the same time, lubricant oil also leaked and caught fire. Within minutes, the fire spread through the entire turbine building. Among the systems affected by the fire were four sets of cables that carried electricity, which led to a general blackout in the plant. One set of cables supplied power to the secondary cooling system, which was consequently rendered inoperable. From the time of the blade failure, it took just 10 minutes for the control room to become filled with smoke, forcing the staff to vacate it.

The operators responded by manually actuating the primary shutdown system of the reactor 39 seconds into the accident. Although the reactor was shut down, since the fuel rods would continue to undergo radioactive decay even after the reactor was shut down, thereby generating heat which could cause a meltdown, some operators heroically climbed onto the top of the building and, under battery-operated portable lighting, manually opened the valves to release liquid boron into the core to slow down the reaction. This instinctive action by the technicians was the fourth and last level of safety protection, and it prevented what would almost certainly have led to a partial core meltdown.

It took 17 hours from the time the fire started for power to be restored to the reactor and its safety systems. Operators who were forced to leave the control room because of smoke could not re-enter for close to 13 hours. An attempt was made to take control of the plant from the emergency control room; but, since there was no power available, even this was not possible. Thus, Narora was almost unique in that for many hours, the operators had no indication of the condition of the reactor!

With the power supply off and the back-up system also down, how did the reactor avert a meltdown? Due to brilliant thinking on the part of the operators. They utilised the diesel generator of the fire engine to keep the circulation of light water going and thus keep removing the heat from the primary heavy water circuit.[66]

Forget about giving rewards, the NPCIL/DAE have not even acknowledged the bravery and quick-wittedness of the Narora plant operators, which prevented a Chernobyl type core meltdown from occurring at the Narora NPP. Why? Probably because they were afraid that if they did so, it would reveal the severity of the accident and therefore the huge dangers of situating a nuclear power plant in a thickly populated state like Uttar Pradesh.

> *Note: In fact, India's atomic energy establishment seems to have forgotten the Narora accident. After the Fukushima accident, our nucleocrats have been claiming that a station blackout is not possible in India!* (See Epilogue, Part V)

(v) Kaiga-1 and 2

Unit 1 of the Kaiga Atomic Power Station located in Karnataka was supposed to achieve criticality in 1996. However, on May 13, 1994, the concrete containment dome collapsed under its own weight. Concrete slabs weighing hundreds of tons came crashing down from a height of about 40 metres, the height of a 12 storey building. The dome had been completed in January 1994, but plumbing, cabling and other such works were going on.

Had the dome collapse taken place after the reactor had commenced operation, it would, in all probability, have led to a Chernobyl-like accident. It would have been a Chernobyl in reverse. In Chernobyl, the reactor core went out of control, and blew away the containment. In Kaiga, the falling debris from the dome would have damaged the reactor core, the coolant pipes and many of the safety systems. With cooling and safety systems inoperative, the reactor would have suffered a loss-of-coolant accident and, with the containment already breached, the reactor core would have been exposed to the environment, resulting in a nuclear catastrophe.

The Kaiga accident did not result in scrapping of the reactor. It only caused a delay in its commissioning, by four years.[67]

(vi) Kakrapar-1 and 2

Kakrapar Atomic Power Station (KAPS) in Gujarat also has two 220 MW (235?) PHW reactors of Canadian design. Unit-1 went critical in 1992 and Unit-2 in 1995.

NPCIL had planned Unit-1 to go critical latest by December 1991, but a major fire broke out in the turbine room of the plant during a routine preliminary test, causing considerable damage and delaying the start up of the reactor.[68]

On June 15, 1994, following heavy rains, flood waters entered the turbine room of the KAPS plant! The flood waters filled the underground room containing pumps and motors which feed the boilers and are necessary for the regular recirculation of the steam from the turbine to the steam generating section in the reactor building. The fury of the floods was such that the waters also breached the waste containment building and lifted and carried some waste canisters out into the open.

Fortunately for South Gujarat, the plant was in a shut down state as the AERB had ordered the shutdown of nuclear power stations for inspection following the fire accident at Narora NPP. Had the unit been operational and producing steam, the flooding of the pumps would have led to steam being pushed back into the reactor, leading to over pressurisation and a massive explosion, a la Chernobyl.[69]

In 2004, an unexplained power surge at KAPS-1 forced NPCIL to shut down the reactor.[70]

The KAPS reactors house the first indigenously-developed microprocessor based control system. However, this has not been tested thoroughly for its reliability; no appropriate facility for such testing exists with the DAE. There have been instances of dangerous and erratic behaviour, such as a control rod coming out when signalled to go into the reactor![71]

(vii) BARC

For all the hype about BARC, this premier nuclear research institution is in an even poorer shape than India's nuclear reactors. There have been numerous accidents at its research reactors, which have led to massive contamination of the premises; some of these also very nearly led to a major disaster. There was an instance of a reactor being started up with an operator inadvertently locked inside. In 1991 Dhruva, a 100 MW research reactor at BARC, operated for almost a month with a malfunctioning emergency cooling system, in complete violation of all safety norms.[72]

Aside from these near catastrophes, there is the more insidious problem of leakage of underground pipes carrying radioactive water in the vicinity of the CIRUS and Dhruva reactors at BARC. An even bigger disaster is the two million tonnes of liquid nuclear waste stored in tanks at the BARC site—these tanks are leaking due to aging, corrosion and faulty welds. The result of these leakages is that cesium-137 has been found in the soil, water and vegetation at the BARC site and the Trombay coast. The leakage of such a dangerous radioactive isotope in an open area outside the reactor complex is a situation unacceptable under any internationally accepted norm. The level of cesium in the soil was found to be 27,000 Bq/gm, which is 30,000 times higher than permissible levels. Considering the long half-life of Cs-137 (over 30 years), this contamination will persist as a threat to the safety of the people and the environment for hundreds of years.

Additionally, the research and reprocessing plants at BARC discharge their nuclear effluents into the Thane creek, which is an extension of the sea at Mumbai port. This has made the bed of the creek highly radioactive. The Thane creek separates Navi Mumbai from old Mumbai, and the radioactive contamination of the creek spells danger to the whole of Mumbai. Thanks to the veil of secrecy surrounding the operations of BARC, the city is oblivious to this danger. The people of Mumbai are going to pay the price for the callousness of BARC officials for centuries to come.[73]

The safety situation at the Kalpakkam Atomic Reprocessing Plant (KARP) run by BARC and located near the MAPS at Kalpakkam near Chennai is no better. There have been numerous cases of workers being exposed to high levels of radiation, including a major accident on January 21, 2003 (discussed in a later section).[74] A report of the United Nations Scientific Committee on Effects of Atomic Radiation (UNSCEAR) says that the routine release of radionuclides from KARP has been high in comparison to the release from facilities in other countries.[75] The reprocessing plants in France and UK are the biggest sources of radioactive pollution in Europe, with radioactive releases from these plants polluting the North Sea as far as the Arctic; one wonders how far has the pollution from KARP spread in the Bay of Bengal/Indian Ocean!

The DAE is totally unconcerned about these terrible radioactive releases from BARC run facilities. Gopalakrishnan writes that during his tenure as Chairman of AERB, the BARC management refused to comply with the procedures and corrective actions ordered by the AERB. Some years later, the DAE ended all possibilities of such disputes by putting safety standards at BARC facilities beyond the purview of this benign regulator. In 2000, Dr R. Chidambaram, then Secretary to the DAE, ordered that the regulatory and safety functions at BARC and its facilities, exercised till then by the AERB, would henceforth be conducted through an internal committee to be constituted by the Director of the BARC. Wow! This should break all records of nuclear safety regulation!![76]

DAE: Terrible Safety Management

In its submission to the IAEA as part of its responsibilities under the 1994 Convention on Nuclear Safety, the DAE stated: 'Safety is accorded overriding priority in all activities. All nuclear facilities are sited, designed, constructed, commissioned and operated in accordance with strict quality and safety standards ... As a result, India's safety record has been excellent in over 260 reactor years of operation of power reactors and various other applications.'[77]

However, the reality is that the DAE is absolutely nonchalant about nuclear safety. We discuss a few examples below, to prove our point.

Fire, Narora, March 31, 1993

Take the Narora accident of 1993. It has been DAE's closest approach to a catastrophic accident. What is most worrisome about it is that the accident could have been foreseen and prevented!

That's because the failure of the turbine blades was avoidable. In 1989, General Electric informed the turbine manufacturer, Bharat Heavy Electricals Limited (BHEL), about a design flaw which had led to cracks in similar turbines around the world and recommended design modifications. BHEL took prompt action and prepared detailed drawings for the NPCIL, the operator of the Narora reactor. However, NPCIL took no action till after the accident!

Secondly, even after the turbine blades had failed, the accident might have been averted if the backup safety systems had been operating, which was possible only if their power supply had been encased in separate and fire resistant ducts. By the time the Narora reactor was being manufactured, this was the established wisdom in the world reactor design industry; this was one of the lessons drawn from the fire at Browns Ferry Nuclear Plant in the US in 1975, and all nuclear plants in the USA had to compulsorily make these modifications. Other countries also adopted this measure. However, even though the Narora plant attained criticality in 1989, this practice was not followed for this plant! The plant was constructed with backup power supply system laid in the same duct, with no fire resistant material enclosing or separating the cable systems. As a result, following the fire in the turbine building, along with the main supply cables, the backup power cables also caught fire and led to a complete blackout in the plant.[78]

Collapse of Dome, Kaiga, May 13, 1994

The accident at Kaiga is unprecedented in the annals of nuclear energy history.

The containment of a nuclear reactor is built strong and built to last. It is designed to withstand not just natural calamities like earthquakes and hurricanes, but even the intense radiation from within in case of an accident in the reactor. But in India, we have a reactor containment that did not even withstand its own weight!

Had the dome collapsed after the reactor had commenced operations, it would in all probability have led to a reactor meltdown, and the entire population of Uttar Kannada and Goa would have been endangered, requiring immediate evacuation. The collapse of the containment in a reactor at any stage is unthinkable—that it should have happened speaks volumes for the safety culture prevailing in our atomic energy establishment. The accident should have led to a complete moratorium on the construction of nuclear reactors in India, and a complete overhauling of the NPCIL/DAE. Yet nothing happened. These august bodies came up with a most 'creative' description of the accident: 'certain sections of the containment got delaminated', and set up committees to whitewash the accident.

Even the devil must be given his due. Like all previous accidents, the NPCIL/DAE have been very efficient in hushing up this accident too—to the extent that till today, nothing is known about the death toll in this accident. At any given time, there would have been at least two hundred workers working inside the reactor building. Immediately after the accident, the reactor building was sealed, no one was allowed to meet the injured, and the NPCIL gave out the statement that at the time of the mishap (11.45 am), no one was in the reactor dome as it was lunch time.[79] Questions cannot be asked of NPCIL, so we'll have to believe the story that on that fateful day: the workers took an early lunch break.

Flooding, Kakrapar, June 15, 1994

The numerous stories about the sloppiness and inefficiency of India's atomic energy establishment would make for hilarious reading, but for the fact that many of these have very nearly led to a 'Chernobyl-like' disaster. The flooding of KAPS on June 15, 1994 due to heavy rains is another such story.

Just behind the turbine room of the KAPS is the Moticher Lake. Outlet ducts of the turbine building connect it to this lake. The lake has gates to control the water level. Following heavy rains on June 15, 1994, the water level in the lake began to rise. The outlet ducts became inlet pipes and water began entering the turbine building on the night of June 15 itself. But such is the level of 'emergency preparedness' of the DAE and its subsidiary, the NPCIL, that even as the flood waters were entering the turbine building to create havoc, the KAPS authorities were soundly sleeping! It was only on the morning of June 16, when the morning shift arrived for work, that the flooding of the turbine room was discovered!!

Even then, the KAPS authorities did not take immediate action. After much dilly-dallying, they finally declared emergency at the 'gentlemanly' hour of 11 am, and evacuated all non-emergency workers from the plant site. They now frantically tried to get the gates of the Moticher Lake opened. But the gates had been neglected for years, and so were jammed! It was only two days later, on June 18, that a large pump arrived from Tarapur and work began to remove water from the turbine building.

Forget big natural disasters, the NPCIL is so incompetent that after more than three decades of experience, it cannot even ensure proper drainage and prevent flooding of its reactors in case of heavy rains![80]

> *Note: Despite such embarrassing accidents, India's nucleocrats claim, after the Fukushima accident, that India's reactors were safer than Japan's because: 'India was uniquely placed as it had a centralised emergency operating centre with well drawn procedures scrutinised by regulators'!* (See Epilogue, Part V.)

Untested ECCS, Kakrapar-1

Unit-1 of KAPS went critical on September 2, 1991. In their hurry to start the reactor, which was delayed by more than a year, the NPCIL/DAE violated their own safety norms. They started the reactor without doing the mandatory 'integrated testing' of the Emergency Core Cooling System (ECCS).

The ECCS is the only back-up system available in case of a 'loss-of-coolant-accident' (LOCA) (an accident wherein the normal reactor coolant drains out for some reason, like breakage of pipes, and so is not available for reactor cooling). In that case, the ECCS automatically comes into play. Therefore the ECCS is a vital safety system, its failure in the case of a LOCA accident could lead to core meltdown.

The ECCS of KAPS-1 had been tested in February 1992, but it did not function to satisfaction. Some repairs were carried out in the following months, but after that, fully integrated testing of the ECCS should have been done to guarantee that it was working satisfactorily. However, the project authorities went ahead with loading fuel and heavy water into the reactor and starting it up, without doing this test. Integrated testing of the reactor is practically possible only once in the lifetime of the reactor, before heavy water is loaded into the reactor core. Once the Primary Heat Transport System of the reactor is loaded with heavy water, integrated testing is no longer possible.

Thus, it is not known if the ECCS of KAPS-1 will function as required in an emergency situation. All that we can do is hope and pray that a loss-of-coolant-accident does not occur in this reactor; in case it occurs, and the untested ECCS fails to perform, it could have disastrous consequences.[81]

As it is, nuclear technology is inherently hazardous. Starting up a reactor, without doing such an essential safety test, is unheard of in the global nuclear industry. The fact that the DAE and NPCIL can take such risks and put the lives of millions of people in danger, just to meet their performance targets, amounts to treason. But we are living in a nuclear dictatorship, where no one dares say that the Emperor is without clothes.

> *Note: Following the Fukushima accident, the Prime Minister ordered a safety review of all our nuclear plants. A few days later, the AERB declared that it had found all our nuclear installations safe. Since integrated testing of the ECCS of KAPS-1 is no longer possible, how did the AERB satisfy itself that all safety systems of this reactor were working satisfactorily?* (See Epilogue, Part V.)

Repeated Mistakes

Not only has DAE's carelessness led to numerous near-misses at Indian reactors, DAE is so casual about nuclear safety that it does not draw lessons even after an accident has occurred, leading to repeated accidents of the same type in different reactors. For instance, the factors which led to the Narora accident were repeatedly present prior to the accident—excessive vibrations in turbine bearings have in fact been common in Indian reactors. In 1981-2, after repeated shutdowns at Rawatbhata-2, it was finally discovered that the problem was due to high vibrations of turbine bearings, and failure of turbine blades was discovered. This led to a prolonged shutdown of more than 5 months. Even after this problem had apparently been fixed, the reactor had to be shut down once again because of high turbine bearing temperatures. Again in 1983, high vibrations were noticed in turbine generator bearings of the reactor and it was revealed that two blades in the second stage of the high pressure rotor had sheared off at the root. Similarly, Rawatbhata-1 had to be shut down in 1985, 1989 and 1990 because of high bearing vibrations in the turbine generator; while Kalpakkam-1 was shutdown repeatedly in 1985 for similar reasons.

Even after the Narora accident of 1993, turbine problems have continued to plague Indian reactors: Narora-2 and Rawatbhata-1 have had to suffer repeated shutdowns due to high turbine bearing vibrations/high bearing temperatures. Not only that, despite the accident in 1993, Narora-1, too, had to be shutdown repeatedly in 1995 because of high vibrations of the turbine generator bearings. Kaiga-2 has also suffered from repeated turbine vibration problems. In 1995, even after repeated shutdowns to mitigate turbine problems, blades failed in the turbine of Narora-2.

Fires have also occurred repeatedly. In Narora-2 in 1996, there was heavy oil smoke from the turbine building. That same year, there was an oil fire in the turbine building of Kalpakkam-2. The following year smoke was observed in Kalpakkam-2, there was a fire in the turbine generator of Kakrapar-1 and smoke was observed from the insulation of the main steam line of the turbine generator in Kakrapar-2. There was a fire due to an oil leak in Kalpakkam-1 in 2000.

Similarly, there have also been numerous heavy water spills. To mention a few instances: there was one such leak in Rawatbhata-1 in 1996, the following year, such leaks occurred at the Kakrapar-1, Kalpakkam-2 and Narora-2 reactors; Narora-2 again leaked in 2000 and 2003.[82]

Yet, the NPCIL Chairperson, S.K. Jain, has the gall to claim that 'India had the best safety record in running nuclear power plants'![83]

Ticking Time Bomb at Tarapur

The Tarapur-1 and 2 reactors are more than 40 years old. The AERB has been periodically extending their operating licenses. S.K. Jain, the Chairman of the NPCIL, recently claimed that the Tarapur reactors undergo a safety audit every 5 years. In the same interview, he, however stated that the last safety audit was carried out in 2004. Implying that the next safety audit, due in 2009, has not been done as of at least March 2011 (when he gave this statement)![84] This itself shows how seriously the NPCIL/DAE takes the safety audit of the AERB—the reactor continued in operation even though its operating license had expired.

That however should be no cause for surprise—even if the AERB had carried out the safety audit as mandated by the rules, it would have been a meaningless exercise as the AERB is not an independent regulator but only a lapdog of the DAE.

The DAE's lackadaisical attitude to giving lifetime extensions to the twin Tarapur BWRs is in keeping with its nonchalant approach towards nuclear safety. This is creating an extremely dangerous situation at Tarapur.

As nuclear plants get older, their various components including critical ones suffer embrittlement due to neutron bombardment, and so they become prone to unanticipated and sudden accidents.[85] While the risk of accidents increases with age for all nuclear reactors, the

Tarapur 1 and 2 reactors are particularly vulnerable as they are of a most vintage design—the Premod design—which is even older than the Mark-1 design of the Fukushima-1 reactor that exploded on March 11, 2011. All other reactors of this design have been shut down long ago![86] Dr Gopalakrishnan, the former chief of AERB, India's nuclear regulator, has repeatedly stated that the two Tarapur reactors have suffered so many safety related problems that they should have been shut down long ago.[87] In fact, according to energy analyst Prabir Purkayastha, the manufacturer (General Electric) had asked the DAE not to extend the life of these reactors.[88]

Yet, the DAE continues to flog these two decrepit reactors—located just 100 kms from Mumbai. It is a form of Russian roulette with millions of lives at stake.

The Growing Problem of Nuclear Waste Disposal

As mentioned earlier, the DAE pursues reprocessing as a way to manage its spent fuel. However, the total reprocessing capacity in the country is very small, and so much of the spent fuel is accumulating in spent fuel ponds near the nuclear reactors (in the case of the Tarapur reactors, it is stored underwater in specially engineered bays in the Spent Fuel Storage Facility at Tarapur).[89]

Even if the DAE had managed to reprocess all its waste, it wouldn't have solved or reduced the problem of safe storage of this waste, as reprocessing does not reduce the total quantity of radioactive waste (instead, in the long run, it only increases the total quantity of waste).[90] However, let us leave aside this debate. The problem brewing at India's nuclear power plants is that since the DAE considers this waste as a resource—from which useful plutonium is to be extracted for use in its 'Three Stage Program'—it has made no effort to even find a temporary solution to the problem of safely storing the growing volume of spent fuel from its nuclear reactors. The waste is simply accumulating at the reactor sites, and will inevitably leak into the environment. (Not that other countries have found a way to safely dispose of this waste. But they are at least trying to find temporary solutions—like storing the waste in underground repositories, or storing it in dry casks—till a more permanent solution is found: which is at least a step ahead.)

The problem is the most acute at Tarapur, the oldest of India's operating reactors, where around 2400 tons of spent fuel waste has accumulated. According to Prof. Matin Zuberi, a well-known nuclear energy expert who has served on several official committees (and so should know), the tanks containing this waste are leaking, threatening to contaminate the environment with their deathly brew for thousands of years.[91]

The spent fuel pools house an enormous amount of radiation. There is about twice as much cesium-137 in a ton of spent fuel as in a ton of reactor fuel. An accident in a spent fuel pool could be catastrophic—much worse than a meltdown in a nuclear reactor. According to a study by Robert Alvarez and others, if a fire in a spent fuel pool released just 10 per cent of the cesium in the pool, the area contaminated would be 5-9 times larger than the area affected to a similar degree by Chernobyl.[92]

The spent fuel pools are not stored in containments as secure as nuclear reactors, and so they are much more vulnerable to terrorist attacks as compared to the latter. Greenpeace commissioned a study that examined the results of an aerial terrorist attack on the nuclear complex at Sellafield. To their horror, they discovered that three and a half million (35 lakh) people could be killed! Other unpredictable events—like earthquakes—can also cause damage to the spent fuel ponds, leading to a loss of cooling water and meltdown.[93]

Storing the waste in dry casks is one possible way to reduce the dangers posed by these spent fuel pools. Several countries around the world are now starting to adopt this temporary solution. However, the DAE has stuck its head in sand, and wished away the problem.

DAE: Attitude towards Worker Safety

In the numerous accidents and mishaps that have occurred at DAE's nuclear plants, thousands of workers have been exposed to high doses of radiation, well in excess of the officially stipulated maximum limits.[94] However, each time such an accident has occurred, the DAE has tried to hush it up, and when it has failed to do so, has tried to downplay the radiation leakage and lie about its medical effects on the workers.

Way back in 1982, Praful Bidwai, the well-known journalist,

writing in *The Times of India*, had documented at least 350 cases of workers being exposed to high levels of radiation at the Tarapur plant alone. Bidwai writes, 'H.N. Sethna, the then DAE Secretary, did not deny the overexposure but blithely declared that it posed no danger.'[95]

KARP Accident, 2003

One of the worst instances of workers being exposed to high levels of radiation in India's nuclear installations took place at the Kalpakkam Atomic Reprocessing Plant (KARP) on January 21, 2003. On that day, some employees at the plant were tasked with collecting a sample of low-level waste from a part of the facility called the Waste Tank Farm (WTF). Unknown to them, a valve had failed, resulting in the release of high-level waste, with much greater levels of radioactivity, into the part of the WTF where they were working. Consequently, six workers were exposed to high doses of radiation.

For one, the accident could have been avoided had radiation monitors or mechanisms to detect valve failure been installed in the WTF, which had not been done even though the plant was five years old!

Let's leave that aside. Much more serious is the attitude of the plant management, that is, the BARC, towards the accident. Despite a safety committee's recommendation that the plant be shut down, BARC's upper management decided to continue operating the plant. The BARC Facilities Employees Association (BFEA) wrote to the director setting forth ten safety related demands, including the appointment of a full time safety officer. The letter also recounted two previous incidents where workers were exposed to high levels of radiation in the past two years, and how officials had always given some or the other excuse to explain away the failure to follow safety procedures. The management gave no response to BFEA's demands. In the face of this intransigence, some months later, the union resorted to a strike. The management's response was to transfer some of the key workers involved in the agitation and give notice to others; two days later, all striking workers returned to work. Finally, in desperation, the union leaked information about the radiation exposure to the press.

Once the news became public, the BARC director, in a press conference six months after the accident, grudgingly admitted that this was the 'worst accident in radiation exposure in the history of nuclear India'. In the same breath, he put the blame for the accident on 'over enthusiasm' and 'error of judgment' on part of the workers! He however refused to reveal anything about the exact medical condition of the workers, including the radiation dosage received by them, except that the workers were 'cheerful'![96]

In 2010, a reporter from the Delhi-based independent weekly, *Tehelka,* tried to trace the whereabouts of the six workers who had been exposed to high levels of radiation due to the accident. The medical superintendent of the DAE established hospital in Kalpakkam told him: 'One of them died, but not due to radiation. The rest are fine.' But his efforts to locate the five surviving workers came to naught.[97]

Tehelka also found evidence of increased incidence of cancer and other diseases among the 30,000 workers living in the five villages located within a 5-km radius from the plant. While the local public health centre denied information to *Tehelka* about cancer-related deaths among workers, saying the information was sensitive, DAE officials maintained that the radiation emission levels were too low to cause problems.[98]

More recently, in the Kaiga incident of 2009 (discussed in Part I of this Chapter), where different reports say that between 35 and 250 workers were affected, the AERB in its press release of November 29, 2010 stated that only two workers received a dose exceeding the 30 millisievert maximum limit stipulated by the AERB. India's nuclear authorities also trivialised the hazards posed by tritium and claimed it was a non-toxic substance. Whereas tritium is a beta-ray emitter and can cause extensive, irreversible damage![99]

It is obvious that the scientists heading the DAE are blatant liars. They lie about accidents at DAE installations, deny occurrence of radiation leakage, lie about the impact of this leakage on workers. Yet, nothing can be done about it. The Atomic Energy Act of 1962 makes them totally unaccountable to the people of India.

Temporary Workers

The DAE's attitude to temporary workers is even more criminal. While the permanent workers have their union to protect their interests, the temporary workers have no such protection. The DAE ruthlessly takes advantage of their poverty and helplessness to make them do the most dangerous tasks, such as cleaning up radioactive materials.

Actually, in doing this, the Indian nucleocrats are only following an old global nuclear industry practice. From America to Japan, the nuclear reactor industry around the world hires large number of poor, untrained 'casual' workers at nuclear plants to reduce the individual doses of the regular staff.[100]

No record is kept of how many such workers are exposed to radiation, and how much radiation they are exposed to. Writing about conditions at the Tarapur nuclear plant in *Business India* in 1978, Bidwai reported that parts of the plant had become 'so radioactive that it is impossible for maintenance jobs to be performed without the maintenance personnel exceeding the fortnightly dose ... in a matter of minutes.' Therefore, instead of regular employees of TAPS, outsiders were employed to carry out maintenance tasks, many of whom did not have knowledge of the hazards they were being exposed to. Obviously, the situation at Tarapur must only have worsened since then!

On September 6, 1992, the *Sunday Observer* reported that temporary workers were used to repair a major radioactive leak from ill-maintained pipelines in the vicinity of the CIRUS and Dhruva reactors at BARC on December 13-14, 1991. These workers were later given a bath, a new set of clothes, and packed off home.[101]

Such anecdotal evidences of poor worker safety culture and workers' health being compromised in DAE establishments are plentiful. The reason why we have only anecdotal evidence is that outsiders do not have access to the health records of DAE workers.

DAE: Terrorising Whistleblowers

If the situation is so bad at DAE's installations, then why don't the workers and scientists reveal what is happening inside the plants? The reason is: they are scared. There is virtually a reign of terror within the

DAE installations, and if someone speaks out, then immediate disciplinary action is taken, as the following example illustrates.

The flooding of the Kakrapar Atomic Power Station in June 1994 due to heavy rains is an absolutely inconceivable accident—it shows criminal negligence on the part of the designers, operators and regulators of nuclear power plants in our country. And yet, nobody has had to suffer the consequences, except Manoj Mishra, the man who blew the whistle.

The KAPS authorities had tried to keep the accident under wraps—so much so that they did not even inform the AERB of the situation. Manoj Mishra, a plant operator and secretary of the employees union of the plant, talked to newspaper reporters and apprised them of the possible serious consequences. It was only after the *Gujarat Samachar* carried news about the accident on June 23 (one week after the accident) that people got to know about it, and the plant authorities dashed to Surat to issue press statements assuring everyone that all was well under control.

Mishra was immediately suspended from work, and later dismissed, for the crime of talking to the press.[102] While all those who displayed singular dereliction of duty continue merrily in their jobs, the one man who put the interests of the country above his selfish interests was hauled over the coals, in order to terrorise others who might develop similar ideas, and ensure that the only 'leaks' from the country's nuclear establishment are official statements.

Part V: India's Nuclear Reactors: Impact on People

India's nuclear reactors are leaking radiation. In 1993, Dr Gopinath, the then Director of the Health Physics Division at BARC, disclosed at a meeting of the United Nations Scientific Committee on the Effects of Atomic Radiation (UNSCEAR) the numerical values of the radioactive discharges from India's nuclear power plants. UNSCEAR was outraged and officially told the Indian government that these discharges were higher than safe limits by about 100 times.[103] That India's reactors are emitting radiation at several times the international norm has also been admitted by S.P. Sukhatme, then Chairman of the AERB, in 2002.[104]

Not much information is available about the impact of these radiation leakages, both the routine releases of radioactivity and radiation released from the innumerable accidents at India's atomic reactors, on the health of people and animal/plant life around these reactors. The authorities have simply not done any studies. The only information we have is based on a survey done by two independent scientists in the villages around RAPS (near Kota, Rajasthan) and some studies done on the impact of MAPS and KARP on life of fisherfolk in the area (around Kalpakkam, near Chennai).

Rawatbhata Survey

Renowned scientists Drs Surendra and Sanghamitra Gadekar of Sampoorna Kranti Vidyalaya, Vedchhi, District Surat, Gujarat did a unique survey of the population living in five villages in the vicinity of the Rawatbhata nuclear power plant in 1991. It is probably the only survey of its kind ever done in the country. The survey found:[105]

- a huge increase in the rate of congenital deformities;
- a significantly higher rate of spontaneous abortions, still births and deaths of new born babies;
- a significant increase in chronic problems like long duration fevers, long lasting and frequently recurring skin problems, continual digestive tract problems, persistent feeling of lethargy and general debility. The young were more affected by these problems. But there were no differences in acute problems like short duration fevers, conjunctivitis, etc;
- diseases of old age prevalent amongst the youth; and
- a significantly higher rate of solid tumours.

The results of the study were published in the journal *International Perspectives in Public Health* [Vol. 10 (1994)] edited by Dr Rosalie Bertell, the eminent environmental epidemiologist. The survey methodology was so good that despite its best efforts, the DAE has not been able to refute its findings. On the contrary, the survey has been such a thorn in DAE's flesh that its scientists spare no effort to denigrate its findings even today, two decades later. Thus, during the public hearing organised by the Maharashtra Chief Minister to reply to fears regarding the Jaitapur Nuclear Park on January 18, 2011 at the Y.B. Chavan Hall in Mumbai, Dr Rajendra Badwe of the Tata Memorial Cancer Hospital dismissed the survey by the Gadekars as being without any foundation since it had not been peer-reviewed and published in reputed scientific journals.[106] A blatant lie! Yet, many prominent Mumbai newspapers published his statement, without verifying the facts.

If the DAE is so sure that its nuclear plants are emission-free, instead of making wild allegations about Gadekars' survey being unscientific, why does it not conduct its own survey and publish its findings? Well, the DAE's scientists may be liars, but that does not mean they are also fools.

Kalpakkam's Forgotten People

The Madras Atomic Power Station (MAPS) at Kalpakkam houses two nuclear reactors, a fast breeder test reactor, a research reactor and a fuel reprocessing plant. The 20-km region around these nuclear facilities, which is the region most affected by the radiation leakages from this complex, has a population of more than 11 lakh people.[107]

Dr V. Pugazhenthi and a team of doctors from Alice Stewart School for Epidemiological Studies, Tellicherry, Kerala and St. Joseph Hospital, Tirunelveli, Tamil Nadu, under the guidance of Dr Rosalie Bertell, the world renowned environmental scientist, did a study of the incidence of goiter and autoimmune thyroid disease (AITD) among the people living in this region in 2007. It is probably the only such epidemiological study in the world. They found a very high incidence of thyroid disorders among women above the age of 14 years living within a distance of 6 kms from MAPS, with the incidence of goiter being an astonishing 23 per cent amongst women in the age group of 20-40, and of AITD being as high as 7 per cent amongst women in the age group of 30-39 years. The average prevalence of these disorders near the plant was around 10 times as high as in the region far away from MAPS.[108] The high incidence of these diseases was obviously due to radiation exposure to routine releases of radionuclides, especially radioactive iodine, from the nuclear reactors and the plutonium reprocessing plant at Kalpakkam.[109]

In another worrying indication, the doctors found several cases of congenital defects and mental retardation in the coastal areas in a radius of 16 kms from the nuclear complex, which are obviously due to exposure of the foetus to radiation. They also detected statistically significant number of cases of multiple myeloma, a rare bone cancer which is linked to nuclear radiation, as well as a case of colon cancer in a young 24 year old worker—it is unusual for people to contract this cancer at such an early age.[110]

The radioactive effluents have badly affected the livelihood of fishermen in the coastal areas surrounding the plant. The area was once rich in lobsters, crabs, shrimp and other varieties of fish, but now the catch has drastically come down. The havoc caused to local life due to the plant is described by the Japanese journalist Tashiro Akira and others who visited several nuclear sites all over the world, including India. Their findings were published in a book titled *Resume*. The Kalpakkam fishermen told them, 'The reason why our catches have declined so drastically is that plant. The warm waste water that comes out of that keeps the fish away, particularly in the area within a few miles' radius of the outlet.' They further stated, 'Lots of dead fish are floating out there. We gather them up and make karuvadu.' *Karuvadu* is a dish made by salting and drying fish for two or three days. But it is not sold locally. The fishermen were blunt about it: 'It all goes to the market. People here won't touch the stuff because they know where it's come from. The villagers take their catch of karuvadu to Madras and sell it there, where it provides a cheap source of protein for the poor people in the city.' When the journalists asked whether it was actually safe for people to eat this fish, the reply was, 'Well, they're probably contaminated, but we can't catch anything else, and there is hardly any money coming in at the moment. We don't have any choice.'[111]

Part VI: India's Fast Breeder Reactor Program

The Worldwide Experience with Fast Breeders

Ever since the dawn of the nuclear age, nuclear energy advocates have dreamed of a reactor that would yield more fuel than it consumes. In the sixty years since then, seven countries—the US, UK, France, Germany, the USSR, Japan and India—established plutonium breeder reactor programs. However, their efforts have failed to produce a reactor that is economically competitive with conventional Light Water Reactors (LWRs). The capital cost per kilowatt of generating capacity of a demonstration sodium cooled Fast Breeder Reactor (FBR) has typically been twice as much as a LWR of comparable capacity.[112] Additionally, plutonium which is the basic fuel for fast breeders is extracted by chemically treating highly radioactive spent fuel at reprocessing plants, and this is an expensive process. This further increases the cost of electricity from fast breeders.[113]

However, more important than the economic aspect is the safety aspect. While all nuclear reactors are susceptible to catastrophic accidents, FBRs are even more so. There are several reasons why accidents involving fast breeders are both more likely and could cause greater damage to public health. Plutonium reactors need fast moving neutrons, and so cannot use water as coolant since water is a moderator. To date, all fast breeders have used liquid (molten) sodium as a coolant. However, molten sodium has serious drawbacks: it is extremely reactive. It burns when exposed to air and reacts violently with water. Therefore, liquid sodium cannot be exposed to air or water, which means operating these reactors is going to be very difficult as even a minor leak can be dangerous. In fact, building and operating even test breeder reactors has been very difficult. Most of the demonstration FBRs that have been built so far have been shut down for long periods due to sodium-water fires caused by leaks.[114]

Another fear with FBRs is that, unlike water-cooled reactors which cease operation if there is a loss of coolant (a safety feature), breeder reactors become even more reactive if there is loss of the sodium coolant. This can result in a core meltdown and a small nuclear explosion. There are fears that if this happens, it can lead to a Chernobyl-type release of radioactivity into the environment.

Repairing an FBR is also more time consuming and difficult as compared to LWRs, as air has to be prevented from coming into contact with the sodium coolant.[115]

Even if success is achieved in building a test reactor, building larger fast breeder reactors will be much more difficult and dangerous as the above problems will multiply in magnitude.

Another kind of problem that plagues breeder reactors arises from their use of MOX fuel (mixed oxide fuel, made from a mixture of uranium and plutonium oxides). Because plutonium is about 30,000 times more radioactive than uranium-235, enormous safety precautions are required during fabrication of this fuel, which makes fabricating MOX several times more expensive than compared to low-enriched uranium fuel. Further, the spent fuel from an FBR typically has a greater buildup of highly radioactive fission products; therefore, the impact of an accident would be much more severe than in the case of a Light Water or Heavy Water Reactor.[116]

This is why, even after six decades and expenditure of $100 billion (in 2007 dollars),[117] the promise of Fast Breeder Reactors remains largely unfulfilled. In the 1970s, breeder advocates were claiming that there would be thousands of FBRs in operation by 2010. Today, the dream is nearly dead. The US, UK and Germany have abandoned their breeder reactor development programs. France still claims that fast breeders have a future, but the country has no operating FBRs, not even demonstration units. Superphénix, the 1200 MW flagship of the French breeder program and the only commercial-size plutonium fuelled breeder reactor in nuclear history, was shut down in 1998 after an endless series of very costly technical, legal and safety problems which rendered it inoperative for the majority of its 11-year lifetime. The other remaining 233 MW demonstration reactor Phenix shut down in 2009. No replacement FBR is planned for at least a decade. The Japanese prototype fast reactor Monju shut down in 1995 after a sodium coolant leak caused a fire. After repairs and many delays, it finally restarted 15 years later on May 8, 2010. This is only a prototype; Japan hopes to build a follow-on demonstration FBR by 2025; only if that succeeds will construction of a commercial FBR begin. It is doubtful if these projections will ever be fulfilled, and Tokyo has been reducing funding for its breeder program for decades.[118]

Apart from India, presently worldwide only Russia is attempting to develop demonstration fast breeders, while China is considering buying two such units from Russia. Russia has one operational fast breeder, the BN-600. But this reactor hardly qualifies as a successful breeder. The Soviet Union/Russia never closed the fuel cycle and has never operated it with MOX fuel. The BN-600 has also suffered repeated sodium leaks and fires. Yet, Russia continues to run the risk of operating it—another example of the extreme callousness of its nuclear establishment.[119]

India's Fast Breeder Program

Despite this worldwide evidence, the DAE continues to persist with its uneconomical and risky Fast Breeder Reactor program.

As discussed in Chapter 7, breeder reactors in India were originall) proposed in the 1950s as the second stage of a three-stage nuclear program. This was seen as a way to develop a large autonomous nuclear power program despite India's relatively small known reserves of uranium ore. However, the DAE began work on FBRs only in 1965, when a fast reactor section was opened in BARC and design work on a 10-MW experimental FBR was initiated. It soon became clear that foreign assistance was required. This project was therefore abandoned and, in 1969, the DAE entered into collaboration with France. It took designs from it for what was to be India's first breeder reactor, the Fast Breeder Test Reactor (FBTR). The DAE sent scientists to France for training, and they formed the nucleus of the Reactor Research Centre that was set up in 1971, at Kalpakkam, to lead the breeder effort. In 1985, this was renamed the Indira Gandhi Centre for Atomic Research (IGCAR). Over the years, this centre has emerged as the main hub of activities related to India's breeder program.[120]

The budget for the FBTR was approved by the DAE in September 1971 and it was anticipated that the reactor would be commissioned by 1976. It was to be a 40 megawatt thermal (MWt)/ 13 megawatt electric (MWe or MW) reactor. However, the reactor attained criticality only in October 1985; and the steam generator began operating only in 1993.

Since then, the reactor has suffered numerous failures and

accidents, which are actually inherent to sodium-cooled FBRs as discussed above and which make it much more dangerous than water cooled reactors. Overall, its performance has been mediocre. It took 15 years before the FBTR even managed 50 plus days of continuous operation at full power (in 2001). In the first 20 years of its life, the reactor has operated for only 36,000 hours, implying that the availability factor has been only about 20 per cent. Despite this chequered history, IGCAR claims to have 'successfully demonstrated the design, construction and operation' of an FBR.[121]

Based on this flawed experience, DAE began making plans for construction of a Prototype Fast Breeder Reactor (PFBR), which would produce 1200 MW of thermal power and 500 MW of electricity. First expenditures on the PFBR were made in 1987-88, and it was reported in 1990 that the reactor would be online by 2000. In 2001, the Chairman of the AEC announced that the PFBR would be commissioned by 2008. Construction of the reactor was finally started in October 2004 and it was then expected to be commissioned in 2010.[122] There is no information about how far its construction has progressed.

Even more worryingly, instead of the mixed carbide fuel used in the FBTR, the PFBR will use plutonium and uranium oxide-based fuel. The DAE has no experience of working with this fuel. Since MOX is thousands of times more radioactive, in combination with liquid sodium as a coolant, it makes the PFBR susceptible to catastrophic accidents.[123]

Construction of a viable FBR is supposed to be the second stage of DAE's ambitious three-stage nuclear program. Given that India has not even built a properly functioning 10 MW demonstration unit more than fifty years after the plan was first announced, the third stage—breeders involving thorium-232 and uranium-233—is unlikely to materialise anytime in the foreseeable future. Yet, the DAE continues to parrot: 'it remains a certainty that thorium-based nuclear energy systems will have to be a major component of the Indian energy mix in the long-term.'[124]

We should actually be heaving a sigh of relief at this failure of the DAE. Breeder reactors are much more dangerous than uranium fuelled nuclear reactors, which is why most countries willing to take the risk of having nuclear power programs have abandoned their fast breeder programs. Furthermore, if they are ever constructed, electricity from these is going to be very expensive. Even assuming that capital costs of FBRs and PHWRs are the same (actually FBRs are costlier), electricity from FBRs would be at least 80 per cent costlier than from PHWRs—mainly because of the high fuel cycle costs associated with reprocessing and the fabrication of MOX fuel.[125]

Therefore, this failure is actually a blessing in disguise!

Part VI: DAE's New Toys: Kudankulam and Jaitapur Nuclear Parks

The DAE/NPC have built and operated India's nuclear reactors so dangerously that it can only be the combined might of the 33 crore Gods in the heavens which has prevented a Chernobyl from occurring in India!

The government of India is so unconcerned about safety at India's nuclear reactors that we don't even have an independent safety regulator, the only country in the world having nuclear power programs where such a situation prevails!!

And now, India's thick-skinned policy makers are planning to set up a string of giant nuclear parks—with reactors three to eight times[126] as big as the ones we have installed at present—all along India's coastline. The first of these is coming up at Kudankulam, in Tamil Nadu, for which Russia is to supply six VVER-1000 nuclear reactors. Construction of the first two units began in 2001, and preliminary agreements have been signed for the construction of another four units.

Preparations for starting construction work at the second nuclear park, in the Jaitapur region of Ratnagiri district (Maharashtra), have reached an advanced stage. This nuclear plant is going to be even bigger than the Kudankulam plant, with six reactors of 1650 MW each, to be supplied by the French nuclear corporation Areva. The project was hurriedly given environmental clearance on November 28, 2010, to please the French President Nicolas Sarkozy, anointed by *Washington Post* as 'the world's most aggressive nuclear salesman',[127] who visited India in early December 2010; on December 6, in his presence, NPCIL and Areva signed a General Framework Agreement and an Early Works Agreement for the construction of the first two reactors.[128]

Routine Impact

The problems generic to nuclear power will of course destroy the environment and health of the people of these areas for centuries to come.

The routine releases of radioactivity from these plants, and the inevitable leakage from the radioactive waste generated by them, will cause the most terrible diseases in the nearby population. In the case of the Kudankulam plant, three large settlements exist within a 5-km radius zone: Kudankulam (population 20,000), Idinthakarai (population 12,000), and a new tsunami (rehabilitation) colony (population 2000-plus). At least 5.7 lakh people live within a 20-km radius around the plant, as per the 2011 census.[129] Similarly, there are many villages within a 5-km distance from the Jaitapur Nuclear Park. According to the 2001 census, the total population staying within a 20-km distance from the plant was 2.6 lakhs.[130]

Both these areas are unique in their ecology. Kudankulam lies at the edge of the Gulf of Mannar, one of the world's richest marine biodiversity areas, with 3,600 species of flora and fauna, 377 of them endemic.[131] Likewise, the Madban area (the site for the Jaitapur Nuclear Plant) lies in the Western Ghats, which is among the world's ten top 'Biodiversity Hotspots' with over 5000 species of flowering plants, 139 mammal species, 508 bird species and 179 amphibian species. More than 325 globally threatened species are found in this region, which also has one of the world's highest concentrations of wild relatives of cultivated plants.[132] The ecology of both the regions is so precious, that only a diabolically destructive mind can make plans to wreck it by building a nuclear plant there.

The cooling systems of these plants will be sucking in and discharging millions of litres of seawater every minute. (The Jaitapur plant is expected to discharge 52 billion litres of seawater a day,[133]

36 million litres every minute.) Billions of fish, fish larvae, spawn, and a tremendous volume of other marine animals will be sucked in and killed by these cooling systems (discussed in detail in Chapter 3, Part IV). This high rate of destruction of fish and fish spawn is going to far exceed the regeneration rate, leading to depletion of fish stocks along both these coastal areas.

Additionally, water discharged into the ocean by their cooling systems will be carrying a terrific amount of heat, and this will dramatically alter the marine environment (we have discussed this issue too in detail in Chapter 3, Part IV).

According to the Ministry of Environment and Forests (MoEF), the temperature of the discharged water should not be more than 7°C for the Kudankulam plant;[134] for Jaitapur, it has imposed the condition that it should not exceed 5°C.[135] Obviously, this condition is going to be 'more honoured in the breach'. For, temperature increases of the coastal waters at India's coastal nuclear reactors already violate these norms: 7.7°C (Tarapur-1 and 2), 8.4°C (MAPS-1 and 2 at Kalpakkam), and 9.5°C (for Tarapur-3 and 4).[136] Tarapur-3 and 4 (2x540 MW) is more than three times as big as Tarapur-1 and 2 (2x160 MW), and temperature rise of the coastal water there is 1.8°C more than at the latter plant. The reactors being built at Kudankulam and Jaitapur are many times bigger—the Kudankulam plant (6x1000 MW) is going to be 13 times bigger than MAPS (2x220 MW), while the Jaitapur plant (6x1650 MW) will be 31 times bigger than TAPS-1 and 2. Therefore, even though the cooling systems of these plants will be sucking in much more water than MAPS and TAPS plants, it is doubtful if temperature rise of the coastal waters at these huge plants will be kept to below MoEF norms. This is going to drive away many indigenous fish species.

Both Kudankulam and Jaitapur are very rich fishing areas. The annual fish catch of Ratnagiri district is around 1,25,000 tons, of which as much as 60-70,000 tons is exported.[137] The three coastal districts of south Tamil Nadu—Tirunelveli, Tuticorin and Kanyakumari—account for 70 per cent of the state's fish catch, and generate over Rs.2000 crores in annual exports.[138] All this is going to be badly affected, destroying the livelihood of tens of thousands of local fisherfolk.

Severe as these effects are, they pale before the most dangerous aspects of these Nuclear Parks.

VVER-1000: A Monster Reactor

The operating experience of the Russian VVER-1000 reactors raises frightening safety concerns! In the last couple of years, in the VVER-1000 reactors at Temelin in the Czech Republic and at Kozloduy in Bulgaria, numerous control rods did not move as designed. That can be catastrophic, as the control rod mechanism is crucial to preventing a runaway fission chain reaction.

The VVER-1000 poses other safety issues too, including the integrity of the pressure vessel (which tends to become extremely brittle with routine neutron bombardment) and the reliability of the steam generator and the auxiliary shutdown system. Even the layout of the plant is problematic. The steam-lines crisscross each other. In an accident, this could lead to broken steam-lines whipping around and hitting electrical supply and control systems, intensifying the accident and its consequences.[139]

These safety issues are so serious that in 1997, the European Bank for Reconstruction and Development cancelled all loans for VVER reactors in Eastern Europe.[140] Dr Alexei Yablokov, chairman of the Russian Federation National Ecological Security Council, and one of Russia's best known experts on nuclear safety, has also admitted in a scientific study that the VVER reactors are unsafe.[141] The IAEA and the US Department of Energy have in fact expressed the opinion that the VVER-1000 reactors cannot meet Western safety standards, even if improvements are made in them![142] (This is not to say that Western standards are very good.)

EPR: Serious Design Problems

The European Pressurised Reactor (EPR) to be constructed at Madban in the Jaitapur region is supposed to be a Generation III+ reactor, that is, it is supposed to belong to the most advanced series of reactors in the world. However, this reactor is of an unproven design, as it is not yet in operation anywhere in the world. The first four reactors of this design are presently in construction in China (two reactors),

Finland and France. The little information available about the latter two reactors makes for scary reading.

As discussed in Chapter 3, the nuclear industry fallaciously claims that these reactors have an improved safety level, whereas the reality is that these reactors are inherently more dangerous as they are of huge capacity (1650 MW, as compared to 400-1000 MW for most present day reactors), and so have much more radioactivity in their core. Additionally, in order to improve the fuel economy, the EPR will use 5 per cent enriched uranium, as against the normal 3.5 per cent in current PWR designs, which will enable its fuel burn-up[143] to reach in excess of 70 GWd/ton as against 30-40 GWd/ton in current LWRs. This is being touted as an advantage of the EPR, but what is not being stated is that such high burn-up leads to much greater radioactivity, and much higher toxicity of the radioactive waste. Consequently, radiation doses to the workers and general public during leakages are going to be correspondingly high. Furthermore, it is reported that the higher burn-up in the EPR will result in greater thinning of the fuel cladding, making it more prone to failure.[144]

Not only are these reactors in no way safer than the present reactors, they also have worrying design problems. Safety regulators in Finland and France have expressed serious reservations about the EPR design, particularly whether there is sufficient independence of the back-up control system—in short, there is the danger that if the main control system fails, there a risk that the back-up system will fail for the same reason. The Nuclear Installations Inspectorate (NII), which is conducting a detailed review of the EPR reactor for the UK, and the US nuclear safety regulator have also expressed the same concerns about the technology. (We have discussed their concerns in Chapter 6, Part III.)

Both the Olkiluoto-3 and the Flamanville-3 reactors have suffered a huge delay in construction and cost escalation (discussed in Chapter 6). Worried, the French government in October 2009 asked Francois Roussely, former chairman of Electricité de France (EDF), to evaluate the EPR and the French nuclear industry in general. The Roussely Report concludes that the difficulties encountered in Olkiluoto and Flamanville are partly due to the complexity of the

EPR model 'including ... the redundancy of safety systems.' For emphasis, we repeat what we have written in Chapter 6:

> This is a damning diagnosis. One of the lessons from Three Mile Island accident was that the design must rationalise the various layers of safety systems, so as to reduce the complexity of the design. This criticism therefore raises questions on one of the most important advancements in design that is supposed to be incorporated in the EPR—that even while having an independent back-up safety system, the complexity of the design is reduced.

On Areva—The EPR Supplier

Areva, the French nuclear corporation and the biggest atomic operator in the world, was voted in 2008 as one of 'the world's most irresponsible companies'. It has resisted cleaning up the radioactive waste from its abandoned mines in France; not only that, its negligence has led to this being used to pave school playgrounds and public parking lots. Its mines in Niger have caused an environmental catastrophe. There have been numerous radioactive leaks from its nuclear plants. Its reprocessing plant at La Hague on the Normandy coast dumps more than 370 million litres of radioactive liquid waste into the English Channel every year and has radioactively contaminated the seas as far as the Arctic Circle. The plant is also one of the world's worst radioactive air polluters. (See Chapter 3, Part III, Section 7 for more details.)

More significantly for India, Areva is failing to implement vital safety measures and has done very shoddy work in the construction of its EPR reactor in Olkiluoto, Finland, in order to save on costs. (See Chapter 6, Part III for a more detailed discussion on this.) The safety and quality standards are so poor that the Finnish nuclear safety regulator has publicly admitted that it may not be able to detect all the problems, and anti-nuclear activists have called for scrapping the construction of the reactor for this reason alone!

Environment Destruction Ministry

We had stated in our concluding remarks in Chapter 3, citing some of the world's most renowned nuclear experts: Nuclear technology 'is a complex technology ... with such high-technology systems involving extremely hazardous materials, it is in the very nature of such systems that serious accidents are inevitable. In other words, that accidents are a "normal" part of the operation of nuclear reactors, and no amount of safety devices can prevent them.'

As if this risk was not enough, the Government of India is now importing giant-sized reactors, whose safety has been questioned even by experts in their home countries! To make matters worse, it has passed a Nuclear Liability Law, indemnifying the foreign equipment suppliers of all liabilities in case of an accident in a reactor supplied by them!!

The Kudankulam-1 and 2 reactors were given approval without an Environmental Impact Assessment (EIA) or public hearing.[145] For the next stage of the project, involving construction of another four reactors, a farcical public hearing was held in June 2007.[146]

For the Jaitapur reactors, an EIA was prepared, a public hearing conducted, and finally the MoEF granted its approval on November 28, 2010. That the FIA was a mere ritual, and was prepared because it was mandated by law to do so, is obvious from the fact that it was prepared by the National Environmental Engineering Research Institute (NEERI), which has acquired a notorious reputation for sloppy work favouring promoters of dubious industrial projects. So far as nuclear reactors go, NEERI, by its own admission, does not have the technical competence to assess radiation related hazards of nuclear reactors.[147] And yet, wonder of wonders, NEERI prepared a 1600-page EIA report for the Jaitapur nuclear power plant!

Despite its bulky size, that the report is a fraudulent exercise is obvious from the fact that it does not deal with any of the serious environmental problems of nuclear power, nor does it deal with the known design problems of the EPR reactor. That, based on this flawed report, the MoEF's approval was going to be a mere formality is obvious from the fact that land for the project was acquired even before the Environment Ministry gave its approval. The MoEF actually fast-forwarded its approval, giving the environmental clearance to the NPCIL just 80 days after it received the EIA report from NEERI, a process which normally takes six months or longer![148] The reason for this hurry was so that the agreement with Areva for supply of the reactors could be signed during French President Sarkozy's visit to India in December 2010. Let us take a brief look at this EIA.

Jaitapur EIA: Phoney Exercise

The EIA does not at all discuss the most important shortcoming of the EPR, its known design problems. Safety regulators of France, Finland, the UK and the USA, all have expressed concern about the design of the Control and Instrumentation (C&I) system of the reactor (discussed in detail in Chapter 6, Part III). The C&I system is the 'cerebral cortex' of a nuclear power station, governing the computers and systems that monitor and control the station's performance, including temperature, pressure and power output levels. While safety regulators of all these countries have asked Areva to rectify this design problem, Indian regulators don't even mention that there is any such problem—even when the problem identified by French, Finnish, UK and US regulators is public knowledge![149] The least they could have done is copy these objections, something they are good at!!

Further, the EIA brushes aside the most important environmental problem of nuclear reactors—the radiation leakages—by making the facile assertion that 'the actual releases will be … far lower than the stipulated limits', without giving any scientific explanation or proof for this postulation. And therefore, since it assumes that the radiation leakages are going to be negligible, the report simply ignores the impact of these radiation releases on the environment and health of the surrounding population![150] One is left wondering, whether the EIA is a horoscope or a scientific document!!

Earthquake Danger Underestimated

The EIA also belittles another potentially serious problem with the Jaitapur plant—its siting in an earthquake-prone zone. The seismic zone map of India divides the country into five zones, from Zone I to V, depending upon the levels of intensity of past earthquakes in that region, with Zone V being the region liable for the most severe earthquakes. The EIA contends that the plant is in Seismic Zone III (Moderate Damage Risk Zone), and that there is no earthquake activity around the Jaitapur site in a radius of 39 km. The implication is that the plant is in a safe zone.[151]

Firstly, what should be remembered is that this classification is only an assessment, and it is possible for a more intense earthquake to occur at a site which has been classified as being in a less intense zone. Therefore, seismic zone classifications are not permanent, and can be revised from time to time, as more understanding is gained of the geology and seismic activity in the area. For example, two major earthquakes, at Koyna (1967) and Latur (1993), occurred in areas categorised as Zone I, supposedly the safest, causing these areas to be revised to Zone IV and III respectively.[152]

Secondly, in contrast to the assertion of the EIA that there have been no earthquakes in the Jaitapur region, data from the Geological Survey of India shows that between 1985 and 2005, there were 92 earthquakes in this region! The biggest of these was experienced in 1993, reaching 6.2 points on the Richter scale![153] This is very disquieting, because it implies that the EIA is lying about simple facts too.

In the light of the above facts, it is possible that the Jaitapur EPR reactor could be stuck with an earthquake of magnitude seven or even more on the Richter scale. If that happens, then it could lead to a major accident, as an earthquake has the possibility of simultaneously affecting many parts of the reactor.

To consider a real life example, a major earthquake of magnitude 6.8[154] stuck Japan on July 16, 2007, severely damaging Japan's largest nuclear plant, the Kashiwazaki-Kariwa Nuclear Power Station (KKNPS). The magnitude of the quake was more than twice as strong as the most extreme cases considered while designing the reactor. It caused at least 50 cases of 'malfunctioning' and 'problems', including damage to the reactor's switchyards, burst pipes, fires, radioactive leakages into the atmosphere and into the Sea of Japan, and the toppling of hundreds of drums of low-level radioactive wastes.

Even more serious is the possibility that an earthquake can cause totally unexpected failures. In the case of the KKNPS accident, underground electric cables were pulled down by ground subsidence, creating a large opening in the outer wall of the reactor's basement—a 'radiation-controlled area' that must be completely shut off from the outside. According to a plant official, 'It was beyond our imagination that a space could be made in the hole on the outer wall for the electric cables.'[155]

The Jaitapur region has already experienced an earthquake of intensity 6.2, and therefore, the possibility of it suffering an earthquake of greater intensity than the one that stuck the Kashiwazaki-Kariwa NPP (intensity 6.8) cannot be denied. Considering this danger, constructing such a large number of high capacity reactors in this area is inviting disaster.

Note: This is precisely what happened at Fukushima too. The Japanese, despite all their skills in planning for earthquakes, had not anticipated and planned for the massive earthquake which hit the plant on March 11, 2011.

No Waste Disposal and Decommissioning Plan

So far as the long-term storage of radioactive waste is concerned, the EIA says: 'The radioactive waste depending upon the activity levels are buried in secured earth trenches, in steel containers which are immobilised in secured concrete vault. The solid waste disposal site is fenced, secured and designed to store waste for sufficiently long time of the order of 100 years.'[156] The EIA thus admits that the plant waste storage system is designed to safely store the waste for only 100 years, which, according to it, is a 'sufficiently long time'. What happens after that? For, the waste is going to remain radioactive for 2.5 lakh years! Well, India's environmental planners are not worried. Why worry about our coming generations, we'll not be there.

The EIA report does not have a decommissioning plan too. It has left this to the future: 'At the end of the operating life of the operating units, which would be around 60 years for EPR-type NPPs proposed to be established at Jaitapur site, a detailed decommissioning plan will be worked out.' No new nuclear plant can be built in Europe or the US without such a plan.[157]

And yet, the MoEF granted environmental approval to the Jaitapur plant!

Fig Leaf: 35 Conditions

To be more precise, the Minister for Environment and Forests Jairam Ramesh gave environmental clearance to the project with 35 conditions attached, of which there are 23 specific conditions and 12 general conditions.[158] Much has been made of these conditions, giving the impression that they would take care of the environmental hazards that may be caused by the plant. Let us examine these.

Take the general conditions. They are actually sanctimonious platitudes. Condition 1 reads: 'The sand for the construction purpose shall be obtained only from the approved quarries.' Condition 7 says: 'Installation and operation of DG (Diesel Generator) sets shall comply with notified guidelines.' Does the MoEF mean to say that all other equipment can violate notified guidelines!

Now for the specific conditions. Some of them are plain stupid. For instance, condition 12 reads: 'During construction of the township and other buildings, it shall be ensured that the buildings confirm to the energy efficiency, water utilisation efficiency as also to the GREHA norms.' Probably Mr Ramesh is also holding the PWD portfolio, and has got confused with his multiple responsibilities.

Most of the specific conditions are actually what the EIA should have tried to prove. Condition 13 reads: 'It shall be ensured that the temperature differential of the discharged water with respect to the receiving water does not exceed 5°C at any given point of time.' Condition 14 says: 'Appropriate safeguard measures shall be taken to ensure that the biodiversities in the sea adjoining Ambolgad are not affected adversely due to the project.' Condition 18 stipulates: 'The radioactive dose apportionment from each unit shall be as per the limits prescribed by the AERB.' And so on ... It was the job of the EIA to assess whether the temperature of the discharged water would exceed 5°C or not, whether the biodiversity of the area would be adversely affected or not, whether the radioactive leakages would be within AERB limits or not, and so on. The EIA has not done any of these assessments! The EIA should have been scrapped, the environmental clearance withheld, and a fresh EIA ordered. Instead, the MoEF gives environmental clearance to the plant, and expresses the pious hope that the plant would fulfill these 'conditions'.

Condition 2 states: 'The following additional details shall be submitted within 12 months: A comprehensive biodiversity conservation plan shall be prepared for Jaitapur ...; a special plan will be made to put in place adequate safeguard measures to ensure that the fisheries in the sea adjoining Ambolgarh are not affected adversely due to the project ...; et cetera.' These are actually mandatory details, and only upon their submission should the environmental clearance have been given. What if it turns out that the fisheries cannot be protected? Will the environmental clearance be revoked?

Actually, while giving environmental clearance for the project, the Environment Minister Jairam Ramesh himself stated the real reason for granting the approval: 'On the one hand there have been many issues raised on the preservation of marine biodiversity, an area in which India has been very weak. But at the same time there are weighty strategic and economic reasons in favour of the grant of environmental clearance now.'[159]

Or rather, despite the adverse environmental impact, it is because of 'weighty strategic and economic' reasons that the project has been given environmental clearance!

The Irresponsible DAE!

Let us for a moment keep aside our main argument about the deathly impact of nuclear power, and accept the government of India argument that the country needs nuclear power as a solution to our energy problems. From the description given above about the VVER-1000 and EPR reactors and the Areva corporation which is going to supply the EPRs to India, it is obvious that these plants need much more stringent supervision during construction, they pose serious safety concerns and so need more exacting management standards during operation, and they are far more risky and so need much greater commitment to safety—an accident at these reactors would be many times more catastrophic than Chernobyl.

Which is the organisation that has been tasked with the responsibility of supervising the construction and subsequently of operating these reactors? The notoriously inefficient and completely untrustworthy DAE, and its subsidiary, the NPCIL ...

... which lie every time an accident takes place at their installations—either they deny it outright, or in case it is not possible to do so, try and play it down by lying about the extent of radiation leakage and its possible impact on their workers and the surrounding population;

... which have built and operated their much smaller 220 MW reactors so carelessly that they are supposed to be the 'least efficient' and the 'most dangerous in the world';

... which are so lackadaisical about the safety situation at their installations that they don't even have an independent nuclear safety regulator!

Yes, it is scary indeed! At the hearing of the PUCL and Bombay Sarvodaya Mandal petitions in the Bombay High Court regarding the safety of India's nuclear power plants (See Part I of this Chapter), the Chief Justice, before summarily dismissing the petitions, remarked that so far, no accident has taken place. To this, advocate for the PUCL, M.A. Rane, asked 'whether the Chief Justice is waiting for an accident to take place from an NPP?'[160] The government of India is apparently of the same opinion, that more and more risks can be taken, till ...

If there is a major accident at Jaitapur, at the very least, Ratnagiri district will have to be permanently evacuated and Western Maharashtra will be radioactively contaminated. If there is a major accident at Kudankulam, at the very least, Southern Karnataka, Southern Tamil Nadu and much of Kerala, along with neighbouring Sri Lanka, will be radioactively contaminated. For 20-30 thousand years. Its consequences will cripple the entire country for many many decades.

Even if there was no alternative, how can we take this risk of damaging the health of our coming generations and rendering large tracts of land uninhabitable for thousands of years, just for meeting our present profligate energy needs?

What is even more stupefying is that we are taking this risk, when there is an alternative safe, green and cheap way of meeting our present and future energy needs! Read on ...

◆◆◆

10

THE SUSTAINABLE ALTERNATIVE TO NUCLEAR ENERGY

PART I: THE OFFICIAL ARGUMENT

The Indian government's argument for embracing nuclear energy in a big way rests on the premise that GDP growth requires a huge increase in electricity generation.

In 2005, the Planning Commission of India appointed a high powered committee to make recommendations on the future of India's energy policy. The Committee submitted its final report, the *Integrated Energy Policy* (IEP) in August 2006. According to this report,

> India needs to sustain an 8 per cent to 10 per cent economic growth rate, over the next 25 years, if it is to eradicate poverty

> and meet its human development goals. To deliver a sustained growth rate of 8 per cent through 2031-2 and to meet the lifeline energy needs of all citizens, India needs, at the very least, to increase its primary energy supply by 3 to 4 times and, its electricity generation capacity/supply by 5 to 6 times of their 2003-4 levels.

The IEP projects that to meet this growth rate, electricity consumption will have to be raised from 553 units per capita in 2003-4 to 2471 units per capita by 2031-2; and the total installed capacity will have to increase to 778,000 MW, implying an increase of close to five times from the current (2010) level of 160,000 MW.[1]

It is in the context of this mammoth future demand projection that the government justifies its massive nuclear energy program. It is expecting around 8 per cent of this demand, about 63,000 MW, to be met from nuclear power[2]—from the 4560 MW at present (March 31, 2010). Justifying this giant leap in nuclear power generation, Prime Minister Manmohan Singh, while speaking at the inauguration of a power plant in West Delhi on March 24, 2008, stated: 'The economy is growing at 8-9 per cent per annum. With growing urbanisation and rising prosperity, the demand for electricity is outpacing existing sources of supply.' And so, he added, India needed to widen its choices for electricity, which should include alternative resources like nuclear power.[3] The same argument is made by many noted intellectuals of the country.

PART II: THE IEP VISION STATEMENT: UNSUSTAINABLE PROJECTIONS

The vision statement of the IEP quoted above estimates that future generation capacity needs to increase by 5 times by 2032.

A number of experts have critiqued the methodology used by India's energy planners to make forecasts of energy consumption. Even assuming that the economy will grow at an average rate of 8 per cent till 2031-2, by extrapolating from recent figures of growth rate of the Indian economy and growth of electricity generation, they show that these projections are exaggerated.[4] *The reason why policy planners make such inflated forecasts is because this then serves as a justification for huge investments in setting up power plants, thereby earning huge profits for the private sector companies who get the construction orders.*

However, let us leave aside this issue and analyse the IEP vision statement on its merit. According to the IEP:

i) a sustained high growth rate of 8-10 per cent for the next two decades is needed for eradicating poverty; and
ii) to deliver such a high growth rate and meet the energy needs of all citizens, would require that the installed electricity generating capacity should increase to between 778 GW (for 8 per cent growth rate) and 960 GW (for 9 per cent growth rate) by 2031-2.[5]

Growth for Whom?

One problem with this argument is that it assumes that GDP growth leads to increased prosperity and better living conditions for the ordinary people. We have argued in detail elsewhere that today, for the common people, the opposite is the case: 'The word "GDP growth rate" has come to have a sinister meaning for the poor. The upper classes measure their increase in wealth by growth in GDP. For the vast masses, it is a measure of the devastation of their lives…' For instance, during the period 2003-07, a period which saw the Indian economy grow at above 8 per cent per year, total employment in the organised sector actually declined in absolute numbers, farmers committed suicides in record numbers, lakhs of people fell below the poverty line and malnourishment in the country increased further from already high levels, and so on.[6]

Likewise, increased electricity generation also does not mean more electricity for the common folk living in small towns and rural areas. This is borne out by past experience—most of the growth in electricity generation during the six decades since independence has gone towards fulfilling the galloping demands of the rich in the big cities in the country. (And in the rural areas, whatever little growth in electricity supply that has taken place has gone to the rural rich and the landowners; most of the rural poor do not have electricity connections.) Thus, the total installed electricity generating capacity in the country has gone up by more than a hundred times since independence, from 1400 MW[7] in 1947 to 160,000 MW in 2010 (Table 10.1). Despite this phenomenal increase, more than 44 per cent of the country's households still have no access to electricity even six decades after Independence. The situation is especially bad in the rural areas, where about 56 per cent of the households have still not been electrified.[8]

Table 10.1: Power Generation Capacity in India (MW)
(as on March 31, 2010[9])

Thermal				Hydro	Nuclear	Renewables[U]	TOTAL
Coal	Gas	Diesel	Total				
84,198	17,056	1,200	102,454	36,863	4,560	15521[Y]	159,398

[U] Renewables includes only wind energy, biomass gassifiers, minihydel turbines (<3 MW), and bagasse fired cogeneration plants; does not include biogas plants, solar PV panels and solar water heaters.
[Y] Based on data as on 30.09.2008

Further, even the 44 per cent villages that have been electrified have very inadequate supply of electricity. Most villages have power supply for durations of less than 12 hours a day on an average. Even this meagre supply is of very poor quality: it is neither regular, nor is it provided when people need it most. Additionally, the low voltage conditions and frequent interruptions make the electrification a cruel joke on the villagers. Anyone familiar with the rural areas of the country is well aware of this.

A recent survey of five states in four regions of the country carried out by Greenpeace also bears this out. In all these five states, there has been a continuous increase in power availability. However, says the survey report, 'most of the additional power available in each state seems to have gone to the cities and towns to meet their insatiable demands ... whereas the villages continue to suffer with inadequate amount of electrical energy even for basic needs.'[10] The government's drive to further add lakhs of MW of additional capacity will also go towards meeting the ever-growing electricity demand of the urban rich, it will not ensure quality power to the rural population and will therefore not lead to their development. That would need an entirely new orientation in our energy policy. We discuss this towards the end of this chapter.

Destructive Projections

Let us leave aside the issue of equitable development, assume that GDP growth is needed, and accept the Planning Commission's argument that for the economy to grow at an average rate of 8 per cent till 2032, it is necessary to increase electricity generation capacity in the country to 778,000 MW over the next two decades through setting up large centralised coal-hydro-nuclear-based power plants. To meet these projections, the IEP assumes a capacity addition of 63,000 MW from nuclear energy and full exploitation of the hydropower potential of 150,000 MW in the country. It also admits that coal shall continue to remain India's most important energy source till 2031-2 and possibly beyond, though the share of coal-based electricity would drop because of added capacity from nuclear and hydro sources.[11] The problem with these projections is that they are simply unsustainable!

We have already discussed extensively in this booklet the disastrous implications of the government's plans to go in for a quantum jump in nuclear energy generation. The IEP has drawn up various possible energy mix scenarios for 2032, and even in the most renewable energy friendly scenario, it expects the share of coal-based electricity to increase to 270,000 MW (from 85,000 MW as of March 31, 2010).[12] Even assuming that the government does pursue this energy mix over the next two decades, the social and environmental costs of setting up coal-based thermal power plants of a total capacity of around 190,000 MW and large dam-based hydropower plants of around 110,000 MW capacity over the next two decades are also going to be huge.[13]

Costs of Coal Power

Even in its most renewable energy friendly scenario (where all options other than coal are pushed to their limits), the IEP projects coal-based power capacity in India to rise to 270,000 MW by 2032; while the Central Electricity Authority (CEA) projects installed capacity of coal-based plants to be 412,000 MW by that year.[14]

The first problem with these projections is: where is the coal going to come from to feed these power plants? The IEP projects the requirement of coal for power generation to increase from 406 million tons (Mt) in 2004-5 to between 1580 and 2555 Mt (for the least and most coal intensive option respectively) in 2031-2. Since the domestic supply of coal is limited, IEP projects that high quality coal import requirement could range from 120 Mt to 770 Mt (for the least and most coal intensive option respectively) by 2031-2.[15] The problem is that, globally, available exportable coal supplies are also running out! A recent study by the Energy Watch Group of Germany predicts that global coal production will increase over the next few years, peak around 2025 and then decline. Clearly then, it is foolhardy to base our future energy security on a resource whose domestic supplies are declining and global availability in adequate quantities beyond 2030 is suspect.[16]

The second, and more important problem with this projection is: the IEP totally ignores the environmental and health impact of such a huge increase in coal-based power generation. It is going to pose a bigger risk to biodiversity in the Indian subcontinent than by all other anthropogenic interferences in the past put together. Each part of the coal cycle—from mining of coal, to burning it in power plants, to disposing of coal waste—causes irreparable damage to the environment and the health of people.

Costs of Coal Mining

The most serious health effect of coal mining is of course on the coal miners—it causes black lung disease, due to the progressive build up of coal dust in the lungs which the body is unable to remove. A much ignored cost of coal mining is the deaths of miners from accidents. Coal mining, especially underground coal mining, is a very dangerous activity, and the list of coal mining accidents is a very long one. Tens of thousands of coal miners have died in accidents during the past century. While, during the early decades of the twentieth century, a large number of these deaths took place in the developed countries (in the US alone, there were more than a 1000 fatalities every year from 1900 to 1945), in recent decades, the most horrifying accidents have occurred in the third world countries, especially China.[17]

Apart from that, coal mining causes displacement of entire communities who are forced to abandon their homes because of the mines. The coal is normally located below thick forests, so mining causes widespread deforestation. It also generates huge waste mountains and blankets surrounding communities with dust particles and debris, seriously impacting their health too.

Costs of Coal Power Plants

The burning of coal in power plants to produce heat and generate electricity leaves a similar trail of destruction in its wake. A groundbreaking medical report, *Coal's Assault on Human Health,* released in November 2009 by the reputed US group, 'Physicians for Social Responsibility', has given in detail the devastating effects of coal-burning on human health. Coal combustion releases sulphur dioxide, particulate matter, nitrogen oxides, mercury and dozens of other hazardous substances into the environment, which damage the respiratory, cardiovascular and nervous systems of the human body. In particular, these emissions contribute to some of the most widespread diseases, including asthma, heart disease, stroke and lung cancer.[18]

Apart from their impact on human health, the sulphur dioxide, nitrogen dioxide and coal dust emissions deposit over large areas, and their synergistic effect is very injurious to vegetation. They adversely affect soil fertility, resulting in a sharp decrease in agricultural yields. When located near forests, like the thermal power plants coming up in the Konkan region, the depositions of these acidic gases put the forests at risk of forest dieback (a condition in which peripheral parts of trees are killed due to factors like acid rain).[19]

Another major problem with thermal power plants is that they require large amounts of water. Given the growing water crisis in the country, the plants being proposed to be sited in inland areas will worsen this crisis. As a solution to this problem, many new thermal power plants are proposed to be set up along the coast, so that they can use seawater. However, this will impact the fish breeding and spawning areas, threatening the livelihoods of fishermen.

Finally, like coal mining, construction of the dozens of thermal power plants required to meet the coal power generation target for

2032 will also mean acquisition of large chunks of land, leading to displacement of lakhs of people. There is no alternate land in the country to be given to them, they will end up in the slums of big cities. True, they will not be annihilated or taken to gas chambers, but the quality of their accommodation is not going to be any better than in any concentration camp of the Third Reich.

Costs of Coal Waste from Plants

The damage caused by coal doesn't end once it's burnt. The combustion waste, also known as coal ash, is very toxic. It contains many chemicals like lead, arsenic, boron and cadmium which can cause cancer and other health effects.[20] While this waste is supposed to be sluiced with water and let into ash ponds, thermal power plants in India invariably discharge ash into nearby water bodies, polluting them and affecting the lives of thousands of people dependent on these water bodies for their water supplies.[21] Not that the ash ponds are any better: most ash ponds are unlined or inadequately lined, and a new official report from the US says that such coal ash ponds have poisoned groundwater or surface water in at least 23 states, and they pose a cancer risk 900 times above what can be defined as 'acceptable'.[22]

Fly ash also contains uranium and thorium. These are present in natural coal in trace amounts. However, when coal is burnt into fly ash, they get concentrated up to 10 times their original levels. When this fly ash is disposed off in landfills, along with other chemicals, the uranium and thorium can also leach into the soil and groundwater from this landfill, affecting cropland and, in turn, food.[23]

Contribution to Global Warming

Given the severity of the global warming crisis which is threatening the very existence of life on earth, probably the gravest problem caused by coal-based power plants is that they are the biggest source of greenhouse gas (GHG) emissions in the world: according to one estimate, they account for one-third of overall global emissions.[24] In the US, the world's second biggest emitter of greenhouse gases (China has now surpassed the US to become the biggest), the electricity sector (meaning mainly the coal fired thermal power plants) is responsible for about one-third of the country's total GHG emissions and 40 per cent of total carbon dioxide emissions.[25] For India, the report *India: Greenhouse Gas Emissions 2007* released by the MoEF in May 2010 says that about 38 per cent of our country's total GHG emissions are due to the electricity generation sector.[26]

To conclude, clearly, even assuming that the government manages to find the coal to fuel its projection of 270-400 GW of coal-based thermal power by 2032, the impact of these plants on the environment and health of the people of the country would be devastating.

Costs of Large Hydropower Plants

Large hydroelectric dams, like coal-fired power plants, also wreak havoc on the ecosystems and communities where they are located. The biggest problem with these plants is that the giant reservoirs of their dams displace huge populations of people, leaving them homeless and destitute. The figures of those displaced so far by large dams are mind-boggling. Arundhati Roy, in her wonderful article *The Greater Common Good* quotes N.C. Saxena, Secretary to the Planning Commission, as saying that nearly 4 crore people have been displaced by dams in the country since independence.[27] That's more than three times the number of refugees created by Partition in India! What about rehabilitation? The government of India does not have a National Rehabilitation Policy. What happened to these 4 crore people, where did they go, where are they now, how do they earn a living now that their lands are gone, no one knows. And now, the government is proposing to set up new hydropower plants to quadruple our present installed capacity!

The second and equally severe problem is environmental: dams submerge millions of hectares of lush forests and large chunks of fertile river valley agricultural lands. The other ecological problems caused by dams are less well known. The World Commission on Dams (WCD), formed in April 1997 to research the environmental, social and economic impact of large dams globally, found that 'large dams generally have a range of extensive impacts on rivers, watersheds and aquatic ecosystems' and 'have led to irreversible loss of species and ecosystems'. Damming of rivers impacts the quantity, quality and pattern of water flow in them, and has caused a huge loss of freshwater diversity: up to 35 per cent of freshwater fish species are estimated to be extinct, endangered or vulnerable.[28]

Contribution to Global Warming

Another myth with regards to large hydropower plants is that they are green, that is, they do not contribute to global warming. The truth is: large dams emit significant amounts of greenhouse gases like methane, carbon dioxide and nitrous oxide. The 'fuel' for these gases is the rotting of the vegetation and soils flooded by reservoirs, and of the organic matter (plants, plankton, algae, et cetera) that flows into dams. According to a study by researchers from Brazil's National Institute for Space Research, the world's large dams emit 104 million metric tons of methane annually, implying that dam methane emissions are responsible for at least 4 per cent of the total global warming impact of human activities.[29] The study also found that more than one-fourth of these emissions, 28 per cent to be more precise, were due to India's large dams! Large dams are in fact responsible for some 20 per cent of India's global warming impact![30]

While the costs of large dams are huge, the benefits are less than projected. Silting of dams leads to a decline in their actual storage capacity, in many cases severely, due to which the area irrigated by them decreases.[31] It also results in decreased electricity generation from their associated power plants. A survey of 208 operational hydel projects in India done by Himanshu Thakkar of South Asian Network for Dams, Rivers, and People (SANDRP) in March 2007 found that 184 of these (88 per cent of those surveyed) were generating less power than their design capacity. And for 90 of these projects (50 per cent of those surveyed), the actual generation of electricity was less than 50 per cent of the design capacity![32]

Therefore, it is not surprising that cost-benefit studies of many large hydropower projects have found the benefits to be less than the costs![33] The evidence is so overwhelming that even the report of the World Commission on Dams, which was sponsored by the World Bank, concluded: 'given the high capital cost, long term gestation period and the environmental and social costs, hydropower is not the preferred option for power generation compared to other options.'[34]

In the light of these facts, the proposal of the government to construct large hydropower projects of nearly three times the present capacity of 37,000 MW over the next two decades is going to be absolutely disastrous, for both the people and the environment.

Conclusion

Clearly then, the government's plans of setting up giant-sized coal, hydro and nuclear power plants to produce the electricity required to power India's future growth:

(i) will not solve the energy crisis of the majority of the Indian people living in the rural areas, who continue to be without electricity even 60 years after independence;
(ii) will have unacceptable environmental, social and health costs.

If that is so, then is there an environmentally friendly way of meeting the genuine present and future electricity needs of all sections of the Indian people? Yes, there is. There exists a *genuinely safe, green, clean and cheap* solution to the energy crisis.

Part III: The Sustainable Alternative: A New Energy Paradigm

It is possible to find a way out of this crisis, but that would call for a totally new approach to energy planning. **Firstly**, we need to reorient our energy planning towards meeting the energy requirements of all sections of the population, and not just the energy needs of the elites living in the cities. The indicator of development must not be statistics showing total energy consumed, but whether the basic energy needs of the people, starting from the poorest sections, are being met. **Secondly**, we need to recognise that what really matters is not how much electricity is generated, but how much of it is finally being converted to work by energy devices (what is called 'useful' energy). That is, we must focus on increasing the services provided by electricity, like lighting, heating, cooling, et cetera, instead of blindly increasing electricity generation. **Thirdly**—and this is important—we need to regulate electricity consumption. Unless we reduce electricity consumption, improving efficiency of the electricity supply network and end-use efficiency will only fuel a rise in total electricity consumption, as has happened in many countries. **Finally**, the electricity supply system must be environment friendly; we cannot ignore its environmental costs.

This new energy paradigm has major implications for the energy system. These include:

A) Demand Side Management (DSM): Increasing useful energy or end-use of electricity does not necessarily mean increasing electricity generation, it can also be achieved by:
   - increasing the efficiency of the electricity generation system and the electricity transmission system, and increasing the efficiency of the devices that convert the electricity delivered to the consumer into the required energy services (heaters, coolers, bulbs, et cetera);
   - curbing demand, by regulating electricity consumption and eliminating wasteful consumption of electricity.

B) Massively increasing the production of electricity from renewable sources like the sun, wind, flowing water (here, we are referring to small hydropower plants and not large hydropower plants) and biomass, for which there is a huge potential in the country.

C) Adopting decentralised energy systems where necessary, as they are often cheaper and more efficient as compared to supplying electricity from a large centralised grid.

We take a look at all these three issues in some detail.

A. Demand Side Management (DSM)

This means adopting policy measures which will increase the availability of useful energy (that is, energy available for consumption) without increasing electricity generation.

1. Improving Generation, Transmission and End-use Efficiency

Even a cursory look at the Indian power sector makes it evident that its overall efficiency is very low as compared to international standards, as table 10.2 attests.

The average Plant Load Factor (PLF) of thermal power stations in the country is reported to be about 77 per cent,[35] while the best run power plants of National Thermal Power Corporation (NTPC), India's largest public sector generating company, have PLF of above 90 per cent, and some of them even have PLF of 100 per cent.[36] By renovating and modernising the plants with low PLF, it should be possible to raise the national PLF. With about 84,000 MW of installed thermal power capacity in the country, every percentage point increase in PLF will save the need for about 840 MW of additional installed power capacity.

Aggregate Technical and Commercial (AT&C) loss includes both transmission and distribution (T&D) losses and also commercial losses like those due to theft, inefficient metering, et cetera. It is thus a better indicator of total losses in the transmission and distribution system. Overall AT&C losses in the country's electricity network were 32 per cent (and T&D losses were 29 per cent) in 2006-07.[38] Bringing

Table 10.2: Power Sector Efficiency in India[37]

Power sector area	*Prevailing level of efficiency / loss in India*	*International best practice*
Generating capacity utilisation (Plant load factor)	Around 77%	More than 90%, to 100%
Aggregate Technical & Commercial losses (AT&C)	Around 32%	Less than 10%
End-use efficiency in agriculture	45-50%	More than 80%
End-use efficiency in industries and commerce	50-60%	More than 80%
End-use efficiency in other areas (domestic, street lights and others)	30-60%	More than 80%

down the losses in the transmission and distribution system to even 15 per cent will reduce the need for additional installed power capacity by about 20,000 MW. This is not a tall order. China's total AT&C losses are 8 per cent; OECD countries' T&D losses are 7 per cent,[39] while South Korea's T&D losses are even lower at just 4 per cent.[40] Assuming a cost of Rs.4.5 crores per MW for setting up new coal power generation capacity, reducing T&D losses by even 15,000 MW would mean savings of Rs.67,500 crores.

Finally, as Table 10.2 indicates, there is huge scope for increasing end-use efficiency in all sectors from agriculture to industry to offices and homes. For instance, agricultural pumping sets or IP sets consume about 30 per cent of all electricity consumed in the country. They consume about 40-45 per cent more energy than required; by investing just around Rs.4000 per set, this wastage can be brought down to less than 15 per cent.[41] Since there were 160 lakh pumping sets in the country in June 2009,[42] this means an investment of roughly Rs.6400 crores will result in a saving of 30 per cent of the electricity consumption in the agricultural sector, that is, a whopping 14,400 MW. Assuming an investment of Rs.4.5 crores per MW for setting up new coal fired power plants, this would mean a saving of (14400 x 4.5) – 6400 = Rs.58,400 crores![43]

Similarly, there is considerable potential for improvement of energy efficiency in Indian industry. According to one recent report, the average Indian cement plant consumes 25 per cent more energy than the global best practice, while the potential for improving energy efficiency of 15 hp and 20 hp industrial motors is between 20-39 per cent. On the whole, for each industry, there appears to be a potential for improvement of energy efficiency that ranges from 15 per cent to 35 per cent.[44]

Another study, by Professor Saifur Rahman, Director, Advanced Research Institute, Virginia Polytechnic Inst & State University, USA, says that India's industrial sector has an estimated energy saving potential of 25 per cent.[45]

Savings from Improving Efficiency of Household Appliances

An important way of increasing end-use efficiency in homes is by promoting the use of efficient electrical devices, which give the same output with lesser consumption of electricity. Most of us are not aware about it, but the potential is huge. A recent study by Prayas Energy Group, a well-known Pune-based research group on policy analysis, has estimated the energy savings that can be achieved by the use of energy-efficient home appliances in the country. The report focuses on nine appliances which contribute to almost all the electricity consumption in Indian households—fans, incandescent bulbs, tube lights, refrigerators, air conditioners, air coolers, electric water heaters, room heaters and televisions (active mode)—and apart from these, stand-by power. Stand-by power is power wasted because appliances are not switched off after use and kept on stand-by mode; the report takes into consideration the stand-by losses of set-top-boxes, DVD players, TVs and computers.

The report takes 2008 as the starting year for its analysis, and calculates the energy savings that would result if all incandescent bulbs were replaced with compact fluorescent lamps (CFLs) and tubelights were replaced with electricity efficient models, and all new purchases of appliances by households were of the most energy efficient model available in the Indian market. The report comes to the astonishing conclusion that after five years, this would result in an annual savings of about 57 TWh (tera watthour = 1000 GWh) in 2013! That is about 30 per cent of the additional annual consumption that would otherwise have happened under a business-as-usual scenario in that year. Retrofitting of lights accounts for about half the savings, while ceiling fans, TVs, refrigerators and reduction in stand-by power account for another 40 per cent of the savings.[46]

These potential savings correspond to saving more than 25,000 MW in generating capacity addition![47] It is more than the total combined capacity of the Jaitapur Nuclear Park, the Kudankulam Nuclear Park and the Mithivirdi Nuclear Park!

Many efficiency measures are easy to implement, and the investment made would be more than compensated by the savings. One such extremely quick and cheap way of bringing about significant energy savings is: replacing incandescent lamps by CFLs. This has the potential to reduce the lighting load of the system by about 80 per cent, and the total cost of lighting to a consumer by about 66 per cent.[48] The energy saving is even more if the bulbs are replaced by LEDs (Light Emitting Diodes), as these consume half the energy of CFLs.

The government of India has taken some tentative steps to improve energy efficiency in the country by setting up the Bureau of Energy Efficiency (BEE), under the Ministry of Power, in March 2002. The agency's function is to develop programs which will increase the conservation and efficient use of energy in India. One of the much publicised schemes of the BEE is a program that aims at the large scale replacement of incandescent bulbs in households by CFLs; it seeks to provide CFLs to households at the price of incandescent bulbs (Rs.15). The scheme was launched in 2009 and was supposed to cover all of India by 2011, but according to newspaper reports, it is being launched in a number of states only this year (2011).[49]

Balance Sheet

All the above mentioned savings resulting from improving generation, transmission and end-use efficiency add up to a whopping 50,000 MW at the minimum. Out of a total generation capacity of 160,000 MW! This means that merely improving the efficiency of the existing electrical infrastructure to even near international standards will reduce the electricity demand by at least 30-40 per cent!

Table 10.3: All-India Power Supply Scenario (2007-08)[50]

Energy demand 739,345 MUs	Energy availability 666,007 MUs	Energy shortage 73,338 MUs (9.9%)
Peak demand 108,866 MW	Peak demand met 90,793 MW	Peak shortage 18,073 MW (16.6%)

Let us now compare the potential of energy savings in India with the power supply deficit in the country (Table 10.3). If we compare the deficit figures with the potential of reducing the electricity demand in the country by improving system efficiency, it is obvious that the entire power sector deficit can be wiped out just by implementing efficiency improvement measures! In fact, there would even be a surplus!! This implies that even without the addition of any new electrical generation capacities, there will be no electricity deficit in the country for the next few years at least.[51]

The cost of implementing these efficiency improvement measures will also be much lower as compared to the cost of setting up new generating capacities—saving a unit of energy costs about one-fourth the cost of producing it with a new plant.[52]

An Observation

While these are absolutely stunning statistics, and most readers would be amazed by the immense potential of energy savings in the country, actually, there is nothing new about this analysis. Some of India's most renowned energy analysts have been writing about the enormous potential of adopting energy efficiency measures in the country since at least the last two decades. For instance, way back in 1991, A.K.N. Reddy critically reviewed the projections made by a committee set up by the Karnataka state government about the energy requirements of the state by the year 2000. He showed that this could be brought down to only about 38 per cent of the conventional demand, by adopting simple efficiency improvement methods like the replacement of inefficient motors with efficient motors, incandescent bulbs by CFLs, and electric water heaters by solar water heaters![53]

The IEP 2006 also admits to the potential of DSM. It says that 'cost effective savings potential is at least 15 per cent of total generation through DSM.' It further says: 'Additional savings are possible on the supply side through reduction in auxiliary consumption at generating plants and lowering technical losses in transmission and distribution.' The IEP concedes that the government has never taken DSM seriously: 'In the 1990s, several studies have estimated the potential and cost effectiveness of energy efficiency and demand side management (DSM) in India. Despite these potential studies, actual implementation has been sluggish.' It also confesses that though the 10th Five Year Plan (2002-2007) has set targets for energy savings at about 13 per cent of the estimated demand, 'there is no specific (financial) allocation to meet the energy savings targets'![54] (That is, the targets are supposed to be musings.)

Astonishingly, even after making all these admissions, the IEP has not factored in the enormous potential of energy conservation while making its forecasts for the energy requirements of the country by 2031-2! This is why it has come to the conclusion that the country must expand its electricity generation capacity by a whopping five times from the present installed capacity of 160 GW (in 2010) over the next two decades.

2. Curbing Luxurious and Wasteful Consumption

However, in practice, a reduction in total energy demand/consumption simply by improving energy efficiency will not occur. That is because of an inherent logic of the capitalist economic system, known as the Jevons Paradox, according to which improving energy efficiency actually leads increase in energy use, to such an extent that the resulting increase in total demand exceeds the savings due to energy efficiency. This concept was first put forth by William Stanley Jevons in the nineteenth century and is considered one of the pioneering insights in ecological economics.

The Jevons Paradox can be seen in the fact that, even though the United States has managed to double its energy efficiency since 1975, its energy consumption has risen dramatically. Over the last thirty-five years, energy expended per dollar of GDP in the US has been cut in half. However, rather than falling, energy demand has increased, by roughly 40 per cent. Moreover, demand is rising fastest in those sectors that have had the biggest efficiency gains—transport and residential energy use. Refrigerator efficiency improved by 10 per cent, but the number of refrigerators in use rose by 20 per cent. In aviation, fuel consumption per mile fell by more than 40 per cent, but total fuel use grew by 150 per cent because passenger miles rose. Similarly, technological advancements in motor vehicles, which have increased the average miles per gallon of vehicles by 30 per cent in the United States since 1980, have not reduced the overall energy used by motor vehicles. Fuel consumption per vehicle stayed constant while the efficiency gains led to the augmentation, not only of the numbers of cars and trucks on the roads (and the miles driven), but also their size and 'performance' (acceleration rate, cruising speed, et cetera), so that SUVs and minivans now dot US highways.[55]

Therefore, in addition to promoting energy efficiency, steps will also have to be taken to curb demand without which total energy demand will not reduce. To give a few examples, steps will have to be taken to push high-end residential consumers into reducing their total consumption, by measures such as sharply increasing cost of electricity for those consuming beyond a certain limit. Curbs will have to be imposed on electricity consumption in offices and institutions. Many are so awfully designed that they need lighting even during daytime in summers, in a tropical country like ours! A particularly bad example is shopping malls and IT companies, which not only have 24-hour lighting, but also 24-hour air conditioning throughout the year. Not only that, even many elite colleges today have air conditioners installed in their classrooms, in order to attract high paying students. To curb such luxurious consumption of electricity, it will not be enough to raise electricity rates, as the rich can afford to consume costly electricity. Restrictions will have to be imposed on such luxurious use of electricity.

Apart from curbing luxurious consumption, wasteful consumption of electricity, like unnecessary illumination of commercial buildings, lighting of roadside hoardings, and the enormous consumption of electricity that takes place to satisfy the recent craze for night time sports, will also have to be curbed.

An Observation

While all the other elements of the Alternate Energy Paradigm are at least accepted in principle by the IEP and the government of India, there is complete silence about this component. This is because it contradicts the basic logic of the present economic system—where the driving force of production is profit making and capital accumulation. Profit accumulation requires: (i) more and more production which therefore means consumption of more and more energy and raw materials; (ii) and more and more sales, even if it means promoting luxurious and wasteful consumption. This is, in fact, the economic reason behind Jevons Paradox too—the capitalist system takes advantage of energy savings to promote proliferation of commodities in order to make more profits.

Therefore, imposing restraint on luxurious and wasteful consumption would mean curbing profit accumulation, something unthinkable under the present economic system.

B. Adopting Renewable Energy

Sources used for electricity generation can broadly be classified into two categories: **non-renewable energy** sources, where the energy source is a finite natural resource that will eventually dwindle, such as coal, gas, oil and uranium, as opposed to **renewable energy** sources, which are naturally replenished in a relatively short period of time, such as sunlight, wind, rain, tides, flowing water (that is, hydro) and geothermal heat. Excluding hydro electricity, the other renewable sources of energy are also called **new renewables**, as they have started being used in a big way only in recent times. The non-renewable energy sources plus large hydro projects (which have also been in use for many years) are also called **conventional sources of energy**, while the new renewables are also called **non-conventional sources of energy**.

A common feature of new renewable energy sources is that they produce little or no greenhouse gases, and rely on virtually inexhaustible natural resources for their fuel. New renewables vary widely in their technical and economic maturity. Some of these technologies are already competitive, and their economies will further improve as they develop technically. In contrast, the price of fossil fuels will continue to rise in future, as reserves get exhausted. The price of electricity from fossil fuels becomes even more expensive if we give the CO_2 emissions by these fuels and the environmental destruction caused by them a monetary value.

1. New Renewables: Global Scenario

Globally, over the last few years, costs of new renewable energies have fallen sharply and production of energy from these energy sources has rapidly expanded. We take a brief look at the global potential of two of these alternatives, wind and solar.

Wind Energy

Harnessing the wind is one of the cleanest and most sustainable ways to generate electricity. Wind power produces no toxic emissions and none of the heat trapping emissions that contribute to global warming. Wind power is also one of the most abundant energy resources. One study, by researchers Christina Archer and Mark Jacobson of Stanford University in California, which collated more than 8000 wind records from every continent, found the global wind power potential to be 72 terawatts, forty times the amount of electrical power used by all countries in 2000. If just 20 per cent of this wind energy potential could be tapped, all energy needs of the world could be satisfied.[56]

However, it is only in recent times, because of growing concerns about global warming, that countries have started investing heavily into research to tap this huge potential. As a result, during the last decade, world wind power generation has doubled about every three years, making wind energy one of the fastest growing sources of electricity in the world. Global wind power capacity increased by a record 31 per cent over the previous year to reach 160,000 MW by the end of 2009, more than triple the 48,000 MW that existed in 2004. Total wind energy production in 2009 was 340 TWh, which was about two per cent of worldwide electricity consumption.[57]

The US has the highest installed wind power capacity in the world (36.3 GW), followed by China (33.8 GW) and Germany (26.4 GW), as of June 2010. China is presently the locomotive of the international wind industry, adding 13.8 GW within one year in 2009, more than doubling its wind power capacity for the fourth year in a row. Several countries have achieved relatively high levels of wind power penetration; wind power was 20 per cent of stationary electricity production in Denmark, 14 per cent in Ireland and Portugal, 11 per cent in Spain, and 8 per cent in Germany in 2009.[58]

The rise in global wind energy capacity has been accompanied by a sharp fall in costs: wind energy today costs only about one-sixth as it did in the 1980s, dropping from about 25 cents/kWh in 1981 to an average of about 4 cents/kWh in 2008—a price that is competitive with new coal- or gas-fired power plants (figures are for the USA). With costs expected to decline further in the coming years, and growing concerns about the environmental costs of conventional sources of energy, wind power generation is expected to grow exponentially in the coming years.[59]

The Global Wind Energy Council projects global wind power generation capacity to reach 332,000 MW by 2013, more than double its current size.[60] While wind power now contributes 1.3 per cent of the global electricity supply, this is projected to increase to 8 per cent by 2018.[61]

The Variability Problem with Wind Energy

The most important question raised about the potential of wind energy is its variability, as variation in wind speed results in variation in power generated. However, electric power generation companies know how to deal with this problem. Even with electric power from conventional sources of energy, electric power companies need to constantly adjust to constant changes in electricity demand, turning power plants on and off, and varying their output second-by-second as power use rises and falls. They also need to meet unexpected surges or drops in demand, as well as power plant and transmission line outages. Therefore, they know how to deal with changes in wind power generation at different wind turbines. In addition, the wind is always blowing somewhere, so distributing wind turbines across a broad geographic area helps smooth out the variability of the resource.

This is also being proved in practice. In the US, which installed a record 8,500 MW of wind power in 2008, capable of producing enough electricity to power more than 2 million typical homes, many electric power companies are already demonstrating that wind can make a significant contribution to their electricity supply without reliability problems. Xcel Energy, which serves nearly 3.5 million customers across eight states, currently obtains eight per cent of its electricity from wind and plans to increase that to about 20 per cent by 2020. There are several areas in Europe where wind power already supplies more than 20 per cent of the electricity with no adverse effects on system reliability. Three states in Germany in fact have wind electricity penetrations of more than 40 per cent.[62]

Solar Energy

In the broadest sense, solar energy supports all life on Earth and is the basis for almost every form of energy we use. The sun makes plants grow, which can be burned as 'biomass' fuel or, if left to rot in swamps and compressed underground for millions of years, take the form of coal and oil. Heat from the sun causes temperature differences between areas, producing wind that can power turbines. Water evaporates because of the sun, falls on high elevations, and rushes down to the sea, spinning hydroelectric turbines as it passes. But solar energy usually refers to ways the sun's energy can be used to directly generate heat, lighting and electricity.

The amount of energy from the sun that falls on Earth's surface is enormous. All the energy stored in Earth's reserves of coal, oil and natural gas together is equal to the energy from just 20 days of sunshine. Once a system is in place to convert it into useful energy, the fuel is free, emission free, inexhaustible, and will never be subject to the ups and downs of energy markets.

The sun's energy when it reaches the Earth's surface is about 1000 watts per square meter at noon on a cloudless day. Sunlight varies from region to region. Deserts, with dry air and very little cloud cover, receive the most sun. Sunlight varies by season as well. Averaged over the entire surface of the planet, 24 hours a day for a year, each square meter collects the approximate energy equivalent of almost a barrel of oil (that is, 159 litres) each year, or 4.2 kilowatt-hours of energy every day. It should be noted that these figures represent the maximum available solar energy that can be captured and used; solar collectors capture only a portion of this, depending on their efficiency.[63]

Though solar power has the potential to provide over 1000 times the present total world energy consumption,[64] it presently provides less than one per cent of it. However, if the rate at which its use has been expanding in the past few years is an indicator, it is poised to become the world's dominant energy source in a few decades from now.

The simplest and most common use of solar energy is using the sun to heat, cool and light buildings. Apart from this passive use of solar energy, mechanical devices can also be used to harness solar energy, the most common being the use of solar heat collectors, solar heat concentrating systems and photovoltaic panels.

Solar Heat Collectors

Apart from using design features to maximise their passive use of the sun, another way in which buildings can use sun's energy is by installing systems that actively gather and store solar energy, called solar collectors. The most common use of solar collectors is for water heating. Solar heat can also be used to power a cooling system using the same principle on which conventional refrigerators and air conditioners work.

Solar water heaters are a very simple way of saving grid electricity. Bangalore city has promoted the use of solar water heaters in a big way, and according to one estimate, it is resulting in a saving of 900 MW peak load![65]

China is the world leader in solar hot water systems, with 60 per cent of the world's capacity. It presently has nearly 27 million rooftop solar water heaters; the energy harnessed by these installations is equal to the electricity generated by 49 coal-fired power plants. In Europe too, rooftop solar water heaters are spreading fast. Cyprus is the per capita world leader, with 92 per cent of the homes having solar water heaters. 15 per cent of all Austrian households now rely on them for hot water. Some 2 million Germans are now living in homes where both water and space are heated by rooftop solar systems. In the United States, heating swimming pools was the dominant application of solar hot water till 2005. In 2006, federal subsidies were introduced, and since then installation of residential solar water and space heating systems has soared.[66]

Solar Thermal Concentrating Systems

By using mirrors and lenses to concentrate the rays of the sun, solar thermal systems can produce very high temperatures—as high as 3,000 degrees Celsius.[67] This intense heat can be used in industrial applications or to produce electricity. One of the greatest benefits of these solar thermal systems, more commonly known as Concentrating Solar Power (CSP) systems, is the possibility of storing the sun's heat energy for later use, which allows the production of electricity even when the sun is no longer shining. Properly sized storage systems, commonly consisting of molten salts, can transform a solar plant into a supplier of continuous baseload electricity. CSP systems now in development will be able to compete in output and reliability with large coal and nuclear plants.

CSP technology is best suited for the desert regions of the world —including desert regions in southern United States, North Africa, Mexico, China and India. Typical CSP plants are of between 50-200 MW capacity. The first commercial CSP plants were built in the 1980s, but it is only in the last few years that capacity has expanded rapidly. The US is the world leader in installed CSP capacity. While it had only 430 MW in operation in 2009, approximately 7,000 MW is in the process of development, of which 3000 MW is expected to be operational by 2011.[68] The European Renewable Energy Council expects total CSP installed capacity to exceed 1000 MW by 2010, and 20,000 MW by 2020.[69] China has also announced plans to set up CSP power plants totalling 2000 MW capacity over the next decade.[70] Large scale CSP plans have also been announced by Jordan, South Africa, United Arab Emirates, Egypt, Morocco, Mexico and several other countries.[71]

CSP costs are declining as technology improves and production increases. Existing CSP plants produce electricity for around 12 cents/kWh. These costs are expected to fall to below 6 cents/kWh by 2015,[72] making it competitive with conventional electricity for decentralised systems. Therefore, CSP-generated electricity is poised for a huge leap in the coming years. A study by 'Emerging Energy Research'[73] projects cumulative global installed capacity of CSP plants to go up to 26,465 MW by 2020.[74] Another study by Greenpeace, the European Solar Thermal Electricity Association and the IEA estimates that CSP has the potential to meet up to 7 per cent of the world's energy needs by 2030, and 25 per cent by 2050.[75]

Photovoltaics

A photovoltaic (PV) cell is a device that converts light into electric current using the photoelectric effect. Though the first solar cell was constructed in the 1880s, due to high costs, its use was restricted to powering spaceships and satellites till the 1960s. This changed in the early 1970s when prices reached levels that made PV generation competitive in remote areas without grid access. PV panels now started being used for off-grid purposes, powering homes in remote locations, cellular phone transmitters, road signs, water pumps, and millions of solar watches and calculators. These off-grid applications accounted for over half of worldwide installed capacity until 2004.

In recent times, due to growing demand for renewable energy sources together with financial subsidies, photovoltaic production has dramatically expanded. Solar PV power stations today have capacities ranging from 10-60 MW, and proposed solar PV power stations will have a capacity of 150 MW or more. Grid-connected solar photovoltaics are the world's fastest-growing energy technology: annual world solar PV installations were 5950 MW in 2008, a 110 per cent increase over 2007. At the end of 2009, cumulative global PV installations surpassed 21,000 MW. Roughly 90 per cent of this generating capacity consists of grid-tied electrical systems. Germany was the world leader in 2009, installing 3800 MW of solar PV in that year.[76]

For solar PV energy to become a dominant source of electricity worldwide, solar costs must become competitive with grid electricity from conventional sources. At present, solar PV (around 30 cents/kWh in the sunniest locations) is a long way from competing with conventional power generation costs (3-5 cents/kWh). But the advantage with solar PV is that decentralised generation is possible with it, meaning the energy source can be located at the consumer's premises, thereby eliminating the transmission and distribution costs. In that case, the solar PV cost needs to be compared with the electricity tariff being paid by the consumer (around 20 cents/kWh), and not the generation cost of conventional electricity (all cost figures are for the US).[77] This gap is not much. Considering the trend of falling solar PV costs over the last many years, solar PV costs (without subsidies) are expected to become equal to or cheaper than grid electricity costs in the sunnier parts of the US, Japan and Southern Europe by 2015. In the more temperate parts of Europe, grid parity is expected to happen around 2020. Grid parity without subsidies is already a reality in parts of California.[78]

Global solar PV generation is therefore all set to surge in the coming years. The US solar PV industry aims to provide half of all new US electricity generation by 2025.[79] Greenpeace and European Photovoltaic Industry Association estimate that by the year 2030, PV systems could be generating approximately 1,864 GW of electricity around the world, or 14 per cent of the global demand.[80]

Conclusion: Renewable Energies Poised for a Leap

Given this huge potential, and with costs poised to fall sharply in the coming years due to technology improvements and economies of scale, it is obvious that it should be possible to meet a substantial part of the global future energy needs by harnessing renewable energy sources. Total global new renewable power capacity (i.e. excluding large hydropower) increased by a healthy 16 per cent over the previous year to reach 280 GW at the end of 2008—of which wind power was 121 GW, worldwide grid and off-grid solar photovoltaic capacity had increased to 16 GW, small hydropower had gone up to 85 GW, biomass power capacity was about 52 GW, and geothermal power capacity had reached over 10 GW. The top four countries in order of installed capacity were China (76 GW), United States (40 GW), Germany (34 GW) and Spain (22 GW); India occupied the fifth position with 13 GW.[81]

Even more significantly, in 2008, for the first time, added power capacity from new renewables in both the United States and the European Union exceeded added power capacity from conventional power (including gas, coal, oil, large hydropower and nuclear).[82]

In March 2007, European leaders signed a binding EU-wide target to source 20 per cent of their energy needs from renewables, including biomass, hydro, wind and solar power, by 2020.[83]

The share of renewable energy in global electricity generation is set to rapidly increase in the coming years. The Energy Watch Group of Germany estimates that 29 per cent of the world's electricity and heat requirements could come from renewables by 2030. The International Energy Agency did a somersault in 2008, and reversing its earlier stand of marginalising renewables, stated that by 2050, if governments support the development of renewables by appropriate policies and incentives, 50 per cent of global electricity supply could come from renewable energy sources.[84]

A report prepared by the European Renewable Energy Council (EREC) and Greenpeace in October 2008 titled *Energy [R]Evolution:*

A Sustainable Global Energy Outlook is even more optimistic. According to this report, if energy efficiency measures are implemented to reduce consumption of electricity, then, new renewable energies (wind, solar, geothermal, ocean and biomass)[85] could provide around 62 per cent of the global electricity generation by 2050! (*We would like to add that along with energy efficiency measures, controls would have to be imposed on energy consumption too, in particular, luxurious consumption of electricity, in order to realise this target, or else, due to the Jevons Paradox discussed earlier, improvement in energy efficiency is only going to fuel a rise in energy consumption.*) The report projects that the global installed capacity of new renewable energy technologies (excluding both small and large hydro) has the potential to grow from 89 GW in 2003 to 5878 GW in 2050, a 66-fold increase in 47 years!!![86]

The Greenpeace-EREC study shows that it is possible to completely phase out generation of electricity from dirty and dangerous nuclear energy all over the world by 2050, reduce worldwide carbon dioxide emissions by 50 per cent below 1990 levels by 2050, and yet meet the global energy needs needed for growth!

Just as this book was going to print, we came across the news that in 2010, for the first time, worldwide cumulated installed capacity of wind turbines (193 GW), small hydro (80 GW, excluding large hydro), biomass and waste-to-energy plants (65 GW) and solar power (43 GW) reached 381 GW, outpacing the installed nuclear capacity of 375 GW prior to the Fukushima disaster. The world is actually moving towards phasing out nuclear electricity and replacing it with renewable energy![87]

2. Renewable Energy Potential in India

India is the only country in the world to have an exclusive ministry for renewable energy development, the Ministry of New and Renewable Energy (MNRE). The website of this Ministry gives a large amount of statistics about the vast renewable energy potential in India. Here is a brief summary:

(i) Solar Energy

Being a tropical country, this is the renewable energy source with the most potential for India. India receives solar energy equivalent to over

5,000 trillion kWh per year. The daily average solar energy incident varies from 4 to 7 kWh per square meter depending upon the location.[88]

According to government figures, more than 35 grid connected solar photovoltaic power plants with a total capacity of 10.28 MW had been installed in the country by the end of March 2010. In addition, total capacity of stand-alone solar PV power plants in rural and other areas to provide power for electrification and running electrical equipments had gone up to 2.46 MWp[89] by the end of financial year 2009-10. The MNRE website also claims that as of March 31, 2010, a total of 5,83,429 PV home lighting systems, 88,297 street lighting systems, 7,92,285 solar lanterns, 7,334 PV pumps and solar water heaters of total collector area of 3.53 million sq. m. had been installed in the country.[90] In January 2010, the government of India announced an ambitious 'Jawaharlal Nehru National Solar Mission' with the aim of generating 20,000 MW of solar energy by 2022; of this, in the first phase, 1300 MW of power is to be added by 2013.[91]

With costs of solar energy systems falling sharply the world over, and given the huge potential of generating usable energy from the sun in India, if the government seriously pursues this option, it should be possible for solar energy to provide an increasing proportion of electricity generation in the country in the coming years.

Note: In July 2011, came the news that solar energy in India has already become cheaper than imported nuclear power. As a part of the first phase of government of India's JN National Solar Mission, seven solar thermal power projects, of total capacity of 470 MW, are being set up by private companies, at an approximate cost of Rs. 12 crores per MW. By early July 2011, these projects had been able to tie up with commercial banks for financing their projects and had informed the government that financial closure had been achieved. This cost of Rs. 12 cr/MW is much lower than the installation cost of the Jaitapur NPP, which is estimated to be at least Rs.21 cr/MW. And the enriched uranium for these EPRs won't come free either—unlike the sunlight.[92]

(ii) Wind Energy

India is now the fifth largest wind power producer in the world, after USA, Germany, Spain and China. As of March 31, 2011, total wind power installed capacity in the country was 14,158 MW. This is slightly less than one-third of the total wind power potential of 48,500 MW as estimated by the MNRE.[93]

The actual wind power potential in India may be many times the official estimate, according to many experts. A new report *Indian Wind Energy Outlook 2009* released in September 2009 in New Delhi by the Global Wind Energy Council says that technological improvements and tapping India's vast offshore potential could result in total installed wind power capacity of 231 GW and power production of 579 TWh by 2030![94] That is huge!

(iii) Small Hydropower

The government of India categorises hydropower projects of up to 25 MW capacity as Small Hydro Power (SHP) Projects, and their responsibility has been vested with the MNRE. These do not have any of the disadvantages of large hydropower plants that have been discussed earlier in this chapter. On the contrary, they are one of the most environmental friendly and cheap ways of providing electricity to remote villages, especially in hilly areas, where providing grid electricity is very uneconomical.

The MNRE estimates the total small hydropower potential of the country to be 15,000 MW. As of December 2009, a total of 700 small hydropower projects aggregating 2,558 MW had been set up in various parts of the country—which is about 17 per cent of the total potential. (In addition, 296 projects of about 936 MW are in various stages of implementation.) The MNRE has set a target of harnessing at least 50 per cent of the SHP potential in the next 10 years.[95]

(iv) Biomass

Biomass, that is, plant and animal waste, is the oldest source of renewable energy known to humans, used since our ancestors learned the secret of fire. A large variety of biomass materials have been used successfully for power generation, including bagasse, rice husk, straw, cotton stalk, coconut shells, soya husk, de-oiled cakes, coffee waste, jute wastes, groundnut shells, saw dust, et cetera; and a number of technologies are available for generating grid quality power from these resources.

The MNRE estimates the surplus biomass availability in the country at about 120-150 million metric tons per annum (covering only agricultural and forestry residues), from which about 16,000 MW of power can be generated. In addition, by modernising sugar mills, about 5,000 MW power could be generated in the country's 550 sugar mills through bagasse-based cogeneration. Thus, the total estimated biomass power potential in the country is about 21,000 MW.

According to the MNRE, as of December 2009, 829 MW of biomass power projects and 1307 MW of bagasse cogeneration projects had been installed in the country for feeding power to the grid.[96]

Biogas

The country also has a huge potential of setting up biogas plants in the rural areas to produce biogas from organic materials like cattle dung. Biogas can be used for providing cooking fuel, for lighting gas lamps and for operating dual fuel engines. According to the MNRE, a potential of setting up 12 million biogas plants exists in the country, which can generate an estimated 17,340 million cubic meters of biogas, apart from providing high quality organic manure. So far, 4.12 million family type biogas plants have been set up.[97]

Energy from Urban Waste: Toxic Energy

Another form of biomass is the rising pile of garbage in India's cities as a result of increasing urbanisation. This has become a big environmental hazard and its disposal a major headache for city corporations. And so, the MNRE is promoting it as another source of non-conventional energy.

The problem is that because the waste is chemically complex, all waste incineration to energy systems (including waste pelletisation, pyrolysis and gasification systems) release toxic emissions into the environment that are detrimental to both human health and the environment. Even new incinerators release toxic metals, dioxins and acid gases. Dioxins are lethal Persistent Organic Pollutants (POPs) that cause irreparable environmental and health damage, including cancer. Far from eliminating the problem of landfills, waste incinerator systems produce toxic ash and other residues. This toxic ash subsequently enters the foodchain.

These waste-to-energy projects being promoted by the MNRE also violate international environmental norms. They violate the Kyoto Protocol, which regards waste incineration as a greenhouse gas emitter; they also violate the Dhaka Declaration on Waste Management adopted by the South Asian Association for Regional Cooperation (SAARC) countries in October 2004, which prohibits member countries from opting for incineration and other unproven technologies.

Therefore, the MNRE must exclude burn-technology-based waste-to-energy programs from qualifying as renewable energy sources and stop promoting it and giving subsidies to it. Instead, appropriate methods such as small scale bio-methanation, composting and proper recycling should be propagated.[98]

(v) Ocean Energy

The ocean can produce two types of energy: mechanical energy from tides and waves (known as tidal energy and wave energy respectively), and thermal energy from the sun's heat.

Tidal power converts the energy of tides into electricity or other useful forms of power. Tidal power is very site specific. For it to work well, there must be an increase of at least 16 feet between low tide to high tide. For India, the most attractive locations for tidal power are the Gulf of Cambay and the Gulf of Kachchh on the west coast, and the Sunderbans in West Bengal. Wave power uses the energy of ocean surface waves to do useful work, like electricity generation, water desalination, or the pumping of water into reservoirs. Ocean thermal energy refers to the solar energy trapped by the ocean. Because of this, different layers of ocean water have different temperatures, and this can be used to generate usable energy.

The theoretical potential of ocean energy is huge, several times greater than the global electricity demand. However, most of the technologies to extract usable energy from the ocean remain in the investigation or demonstration phase. Right now, there are very few ocean energy power plants operating commercially in the world, and those operating are fairly small. Among the best examples are the tidal energy plant on La Rance River in northern France, which can produce up to 240 MW per year, and provides enough energy to satisfy demands of 240,000 homes in France. Experimental wave farms are being developed or are in operation to collect energy from ocean waves in Scotland, England and Australia.

The identified economic tidal power potential in India is of the order of 8000-9000 MW; while India's 7500 km long coastline (including coastline of islands) has a wave energy potential of 40,000 MW. The potential of ocean thermal energy is even more. Of course, it is going to be many years before this potential can be realised, as much technological development still needs to be done.[99]

(vi) Geothermal Energy

Below the Earth's crust, there is a layer of hot and molten rock called magma. Heat is continually produced there, mostly from the decay of naturally radioactive materials such as uranium. The amount of heat within 10,000 meters of Earth's surface contains 50,000 times more energy than all the oil and natural gas resources in the world.

This heat energy, known as geothermal energy, can be tapped in many ways, from large and complex power stations to small and relatively simple pumping systems. Many regions of the world are already tapping geothermal energy as an affordable and sustainable solution to reducing dependence on fossil fuels.

The most common current way of capturing geothermal energy is to tap into naturally occurring 'hydrothermal convection' systems where cooler water seeps into Earth's crust, is heated up, and then rises to the surface. When heated water is forced to the surface, it is a relatively simple matter to capture that steam and use it to drive electric generators. Worldwide, about 10,715 MW of geothermal power is online in 24 countries. Geothermal plants produce 25 per cent or more of electricity in the Philippines, Iceland and El Salvador. The United States has more geothermal capacity than any other country, with more than 3,000 megawatts in eight states.

Geothermal springs can also be used directly for heating purposes. In Iceland, virtually every building in the country is heated with hot spring water. In fact, Iceland gets more than 50 per cent of its energy from geothermal sources.

A much more conventional way to tap geothermal energy is by using geothermal heat pumps, which take advantage of the constant year-round temperature of about 50°F that is just a few feet below the ground, to provide heat and cooling to buildings. Either air or antifreeze liquid is pumped through pipes that are buried underground, and re-circulated into the building. In the summer, the liquid moves heat from the building into the ground. In the winter, it does the opposite, providing pre-warmed air and water to the heating system of the building. In regions with temperature extremes, such as the northern United States in the winter and the southern United States in the summer, ground-source heat pumps are the most energy-efficient and environmentally clean heating and cooling system available. More than 600,000 ground-source heat pumps supply climate control in US homes and other buildings, with new installations occurring at a rate of about 60,000 per year.[100]

In India, the Geological Survey of India has identified 350 geothermal energy locations in the country. The most promising of these is in the Puga valley of Ladakh. The estimated potential for geothermal energy in India is about 10,000 MW. However, so far, the government has not taken this energy source seriously. Only in recent years have some private sector companies evinced an interest in tapping this energy. India's first fully-operational commercial geothermal power plant of an initial capacity of 25 MW is likely to come up in 2012 in the state of Andhra Pradesh.[101]

C. Adopting Decentralised Energy Systems

The current energy paradigm in India is to build large centralised power generation systems, mainly thermal plants (coal, gas), large dams, and now nuclear power plants as well. Inherent within such a generation system are very long transmission lines, a hugely complex distribution system, and a network of transformers to step up and step down the voltage of electricity being transmitted. Each of these adds to the complexity, reduces the efficiency, increases the electricity losses, and results in increased capital and operational costs. These factors make centralised generation systems based on large power plants an economical option only for concentrated loads.

Indian villages are spread over wide distances. Supplying them electricity from a centralised electricity generation system requires long transmission lines; this in turn implies huge transmission losses. Therefore, supplying electricity to villages from centralised generation systems is expensive. But at the same time, most villages do not have large populations, and development levels are low; so, they cannot provide the substantial loads that towns and cities can. This therefore means that the per unit cost of supplying electricity to India's far flung villages from centralised electricity generation systems is very high.

A very simple, efficient and cost-effective solution to this problem is making use of decentralised power generation systems (meaning electricity generated at or near the point of use), based on renewable sources of energy. These can be a mix of wind (especially wind mills in preference to wind turbines), micro hydel, solar and biomass, depending on the location and availability of local resources. Since a decentralised generation system is connected to a local distribution network, instead of a high voltage transmission system, the losses are very low. Even if the cost of electricity from this decentralised system is more than the generation cost of conventional grid electricity, because of the huge costs and losses involved in transmitting the latter to remote villages, for the rural consumer decentralised electricity would be cheaper than the real end cost of conventional electricity. Further, as we have seen above, costs of decentralised electricity from renewable sources are rapidly falling, while that of conventional electricity are bound to increase as global coal-gas-oil-uranium resources become more scarce. Renewable energy systems also produce less carbon emissions and have none of the environmental, health and social costs associated with large conventional power plants. Finally, the decentralised electricity supply system also has the benefit that the technology being simple, it empowers local people as they can easily control and manage it. They can then get electricity as per their requirements, instead of having to wait for hours for unreliable electricity from the grid which often comes at the most odd hours.

An Observation

The potential of renewable energy sources to meet a substantial part of the country's future energy needs, and within this, the potential of decentralised renewable energy sources to meet a large part of our rural energy needs, making it possible to phase out deathly nuclear energy, is well established. Many European countries which do not operate or are phasing out nuclear plants are going in for renewable and decentralised energy in a big way to meet their energy requirements.

Bewilderingly, despite being the only country in the world to have an exclusive ministry for renewable energy development (the Ministry of New and Renewable Energy or MNRE), the Integrated Energy Policy (2006) of the government of India does not take into consideration the huge scope of renewable energy sources in meeting a significant part of the future electricity generation needs of the country! On the contrary, it gives renewables only a 5 per cent share in total power generation by 2031-2, ignoring even the estimates made by the MNRE about the huge potential of renewable electricity generation in the country! It further goes on to say that their role would be marginal even up to 2050!![102]

PART IV: THE POTENTIAL OF THE ALTERNATE ENERGY PARADIGM

The above analysis shows that it is possible to solve India's energy crisis with an Alternate Energy Paradigm, whose basic elements are:

1. maximising energy efficiency, including efficiency of the electricity delivery system and end-use efficiency;
2. curbing growth of demand by imposing restrictions on luxurious consumption of electricity by the rich, and eliminating wasteful use of electricity;
3. making the maximum possible use of renewable energy sources; and
4. reducing load on the grid by promoting decentralised renewable energy supply systems.

Given the huge scope of improving energy efficiency in the country, if the government indeed implements the energy efficiency measures outlined in Part III-A-1 above, imposes restrictions on luxurious consumption of electricity, takes steps to eliminate wasteful consumption of electricity, and promotes the use of decentralised energy systems to meet the energy needs of India's far-flung villages—over half of which have still to be electrified sixty years after independence—then the additional grid electricity generation required for meeting our future growth needs is substantially reduced; in fact for a few years we may even be in surplus.

In that case, a major portion and possibly all our future electricity needs can then easily be met from renewable energy sources, whose potential in the country is huge as discussed earlier. To summarise the potential of grid connected renewable electricity generation in the country as estimated by the government:

- 48,500 MW of Wind Energy;
- 15,000 MW of Small Hydro Power;
- 21,000 MW of Biomass Energy; and
- at least 50,000 MW of Solar Energy.[103]

According to the World Institute of Sustainable Energy (WISE), the well-known not-for-profit non-governmental organisation that promotes sustainable energy, the actual grid connected renewable energy potential in India is much more than this:[104]

- Wind Energy – 100,000 MW;
- CSP-based power generation – 200,000 MW; and
- Solar PV-based power generation – 200,000 MW.

Therefore, if serious efforts are made to harness this massive renewable energy potential:

- there would be no need to set up the giant sized nuclear power plants—with all their deathly consequences—being planned by the government; and the operating nuclear reactors can also gradually be phased out;
- there would also be no need to set up large centralised coal and hydro-based power plants on the scale visualised by the government.

◆◆◆

11

UNITE, TO FIGHT THIS MADNESS!

PART I: WHY THIS MADNESS?

When such a cheap, clean, green and safe alternative energy paradigm is available, why are India's rulers indulging in this mindless spree of constructing giant foreign-supplied nuclear parks and indigenous nuclear plants? And not just nuclear power plants, but also ultra mega coal power plants and giant hydroelectric projects!

So That Foreign Corporations Can Party Through the Night...

It's obviously not for meeting the energy crisis of the country; as we have seen above, there are safer, environment-friendly and cheaper options to mitigate the energy crisis. The real reason is: to provide US, French, Russian and other foreign corporations, and apart from them, the big Indian business houses, a fantastic investment opportunity, so that they can make huge profits. This was in fact the real 'deal' behind the Indo-US Nuclear Deal: the US signed the Nuclear Deal in return for India agreeing to buy $150 billion worth of US nuclear reactors, equipment and materials.[1] And not just nuclear reactors, Indian Prime Minister Manmohan Singh's special envoy Shyam Saran also promised that US companies would "benefit for decades" from Indian orders for military hardware, ranging from fighter jets and aircraft carriers to anti-nuclear missile shields.[2]

The nuclear deal had thus nothing to do with India's growing strength, with the US recognising India's growing clout in the world, etcetera, etcetera. It was all about big business. This is why both US and Indian big corporations lobbied hard to get the deal approved by the US Congress. Ron Somers, the president of the US-India Business Council, put it very straightforwardly in July 2007: '[The US-India nuclear deal] will present a major opportunity for US and Indian companies.' He added that the deal would create up to 27,000 'high-quality' jobs per year over the next decade in the US nuclear industry. The Confederation of Indian Industries, a lobbying group of big Indian business houses, funded numerous trips to India for US congressional delegations. Modest estimates place the total cost at about $550,000.[3]

Even before the deal was finally approved by the US Congress in October 2008, several US multinational energy firms, including General Electric, Bechtel, Edlow International, Nukem, Thorium Power and Westinghouse, sent representatives to New Delhi for discussions on future contracts. (Westinghouse, although a subsidiary of Toshiba since 2006, is based in Pennsylvania.) WM Mining, a uranium mining firm, even negotiated an agreement with Nuclear Fuel Complex, Hyderabad, to supply 500 metric tons of uranium annually with an expectation of $1.3 million in profits.[4] And within months of the deal being signed, GE Hitachi Nuclear Energy and Westinghouse signed memorandum of understandings with NPCIL regarding deployment of their 1350 MW Advanced Boiling Water Reactor and AP-1000 reactor respectively.[5]

Similarly, the 45 member countries of the Nuclear Suppliers Group (NSG) also gave their approval to ending the embargo on nuclear trade with India on the promise of lucrative business opportunities. In early 2007, anticipating a change in NSG rules, Russia and India signed an agreement for Russia to supply four 1000 MW nuclear reactors to India, a deal potentially worth $10.35 billion.[6] All 27 EU countries are members of the NSG; to win their approval, India promised to begin construction of six EPRs designed by French and German utilities as soon as the NSG permitted it to do so.[7] Two days after the NSG gave its green signal for nuclear commerce with India, a British nuclear power industries delegation arrived in India to explore the market (on September 8, 2008).[8] The same month, France signed a bilateral nuclear cooperation agreement with India allowing for the sale of French reactors, as well as other civilian nuclear material. The agreement was ratified in January 2010,[9] and the framework agreement for the supply of the first two reactors was signed between the two countries in December 2010. Though the final price has not yet been announced, this deal should be worth at least $14 billion;[10] and France is to supply four more such reactors.

Not to be left behind, in November 2009, after a break of three decades, Canada also signed an agreement with India paving the way for Canadian firms to supply nuclear equipment to India, including the ACR-1000 reactor.[11]

Indian Big Business Joins the Tango ...

India's big business houses are expecting to get subcontracts from these foreign corporations worth thousands of crores of rupees. Speaking to reporters after news came in of NSG granting permission for nuclear trade with India, Venugopal Dhoot, head of the consumer goods-to-energy firm Videocon Group, stated that it was a huge business opportunity for India and that some 40 companies were negotiating nuclear power joint ventures with foreign firms.[12] Amongst the companies that have announced plans to enter this sector are Tata Power, Reliance Power, JSW, GMR and Lanco.[13] Many Indian companies have already concluded agreements with foreign nuclear corporations. In 2009, the Mumbai-based Indian conglomerate Larsen & Toubro (L&T) signed four agreements with foreign nuclear reactor vendors. In January 2009, it signed an agreement with Westinghouse Electric Company to produce component modules for its AP-1000 reactors. The second agreement was with Atomic Energy of Canada Ltd. (AECL) 'to develop a competitive cost/scope model for the ACR-1000'. In April, it signed an agreement with Russia-based Atomstroyexport primarily focused on manufacturing components for the next four VVER reactors at Kudankulam, but extending beyond that to other Russian VVER plants in India and internationally. Then, in May 2009, it signed an agreement with GE Hitachi to produce major components for its ABWRs, including the supply of reactor equipment and systems, valves, electrical and instrumentation products. Early in 2010, L&T signed an agreement with Rolls Royce to jointly make components and provide services for light water reactors in India and internationally.[14]

Similarly, Hindustan Construction Company (HCC), which has done civil work for eleven of India's nineteen existing reactors, has tied up with the UK-based engineering and project management company AMEC Plc to provide design consultancy and engineering services for nuclear power plants in India. It has already completed civil work for the Russians at the Kudankulam Nuclear Plant.[15] Meanwhile, Areva has announced that it will be looking at Tata Engineering, L&T and Bharat Forge to provide nuclear components for the Jaitapur reactors.[16] In January 2009, even before it signed the formal agreement with the Indian government for supply of its EPR reactors, Areva signed an agreement with Bharat Forge, India's biggest forging company, to build a manufacturing facility for heavy forgings in India by 2012.[17]

Presently, nuclear power production in India is under government control and only NPCIL can set up and run nuclear power plants. However, the law is very likely to be amended soon to allow the entry of private players. The *Economic Survey* of 2008-09 has already aired the view that the Atomic Energy Act needs to be amended to permit private corporate investment in nuclear power.[18] In anticipation, big Indian conglomerates like Reliance, Tata, GMR and Essar have begun preparations to set up and own nuclear plants.[19]

In the mad world of capitalism, profit is all that matters, even if it means afflicting and killing tens of thousands of people with the most terrible diseases for centuries. In Charlie Chaplin's epic black comedy film *Monsieur Verdoux* made in 1947, Henri Verdoux is accused of making a business out of robbing and killing unsuspecting women. Verdoux, in his reply, says: 'As for being a mass killer, does not the world encourage it? Is it not building weapons of destruction for the sole purpose of mass killing? Has it not blown unsuspecting women and little children to pieces? And done it very scientifically? As a mass killer, I am an amateur by comparison.' Chaplin in this film was referring to the weapons of mass destruction used by both sides during the Second World War. Nuclear reactors are even bigger weapons of mass destruction than the biggest bombs used during the Second World War!

Part II: Nuclear Madness: Part of Globalisation

For their narrow interests of profits and commissions, India's rulers are willing to condemn people living in the area around nuclear plants to suffer from all kinds of terrible diseases and give birth to deformed children for thousands of years! Their greed has made them so shortsighted that they are willing to even risk a nuclear accident – which can render huge areas of the country uninhabitable for centuries!!

Why is the Indian government mortgaging the interests of the people of the country to benefit big foreign and Indian corporations? It has actually been happening for the last two decades, since 1991 to be more precise, when under World Bank-IMF pressure, the government of India decided to restructure the Indian economy. The Indian economy was trapped in an external debt crisis. Taking advantage of this, India's foreign creditors, that is, the USA and other developed countries—also known as the **imperialist countries** because they are seeking to re-establish their control over the economies of the third world countries that they had once colonised—through the World Bank and the IMF (which are controlled by them), arm-twisted the Indian government into agreeing to this restructuring.[20] The basic elements of this so-called 'Structural Adjustment Program' were:

(i) opening up the economy to unrestricted inflows of foreign capital and imports;
(ii) privatisation of the public sector, including welfare services; and
(iii) removal of all controls placed on profiteering, even in essential services like drinking water, food, education and health.

This restructuring of the Indian economy at the behest of India's foreign creditors has been given the high-sounding name of 'globalisation'. Since then, governments at the Centre and the states have continued to change, but globalisation of the economy has continued unabated.

Unbridled Corporate Plunder

The essence of globalisation is that the Indian government is now running the economy solely for maximising the profits of giant foreign corporations—also known as Multinational Corporations (MNCs)—and India's big business houses. These corporations are on a no-holds barred looting spree. They are plundering mountains, rivers and forests for their immense natural wealth. They are seizing control of public sector corporations, including public sector banks and insurance companies, created through the sweat and toil of the common people, at throwaway prices. Privatisation is also enabling them to enter essential services—including education, health, electricity, transport, even drinking water—and transform these into instruments of naked profiteering. Because these are essential services, the profits are huge.

In this plunder, India's business houses can only be junior partners of the foreign MNCs, as the latter are gigantic: just 200 MNCs control a quarter of the global economic activity;[21] of the world's 100 largest economies, 51 are MNCs, and 49 are countries.[22] But that is all right with them, as their profits too are increasing. They are not anymore bothered about who is controlling the Indian economy, but only about filling their coffers.

Hoarders and blackmarketeers are having a field day, as laws controlling their activities have been relaxed in the name of freeing the markets. The speculators are ecstatic; they have never had it so good. The swanky upper middle classes are also in raptures over globalisation; the world's most trendy consumer brands are now available in the country.

In sum, globalisation has become the consensus policy of India's elites. And since it is the elites, especially the big business houses, who finance and thereby control the political parties, all the major parties, irrespective of their colour, are implementing the economic 'reforms' wherever they are in power (in the Centre and the states). The top Indian intellectuals and the media—faithful servants of the capitalist classes—have launched a massive propaganda offensive to convince the Indian people about the benefits of globalisation.

The government of India has given up all concern about the future of the country, about conserving the environment for our future generations, about the livelihoods of the people of the country, about making available essentials like food, water, health and education to the people at affordable rates so that they can live like human beings and develop their abilities to the fullest extent. It is now only concerned about how to provide new and profitable investment opportunities for foreign multinational corporations and their Indian collaborators. *The invitation to foreign nuclear power corporations to set up giant nuclear parks in the country is just another of these policies, though it is undoubtedly amongst the most disastrous with consequences that will plague us for thousands of years.*

Part III: India on SALE

Let us take a brief look at the kind of policies being implemented by the government of India to enable MNCs and Indian corporate houses to earn multi-billion dollar profits.[23]

Robber Capitalism

The tribal districts of the states of Orissa, Jharkhand and Chhattisgarh are home to much of the country's forests and minerals. With the naked connivance of the politicians, police, bureaucracy and the courts, corporations have unleashed a fascist reign of terror on the people of these states, in order to drive them out and seize control of their lands, cut down the forests, and set up mining projects, huge steel, iron and aluminium plants, and ultra mega thermal power plants. Real estate speculators are also participating in this huge land grab, to set up IT parks, golf courses, five star hotels and ultra luxurious residential complexes.

Urban Real Estate Loot

The country's metropolises have the potential of attracting billions of dollars of investments in hotels, airports, malls, stadiums, metro rails, flyovers and other urban infrastructure. But for that, land is needed. Where does that come from? The palatial houses of the rich cannot be touched. The only alternative is to evict the poor from the slums, which occupy a considerable portion of the land in the cities. So they are being bulldozed out.

Great Land Grab

In one of the greatest land grabs in modern Indian history, hundreds of thousands of hectares of agricultural land is being transferred to private industry to set up Special Economic Zones (SEZs). Investors in the SEZs are being given the most amazing concessions: no import duties, no controls on imports and profit repatriations, 100 per cent tax holiday for 5-10 years, and what not. Labour laws and environmental laws will not be applicable to these zones. The Development Commissioner of the SEZ will function like a virtual dictator of the area—Indian democracy will end at the border of these zones! The government has even declared that these areas will be considered as foreign territories for the purpose of trade operations, duties and tariffs —the foreigners now no longer have to come with arms to win trade concessions!!

Global Garbage Dump

In its feverish desire to promote foreign investment, the government is allowing the country to be transformed into a toxic and garbage waste dump of the developed countries. Since e-waste recycling is a very hazardous industry which contaminates the soil and groundwater, and causes severe health problems to the workers, the government is allowing developed countries to ship their waste to India—to the extent that over 70 per cent of the electronic waste generated in the developed world is now coming to India. India is also the toxic waste dump of world shipping: toxic ships from around the world, contaminated with thousands of tons of deadly chemicals, are brought to Alang in Gujarat, the world's largest ship-breaking yard, to be broken up. As if this was not enough, the country is also becoming the household waste dump of the developed countries, recycling rubbish in the developed countries is a complicated and costly affair as it is environmentally very polluting, and they find it cheaper to ship it to India where regulations are practically non-existent.

1000 More Bhopals

In the name of development, foreign and Indian corporations are being allowed to set up the most polluting industries in India. Even industries banned in the West are being allowed to operate in India: asbestos, banned in the European Union and USA, is a Rs.2000 crore industry in India, with annual consumption of over 125,000 million metric tons a year; endosulphan, banned in 62 countries as it causes appalling birth deformities, continues to be used as a pesticide in India; the list of such chemicals is very long. The result is that India is home to some of the world's 'top toxic hotspots': the Eloor industrial estate near Cochin, Kerala, which has polluted the Periyar River and nearby villages with persistent organic and inorganic compounds and heavy metals; Sukinda Valley in Orissa, where one-fourth of the people suffer from pollution-induced diseases; the chemical industry belt between Vapi and Ahmedabad in Gujarat, where mercury levels in groundwater are 96 times higher than safety levels ... A survey by India's Central Pollution Control Board found that groundwater was unfit for drinking in every one of the 22 major industrial zones surveyed by it!

GM foods—Self-destroying Food Security

The US agribusiness corporation Monsanto has been desperately trying to push the government of India to permit the growing of Bt Brinjal, a genetically modified (GM) variety of Brinjal developed by it, in the country. Once Bt Brinjal is approved, it will pave the way for introduction of other GM food crops too. GM seeds have to be purchased from the seed company for each sowing. Therefore, as their use spreads in the country, Monsanto which has patented these seeds will be able to charge monopoly prices for them and earn stupendous profits! It will also result in spread of monoculture. With the control of seeds passing from farmers into the hands of foreign agribusiness corporations, it will destroy our food security. There is extensive evidence demonstrating that GM crops and foods have adverse impacts on human and animal health, and also on the environment. Further, once released into the environment, GM seeds can never be recalled, because seeds have a life of their own, and propagate themselves in uncontrollable ways. Because of these potentially very dangerous effects, more than 180 countries in the world have banned the growing of GM foods. Nevertheless, the Government of India is keen to grant approval to Bt Brinjal, without rigorously ascertaining by scientific tests as to whether it is safe or not! It nearly succeeded last year, but massive protests all over the country forced the Environment Minister to announce a moratorium on its introduction in February 2010. Now, to get around this suspension, the government is planning to move a new bill in Parliament which will enable it to fast track approval to the growing of GM food crops in the country! There is no limit to the betrayal by India's rulers.

Health, Education, Drinking Water ... all on SALE

The government is gradually withdrawing from providing welfare services to the poor at subsidised rates; they are being privatised and handed over to private corporations for their unbridled loot. Government hospitals and municipal schools are being privatised; medicine prices have zoomed; college fees have gone through the roof; electricity prices are rising; bus fares are rising; the public distribution system designed to check speculation in prices of foodgrains is being eliminated; and now, drinking water supply in cities is also being handed over to these corporations, who will then hike its price by 10-15 times.

Its Consequences

Obscene Inequality

Globalisation has led to a sharp increase in polarisation in the country. The rich have grown enormously richer: in 2010, the number of dollar billionaires increased by 17, driving the total to a record 69;[24] those with disposable assets of over $1 million (Rs.4.6 crores) increased by 51 per cent over the previous year to an estimated 1.27 lakh during 2009.[25] And so they have declared: 'India is shining', India is becoming an 'economic superpower'; and now after the 1,2,3 Nuclear Deal, India is on its way to becoming a 'nuclear superpower' too.

On the other hand, globalisation has also led to crores of people being pushed below India's already shamefully low poverty line. This is evident from a host of government data. The latest official survey carried out in 2004-5 shows that an appalling 87 per cent of the rural population are unable to access the minimum recommended 2400 calories per day; the corresponding percentage for urban India, where the nutrition norm is lower at 2100 calories, was 64.5.[26] Another official report submitted to the Prime Minister says that an overwhelming majority of the population—77 per cent, or 83.6 crore people—are living on Rs.20 or less a day.[27] According to the National Family Health Survey of 2005-6, nearly half (46 per cent) of India's children under the age of three are underweight, an indicator of malnourishment.[28] Then how come the government is claiming that poverty in the country has reduced after globalisation? It has managed this by a simple trick of lowering the country's already low poverty line![29]

Heading Into a Financial Collapse

As the country's ruling classes ruthlessly push ahead with economic reforms, the economy is getting more and more entrapped in a financial crisis. In fact, the country's present external financial situation is worse than the foreign exchange crisis of 1991 that pushed the government into globalising the economy! Discussing this in detail is beyond the scope of this book, here we restrict ourselves to giving a few statistics:

- The external debt of the country has gone up to $273 billion as of end-June 2010,[30] which is more than three times the debt of $83.8 billion at the end of March 1991![31]
- The trade deficit zoomed to $117 billion in 2009-10,[32] which is more than 40 times the deficit in 1991-92 of $2.8 billion!![33]
- The current account deficit rose sharply in the quarter ending June 2010 to $13.7 billion, from $4.5 billion a year ago. Even the Finance Minister Pranab Mukherjee has admitted that it is worrying!!![34]

This deepening crisis has made the Indian economy more and more dependent on inflows of foreign capital—both Foreign Direct Investment (FDI) as well as speculative capital investments in the stock markets euphemistically known as Foreign Institutional Investor (FII) capital—to keep it afloat. If the foreign investors stop investing, and start withdrawing their speculative investments in the stock markets, the economy will suffer a recurrence of the financial collapse of 1990-91. This has pushed the country firmly into the clutches of foreign corporations and speculators and their governments; they can now impose any conditions they want, and the Indian government has no option but to obey.

Worsening Environmental Crisis

With the Indian government desperate to attract FDI inflows, it is willing to allow foreign investors to set up the most environmentally destructive projects. The result is: the country is heading into an ecological catastrophe. We have extensively discussed the terrible environmental destruction caused by nuclear plants in this book. But apart from that, the numerous other projects being implemented in the country are also going to cause enormous environmental ruin, the signs of which are already very evident: the increasing contamination of air and surface waters with industrial pollutants, the contamination of fresh water fish and ocean fish with mercury and numerous industrial organic chemicals, the pollution of groundwater with pesticides and other contaminants of chemical intensive agriculture, the accelerating depletion of groundwater levels all over the country, the extensive destruction of forests, widespread soil degradation which is threatening agricultural productivity in large areas in the country, et cetera.[35] (These problems are actually a part of the growing ecological crisis gripping planet Earth itself, resulting from similar policies being implemented by most countries in the world. But we confine our discussion to India, because that is where we can act to bring about a change.) As these problems worsen, they are going to affect survival of life itself—and the timetable is shorter than earlier thought!

The fundamental reason for this growing ecological disaster is the inherent logic of capital which is playing itself out as capitalist globalisation advances. For capitalists, whether they be foreign or Indian, environment is not a place with inherent boundaries where human beings live together in harmony with other species and which is to be conserved for future generations. Capitalists are not bothered about the future, they only live in the present. For them, the environment is a just another realm to be exploited as they go about seeking to maximise their profits. If, in this process, the environment is destroyed, so be it.

Part IV: Unite and Fight, to Save the Future

The solution to this growing economic and environmental crisis is not to replace one political party with another, for they are all the same, all are lackeys of the country's elites, all are in agreement on the issue of globalisation. We, the ordinary people, will have to overcome our fears, take the initiative, unite, come out on the streets, and assert that we are not going to be passive spectators anymore to this unbridled loot of our country.

People Are Beginning to Stir ...

People all over the country have already begun waging heroic struggles against the destructive projects being implemented in the country. The tribals and small farmers of Orissa, Chhattisgarh and Jharkhand are waging heroic struggles against giant corporations like Vedanta, POSCO (South Korea, actually USA) and the Tatas who are seeking to take over their lands and destroy their livelihoods. The people of Goa got together to force the state government to cancel all the SEZs in the state. The people of 22 affected villages of Maha Mumbai SEZ which was going to be set up by Reliance in Raigad district of Maharashtra launched a determined struggle, refusing to part with their lands, forcing the government to ultimately cancel the project and de-notify the 16,000 acres of land earmarked for the project. The Warkaris of Western Maharashtra rallied in thousands to force the state government to cancel permission given to the notorious Dow Chemical Company to build a dangerous research centre near Pune. The people of Uttarakhand have launched an intense agitation against the mad scheme of the state government to build bumper-to-bumper dams all along the course of the Ganga River, forcing the government to cancel many of these devastating projects. In Gujarat, thousands of farmers from the Mahuva area in the Bhavnagar district of Gujarat have repeatedly courted arrest to protest the government's sanction for a cement factory and limestone quarry in their area, which would destroy a huge waterbody that is the lifeline of many villages. In Andhra Pradesh, people fought hard against plans to build a 1000 km long coastal industrial corridor by acquiring 50 lakh acres of land displacing about 2 crore people, eventually forcing the state government to scrap the project; now, they are organising against another such coastal corridor project in the Vishakhapatnam and East Godavari districts! Farmers' organisations of 14 states fought a fierce struggle to force the Environment Minister to impose a moratorium on the introduction of Bt Brinjal in the country. Amongst the most inspiring of these struggles has been the heroic struggle of the people of Nandigram, who battled the entire might of West Bengal police and the goondaism of the ruling party and eventually forced the state government to back off from setting up a chemical SEZ on their fertile lands.

And of course, as we have mentioned in the Introduction to this book, powerful struggles by people of Meghalaya and Nalgonda (Andhra Pradesh) have forced the respective state governments to put on hold proposals to start uranium mining in these areas. Local people are also waging fantastic struggles against DAE plans to build nuclear power plants in Jaitapur (Maharashtra), Haripur (West Bengal), Gorakhpur (Haryana), Mithivirdi (Gujarat) and Kudankulam (Tamil Nadu).

Yes, people are beginning to stir all across the country ...

Lot More Needs to be Done

However, these struggles have not prevented the ruling classes from going ahead with their sordid agenda. For every struggle won and project cancelled, the ruling classes have been able to implement ten other projects. Often, if a project is cancelled due to strong protests by the local people, it is simply shifted to another region. Therefore, at the most, these struggles have slowed down the pace of globalisation of the Indian economy. We need to do a lot more.

Advance, to Build a New World

We need to involve more people in our struggles. We need to unite our different struggles. We need to deepen our struggles, and advance from opposing this or that project, to challenging the entire project of capitalist globalisation being implemented by the ruling classes of the country at the behest of the imperialists.

The ruling classes are claiming: 'TINA'—there is no alternative to globalisation! In order to see through their propaganda, in order to be able to understand their real agenda, we must read, think, analyse. We must develop our political consciousness. Only then will we develop the realisation that actually, **there is no alternative to fighting globalisation**!

The juggernaut of capitalist globalisation is threatening the very existence of life (not just in India, but on planet Earth itself)! We need to take our struggles beyond the boundaries of globalisation, and advance them to fighting for building a new society whose basic logic does not promote individual selfishness and aggrandisement but cooperation and collective well-being, where production is oriented not for the profit maximisation of a few but for fulfilling the basic needs of all human beings—healthy food, best possible health care, invigorating education, decent shelter, security in old age, clean pollution-free environment. Only such a society will implement the alternate energy paradigm discussed in the previous chapter, which is oriented towards meeting the energy needs of the poorest sections of the society in the cheapest and most environment friendly way.

It is possible to build a new world! The people of Madban (Jaitapur), Gorakhpur, Mithivirdi, Kudankulam, Nandigram, Mahuva, Niyamgiri, Kashipur ... are showing the way with their inspiring struggles, braving police *lathi*-charges, false cases, arrests. Let us DARE to support them, and take our own initiatives!!

Epilogue

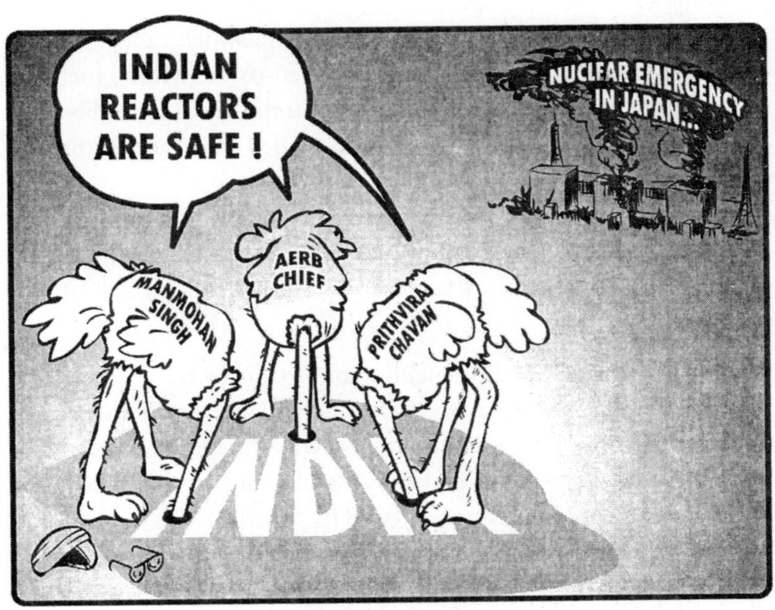

THE FUKUSHIMA CATASTROPHE IN JAPAN

The Fukushima Dai-ichi Nuclear Power Plant (also known as Fukushima One NPP—in Japanese, Dai-ichi means 'number one'), constructed and run by the Tokyo Electric Power Company (TEPCO), is located on a 3.5 square kilometer site in the Fukushima prefecture (a prefecture is somewhat similar to a state in India) of Japan. The plant has six Boiling Water Reactors. The oldest, Unit 1, of 460 MW, was connected to the grid in 1971; it was initially scheduled for shutdown in early 2011, but in February 2011, Japanese regulators granted it a lifetime extension of ten years. Units 2 and 3, both of 784 MW, began commercial operation in 1974 and 1976 respectively, while Units 4 and 5, also of 784 MW each, came online in 1978. Unit 6, of 1100 MW, was connected to the grid in 1979. These six reactors, with a combined installed capacity of 4700 MW, made the plant one of the 15 largest NPPs in the world.[1]

All reactors use low enriched uranium as fuel, except Unit 3, which also contains a small amount of mixed-oxide (MOX) fuel, that is, fuel made from both uranium and plutonium oxides rather than just uranium. This MOX fuel was first loaded into the reactor only recently, in September 2010.

Reactors 1 to 5 are of Mark 1 design, while Reactor 6 is of Mark 2 design. Like in all reactors, in these reactors too, there are four levels of shielding to prevent the fission products from leaking out into the environment:

1. The thick Zircaloy shielding of the fuel rods;
2. The reactor pressure vessel, generally made of 6 inch steel;
3. The primary containment, generally made of 1 inch thick steel and surrounded by a steel-reinforced, pre-stressed concrete 1.2–2.4 meters (4–8 ft) thick;
4. The secondary containment, the reactor building, also made of steel-reinforced, pre-stressed concrete 0.3 m to 1 m thick.

A special feature of these designs is that the spent fuel is stored within the reactor building in a swimming pool like concrete structure

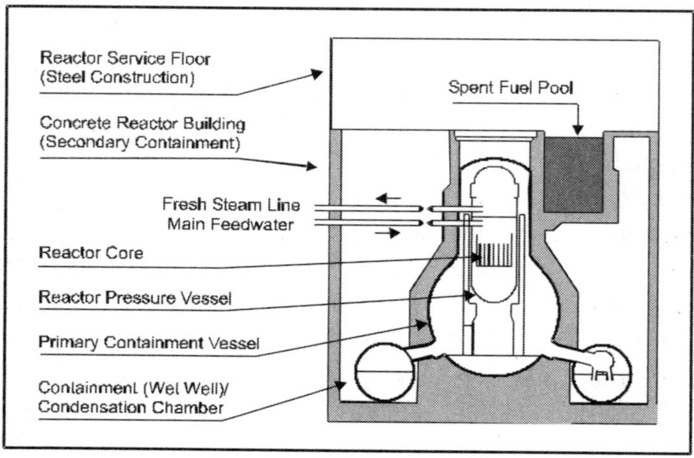

Figure: Fukushima Reactor

near the top of the reactor vessel. The pool water is cooled continuously to remove the heat from the spent fuel.

Part I: The Beginnings of the Accident

On March 11, 2011, a massive earthquake measuring 9.0 on the Richter scale struck Japan. The Japanese had planned well, and the Fukushima Dai-ichi reactors were shut down by sensors the moment the earthquake struck.

However, nuclear technology is a uniquely hazardous technology. In all other industries, however hazardous the industry might be, the moment the plant shuts down, the risk virtually ends. Barring a natural disaster, or an act of sabotage or terrorism, its toxins will remain contained and isolated from the environment. They will not on their own and spontaneously begin to cause havoc. However, in the nuclear industry, even after the reactor is shut down by bringing the fission reaction to a stop by inserting control rods into the reactor fuel, the danger of things going out of control does not end. Within the reactor core, the deadly radionuclides created during the fission reaction continue to give off heat. Therefore, even after the reactor has been shut down, the reactor needs to be continuously cooled for a very long time. If this cooling is disrupted for some reason, then the reactor temperature starts to rise. If it rises significantly, it could lead to a melting of the zirconium tube or 'cladding' that encases the uranium fuel pellets, leading to leakage of radioactivity from the nuclear fuel and its possible dispersal into the environment.

This is precisely what happened at Fukushima, the cooling got disrupted.

What Happened at Fukushima

The Fukushima plant had multiple backup features to prevent the cooling from getting disrupted and thus prevent what is called a loss-of-coolant accident. In case the power supply failed for some reason like an earthquake, they had a backup supply from diesel generators. If these failed, then there was also an eight-hour battery supply. To meet the danger of a tsunami, they had built a six metre high wall around the plant. However, on March 11, 2011, all their planning failed.

The Earthquake

For nearly two months after the accident, TEPCO claimed that its reactors had successfully withstood the massive earthquake of magnitude 9; it was only because the earthquake triggered a massive tsunami barely an hour later, with waves as high as 14 metres, that the accident occurred. TEPCO argued that it can't be faulted for the accident as it could not have possibly foreseen this double natural calamity, which was a one-in-a-million chance occurrence.

On May 16, 2011, TEPCO finally admitted that the accident in Unit 1 had started immediately after the earthquake struck Japan, that is, before the tsunami. A radiation alarm—which was set to go off at high levels of radiation—went off minutes before the station was overwhelmed by the giant tsunami. This implies that the coolant was lost and fuel melting began almost immediately after the earthquake struck the plant. Most likely, a major pipe carrying cooling water to the core was damaged by the earthquake. The ensuing tsunami and loss of back-up power exacerbated the accident.[2]

Tsunami Worsens the Accident

At the time of the quake, Reactor 4 was in a de-fuelled state, all its fuel rods had been shifted to the spent fuel pool on an upper floor of the reactor building; Reactors 5 and 6, though fuelled, were in shutdown state for planned maintenance; while Reactors 1, 2 and 3 were operating.

As soon as the earthquake struck, Reactors 1, 2 and 3 were immediately shut down by sensors. The earthquake knocked off the primary AC supply; an hour later, the tsunami tore through the wall like a knife cuts butter, to smash the backup diesel generator sets. Power couldn't be restored for days, and since the battery backup was only of 8 hours, the plant suffered a power blackout.

At this stage, the reactor operators were confronted with two separate set of problems. One problem was to cool the reactor cores in Reactors 1, 2, 3, 5 and 6. The second problem was that the spent fuel pools of all six reactors also needed to be continuously cooled. If the cooling is disrupted and temperature rises, the zirconium cladding of the fuel assemblies, whether in the reactor core or in the spent fuel pool, would melt, releasing its contents into the atmosphere.

As mentioned above, these reactors had the spent fuel bundles tucked in a room near the top of the reactor vessel. While the reactor core is encased in a steel vessel inside the primary containment, the spent fuel is outside this containment. All that shields the radioactivity from the spent fuel from getting dispersed into the environment are the thick outer walls of the reactor building—the so-called secondary containment. The spent fuel contains even more radioactivity than the reactor core, and so is potentially far more harmful to the environment than the fuel inside the reactor core. Of particular concern was the Reactor 4 spent fuel pool as it contained a large amount of fresh spent fuel that was moved to the cooling pond from the core a few months ago, and so was highly radioactive.

At Fukushima, once the batteries got discharged and there was no power to run the cooling systems, temperatures of the fuel assemblies began to rise. As a result, more and more water started boiling off. As long as the fuel assemblies were covered with water, all was well, but eventually, in Reactors 1, 2 and 3, the tops of the fuel assemblies became bare and their temperatures passed the take-off stage. (In Reactors 5 and 6, this was prevented due to restoration of power to the cooling system in time.)

Hydrogen Explosions

Fuel assemblies are composed of pellet-sized slightly enriched uranium fuel enclosed in a zirconium alloy cladding. Zirconium is used as cladding material because not only does it have good thermal properties like any metal, but more importantly, it does not capture too many neutrons and thus helps the neutron economy inside a reactor. However, it has a serious drawback. At high temperatures, it reacts with steam to produce zirconium oxide and hydrogen gas. Moreover, the reaction is exothermic—that is, it releases a great deal of heat—further raising the temperature, thereby aggravating the problem. The same phenomenon can occur in a spent fuel pool in the case of loss of cooling water.

This is precisely what happened at Reactors 1, 2 and 3 at the Fukushima Daiichi plant. With the cooling disrupted and temperature of the fuel assemblies rising, the zirconium cladding started reacting with steam to produce hydrogen gas. As more and more gas was produced, the pressure inside the reactor vessels of these three units started increasing. In desperation, the operators decided to use sea water to cool the reactors—a decision which meant the writing-off of the reactors since sea water is highly corrosive.

However, with high pressure inside the core vessels, and the pumps pumping seawater into the vessels operating at low pressure, this water just wouldn't go in or at least not in sufficient quantities. Periodically, workers opened valves to vent steam and gas from the reactor vessel into the primary containment. This, in turn, increased the pressure inside the primary containment. When the pressure here became very high, workers vented the gas from the primary containment into the reactor building, that is, the secondary containment. As pressure rose inside the reactor building, operators vented the secondary containment too, into the atmosphere. However, this would also mean releasing radiation into the atmosphere, so they sought to minimise the amount of gas vented from the reactor building into the atmosphere, by not having too frequent venting.

For reasons not fully understood, the hydrogen gas within the reactor building exploded, demolishing the roof of the building, at Reactor 1 on March 12 and Reactor 3 on March 14. In both the units, the upper structure of the building, where the spent fuel pool is housed, got destroyed, exposing the spent fuel pool to the atmosphere.[3] According to Arnie Gundersen, an eminent US nuclear engineer with 39 years of experience,[4] the explosion in Reactor 3 may have been more serious than has so far been admitted. According to him, an initial hydrogen explosion probably caused a 'prompt criticality' in the spent-fuel pool at the top of the Reactor 3 building, that is to say, a runaway nuclear chain reaction may have taken place in the spent fuel rods. The upward thrust from the explosion threw spent fuel rods from the pool into the atmosphere and scattered them for miles, and also blew aerosolised plutonium and uranium into the atmosphere.[5]

On March 15, a third hydrogen explosion rocked the plant, this time in Reactor 2. Simultaneously, the spent fuel pool in Reactor 4 caught fire. Soon after, a massive hydrogen explosion damaged the upper portion of the reactor building in this unit too. According to TEPCO, the hydrogen that caused this explosion did not come from the overheating of spent fuel assemblies in Reactor 4, but from Reactor 3; it flowed through a gas-treatment line and entered Reactor 4 because of a breakdown of valves.[6]

Reactor units 5 and 6 survived the accident without significant damage. In the first few days after the quake, the water level in the spent fuel pool of Reactor 5 had started falling; there was also some overheating of the reactor cores of both these reactors. Fortunately, however, before things could get worse, operators managed to restart an emergency diesel generator of Reactor 6, and it became possible to supply power to the cooling units of these two reactors. By March 20, these two reactors had reached cold shutdown conditions, and their spent fuel temperatures too were decreasing.

Clearly, the Fukushima accident is worse than Chernobyl. The Chernobyl accident involved a single reactor. The Fukushima accident involves three reactor cores (of Units 1, 2 and 3) and four spent fuel pools (of Units 1, 2, 3 and 4). Each spent fuel pool has fuel rods equivalent to several cores each. In all, that's the equivalent of as many as twenty reactor cores![7] Fukushima is clearly the biggest industrial catastrophe in the history of mankind.

Part II: The Present Situation

Meltdown in Reactors 1, 2 and 3

In the reactor core, once cooling gets disrupted and fuel rod temperature increases, the gas pressure inside the fuel rod also increases and eventually can cause the cladding to balloon out and rupture. This would result in the release of radioactive fission gases and some of the fuel's radioactive material in the form of aerosols into the reactor pressure vessel and, in the case of the spent fuel assembly, into the building that houses the spent fuel.

If the temperature increases further, the cladding melts, and eventually the fuel pellets themselves melt, to release even larger amounts of radioactive gases. This lava-like molten mixture is known as **corium** and reaches temperatures of as high as 2200 degrees on the outside and 3500 degrees Celsius inside.[8] If it relocates within the reactor core, it can cause additional problems: for instance, if it falls to the bottom of the reactor pressure vessel, it can burn through it and flow out.

On May 12, 2011, two months after the earthquake and tsunami, Japan finally admitted to what nuclear scientists had long since suspected, that the entire nuclear fuel in Reactor 1 had melted! According to TEPCO, within six hours after the earthquake hit the reactor, temperatures reached 2,800°C, causing the fuel pellets to start melting. Within 16 hours, the reactor core melted, and dropped to the bottom of the pressure vessel in a clump.[9] In early June, the Japanese admitted that Reactors 2 and 3 had also suffered full nuclear meltdowns. In a report prepared for the IAEA, they also admitted that in all three reactors, the fuel had probably melted through the bottom of the reactor vessel to their outer steel containments.[10]

Containments Damaged: Cooling Water Leaking

This essentially means that the three reactors no longer exist. The molten uranium is lying at the bottom of their containments.

There are also reports that the primary containments of the three stricken reactors are most probably damaged. A report prepared by TEPCO for the Economy, Trade and Industry Ministry's Nuclear and Industrial Safety Agency (NISA) on May 23 says that not just the reactor pressure vessels but also the containment vessels of all three reactors were probably damaged within 24 hours of the earthquake and tsunami. Other reports say that the primary containment of Reactor 1 is definitely damaged, and the situation of the containments of the other two reactors is not known.[11]

And so, the water being poured in to cool the molten fuel in the reactors is flowing out through the cracks in the containment, after directly touching plutonium and cesium and strontium and all the intensely radioactive products of a nuclear fission reaction, and is carrying all those radioactive isotopes out as liquids and gases into the environment. Which is why TEPCO discovered that its radiation detectors had all maxed and become non-functional! No wonder TEPCO selectively stopped reporting radiation releases—it was in the middle of not one, but three Chernobyl-like core fuel meltdowns![12]

An important fact being played down by the Japanese media is that Unit 3 contains MOX fuel. Even though it was only 6 per cent of the fuel (32 out of 548 total fuel assemblies), MOX is exceedingly dangerous because it contains plutonium—a single milligram (mg) of MOX is as deadly as 2,000,000 mg of normal enriched uranium, according to a report by the internationally renowned anti-nuclear group Nuclear Information and Resource Service (NIRS).[13] Furthermore, plutonium has a half-life of 24,000 years; even if some of this terrible material escapes from the reactor, it is going to contaminate the environment for tens of thousands of years. And with Reactor 3 also leaking water, this is obviously happening.

TEPCO has poured in millions of litres of water to cool the reactors and the spent fuel pools. This water, which has become highly radioactive after coming into contact with the nuclear fuel, has flowed out into basements, connecting tunnels and service trenches at the plant. According to a newsreport published in the magazine *Bloomberg Businessweek* issued from New York on May 19, more than 10 million litres of this radiation-contaminated water has leaked or been released into the sea (before the leaks were discovered by TEPCO and plugged).[14] This water is emitting radiation of as much as 1 sievert (or 1000 millisieverts) per hour—a level high enough to cause acute radiation sickness after a short exposure.[15]

On June 4, 2011, a robot sent into the Reactor 1 building reported that radiation levels in the air around the reactor were at 4000 millisieverts per hour—a level so high that it is 100 per cent lethal within 1.5 hours of exposure![16] There are as yet no reports of robots entering similar areas inside Units 2 and 3, but there would be similar or higher measurements in those much larger reactors.

Due to these high radiation levels, it is not going to be possible to repair these reactors and stop these radiation leakages anytime soon. According to many experts, the melted fuel will have to be cooled for at least one to two years![17] Till then, the plant will continue to leak cesium and strontium and plutonium into the atmosphere.

On June 3, 2011, *Bloomberg* reported that the amount of

contaminated water has risen to about 105 million litres. TEPCO has warned that with the storage capacity fast filling up, the water may start overflowing by June 20 (2011), sooner if there are heavy rains. It estimates radiation in the water at an astronomical 720,000 trillion becquerels—this is roughly equal to Japan's latest estimate of the amount of radiation that escaped into the air from the plant in the first week after the accident.[18] This means that the danger from the overflow is as serious as that from the melted fuel!

Arnie Gundersen points out that the big problem before TEPCO is going to be what to do with this huge volume of highly radioactive water? It is not going to be easy to filter it, as the filters are made of plastic material and might melt, and workers will not be able to go near the filters to replace them because of the high radioactivity.[19]

The debris from the hydrogen explosions at the plant is also highly radioactive, and is lying within and outside the plant compound. This debris is emitting 900 to 1,000 millisieverts of radiation per hour.[20]

The blob of melted nuclear fuel lying at the bottom of the reactor containment in all the three reactors is at a very high temperature—as much as 5000 degrees Celsius in the inside of the molten mass (the water is only able to touch and cool the outside of the melted fuel).[21] It could fission its way through the containment vessel, melt through the basement of the power plant and enter the soil and water table, causing huge contamination of the crops and groundwater around the power plant ... for tens of thousands of years. And there is not much that the operator can do to prevent this nightmare scenario from unfolding, except keep pumping in thousands of litres of water every day to cool the molten fuel. Workers cannot be sent in to do major repair works due to high radiation levels. All that can be done is to pray, and hope for the best!

Spent Fuel Pools Damaged

The situation at the spent fuel ponds of Reactor units 1 to 4 is also very bad. Indications are that at least in one of the pools (or if not there, then in the reactor core of Unit 3), the fuel is reaching criticality. However, once this starts to happen, the fuel heats up, and gets to a point where it gets so hot that it shuts itself down; as it cools down, the cycle starts again.[22]

The situation at the spent fuel pools of Reactors 3 and 4 is of particular concern. As discussed earlier, on March 14, there was an explosion in the spent fuel pool of Reactor 3 (along with the hydrogen explosion), which badly damaged its spent fuel pool; a report by the US NRC says that fuel may have been ejected from the pool up to one mile from the plant. According to Arnie Gundersen, chief nuclear engineer with the energy consulting firm, Fairewinds Associates, shockwaves from the explosion caused the fuel rods to be launched into the air out of their containment vessels 'like the muzzle from a gun'.[23] This would account for the discovery of plutonium contamination as far away as 50 kms from the plant.[24]

In Reactor 4, the big problem is that the building housing the spent fuel pool has partially sunk and is threatening to collapse. It is probably because the leakage of radioactive water has softened the ground, and the hydrogen explosion has weakened the structure. Another earthquake, or even an aftershock, could mean catastrophe.[25]

A Nuclear Accident Never Ends ...

How bad is the situation at the plant? What is the situation of the molten fuel lying on the reactor floor in Reactors 1-3? What is the extent of damage to the four spent fuel pools? None of this is known, and will probably not be known for years. Eight months after the Three Mile Island accident, an Oak Ridge National Laboratory scientist declared, 'Little, if any, fuel melting occurred, even though the reactor core was uncovered. The safety systems functioned reliably.' A few years later, robotic sorties into the area revealed that half the core—not 'little, if any'—had melted down.[26]

On April 17, TEPCO announced plans of reducing radiation leaks from the reactor within three months and bringing the plant to a cold shutdown within six to nine months.[27] Cold shutdown means that the fuel has cooled sufficiently for the coolant water to be at less than 100°C. However, with fuel having melted down and lying on the floor in all three reactors, this timeline has obviously gone through the roof. It may take TEPCO as much as two years to bring the melted reactor cores to a cold shutdown, according to Dr M.V. Ramana, a physicist at Princeton University who specialises in issues of nuclear safety.[28]

Even after that, the problem is, what do you do with the melted fuel? How do you remove it from the environment for hundreds of thousands of years? According to Gundersen, 'Somehow, robotically, they will have to go in there and manage to put it in a container and store it for infinity, and that technology doesn't exist. Nobody knows how to pick up the molten core from the floor, there is no solution available now for picking that up from the floor.'[29] Many eminent scientists are now of the opinion that the only way to stop the release of radioactive materials and further contamination of air, soil and water at Fukushima is to entomb the reactors, like at Chernobyl.[30]

However, that too is not going to be easy.

Following the Chernobyl accident, a huge sarcophagus or coffin made from more than 400,000 cubic metres of concrete and 7,300 tons of metal framework was built over the destroyed reactor in order to prevent the release of radioactive materials from the melted fuel. It locked in 200 tons of radioactive corium, 30 tons of highly contaminated dust and 16 tons of uranium and plutonium. This sarcophagus is now cracking up and leaking radiation, and needs to be urgently replaced, or else it could come crumbling down, unleashing a catastrophe on the same scale as the original accident 25 years ago! Experts had recommended that the structure be replaced by 2006; the Ukrainian government, with financial help from the European Union and several countries, started work on building a gigantic new shell to cover Chernobyl's exploded reactor and the existing steel sarcophagus only in 2010. The new structure, an arch more than 100 metres high, 250 metres wide and 160 metres long, and expected to weigh 20,000 tons—the largest such structure in the world—will be assembled close to the Chernobyl site and then slid on rails over the existing sarcophagus, before the ends are blocked up. It is expected to be completed in 2015, ten years later than required.[31] Let us keep our fingers crossed and pray that the sarcophagus will cooperate and won't cave in on itself before then.

This new structure is expected to last for at the most 100 years.

That isn't a very long time. Additionally, even after the new shelter is built, the danger from the destroyed reactor will not be over. How far down into the concrete foundation has the nuclear lava penetrated? How great a threat is it to groundwater? No one knows the answers.[32]

Entombing the Fukushima reactors is going to be an even more difficult task than Chernobyl as there are four reactors here which would need to be encased. Moreover, it cannot be done immediately, as the cores are still hot. It is going to take at least a year, even two years, to cool the reactors sufficiently for it to become possible to fill them up with concrete and let them lie there, like a giant mausoleum. However, this is possible for only Reactors 1, 2 and 3. This cannot be done so simply for Reactor 4, as here, all the fuel is at the top of the reactor building. Concrete can't be poured into this reactor from the top because it will collapse the building. And the fuel cannot be lifted out because it is highly radioactive. According to Arnie Gundersen, the world renowned nuclear engineer, the solution is not going to be easy. The Japanese will need to use massive cranes, cranes that lift a hundred and fifty tons, and put the nuclear fuel into canisters, which can then be removed. But this cannot be done in air; it has to be done under water. So a building will have to be built around the reactor building to provide enough shielding and water, and then the cranes can be sent in to put the fuel into canisters. The whole process is going to take many years.[33]

But what if the building housing Reactor 4 collapses before the fuel is shifted into canisters? Gundersen, a kind of living legend in the field of nuclear engineering, recommends that if this reactor topples due to an earthquake or some other reason, the people of Tokyo should simply get on a plane and get out of there, irrespective of whatever the authorities say![34]

Part III: Nuclear Pinocchios

A History of Downplaying Nuclear Tragedies

Ever since the start of nuclear technology, those behind it have made heavy use of deception, obfuscation and denial to downplay its potential impacts, often with the complicity of most of the media. *New York Times* reporter William Laurence wrote a widely-published press release covering up the first nuclear test in New Mexico, which was conducted by the United States Army on July 16, 1945, claiming that it was nothing more than an ammunition dump explosion.[35] For 25 years after the atomic bombing of the cities of Hiroshima and Nagasaki on August 6 and 9, 1945, the United States engaged in airtight suppression of all photographs and film shot in Hiroshima and Nagasaki after the bombings, in order to hide from the US public the horrifying impact of the atomic bomb explosion on human life.[36]

After the Three Mile Island accident in 1979, the US nuclear industry and government agencies claimed that relatively small amounts of radiation had escaped from the plant, and that no one was even injured. This is still the official version of the accident, even though several books have given details of adult and infant deaths as a result of the accident; there is also a TV documentary, *Three Mile Island Revisited*, that focused on the cancers and deaths in the area around the plant, and how its owner has quietly given pay-outs, of as much as $1 million per person, to settle with people who suffered health impacts or lost family members because of the accident.[37]

Following the catastrophe at Chernobyl, spokespersons of the global nuclear industry and international nuclear agencies have made the most outrageous (or, should we say lunatic) claims in an attempt to downplay the accident. A few months after the accident, M. Hans Blix, long time Chairman of the IAEA, declared that 'due to the importance of this source of energy, the world could support one accident of the Chernobyl scale every year ...'[38] Anil Kakodkar, the former Chairman of India's Atomic Energy Commission, has gone to the extent of reducing Chernobyl to smaller than a road accident. He claims that 'the Chernobyl accident caused 47 deaths till the year 2004 among firemen and severely exposed persons'![39] These salespersons have deliberately chosen to ignore the findings of the most comprehensive study done on the effects of Chernobyl by a team of eminent scientists from Russia and Belarus and published by the New York Academy of Sciences in 2009. They studied health data, radiological surveys and scientific reports—some 5,000 in all—

and estimated that the accident caused the deaths of 985,000 people worldwide from 1986 to 2004. More deaths, they wrote, will follow. The mainstream media, everywhere, from the US to India, has blacked out this report.[40]

Downplaying Fukushima

The multi-trillion dollar nuclear industry knows that if the full scale of the tragedy at Fukushima becomes known to the people, a cry would emanate from all corners of the world, people would refuse to passively wait for their death from its impacts, and it could well sound the death-knell for the industry. And so from the beginning of the accident, the global nuclear industry and its accomplices—the governments of pro-nuclear countries like the United States and France, their bootlicking Indian government, and of course the Japanese government—have tried to downplay its potential impact.

Initially, the Japanese government categorised the Fukushima accident as a Level 4 accident on the INES scale.[41] A week later, on March 18, it elevated the disaster to Level 5, the same level as the Three Mile Island accident of 1979. Finally, a month after the accident, it reluctantly raised the severity level of the accident from five to seven —the maximum on the INES scale, on par with the Chernobyl accident of 1986. Nevertheless, the Japanese government and TEPCO, the plant operator, continued to emphasise that the accident was much less serious and radiation leaks from the plant were far less than the Chernobyl accident.[42] Only two months after the accident did they admit that there was a meltdown in Reactor 1. This, despite the fact that they knew at least from late March that this reactor had suffered a meltdown within hours of the earthquake![43]

Yayatis All

What's worse, from the beginnings of the Fukushima tragedy, instead of taking serious measures to protect its people from the terrible health effects of radiation, the Japanese government has been more keen on downplaying the contamination levels and falsely assuring its people that the health effects are going to be minimal. Thus, despite being aware that radiation levels at distances of more than 30 km from the plant exceeded safe levels, it withheld this data and advised only those people staying up to 30 kms from the plant to stay indoors.[44] When soil samples outside the 20 kms evacuation zone declared by the Japanese government were found to have very high levels of radiation, the chief government spokesman, Yukio Edano, declared: 'At the moment, we have no reason to believe that the radiation will have an effect on people's health.'[45] When strontium was found in soil samples in the city of Fukushima, 62 kms away from the crippled nuclear power plant, Japan's Nuclear Safety Commission said it was unlikely to pose an immediate threat to human health.[46] But what probably deserves a 'Nobel Prize for Obfuscation' is this statement from TEPCO after plutonium was found in five different soil samples outside the Fukushima nuclear plant: 'It is not a health risk to humans'![47]

On May 11, the Japanese government abolished the 50-millisievert yearly limit of maximum permissible amount of radiation exposure for workers at the troubled Fukushima plant.[48] Before that, on April 19, it had sharply ramped up its radiation exposure limit for school children in Fukushima prefecture by twenty times, from 1 mSv/year to 20 mSv/year; only after huge protests by angry parents did it partially reverse its decision towards the end of May.[49]

Initially, on March 23, 2011, the Japanese government barred Fukushima prefecture from distributing various greens and turnips, including broccoli, cabbage, cauliflower, parsley, and many others; Ibaragi prefecture was stopped from shipping spinach, *kakina* and parsley; while Tochigi and Gunma prefectures were disallowed from shipping spinach and *kakina*. However, less than a month later, by mid-April, it did a 180-degree turnaround and launched an initiative —in the name of nationalism—to promote the consumption of greens from Fukushima and other prefectures near it! Prime Minister Naoto Kan assured the people that produce from the region around the damaged Fukushima facility was safe to eat, despite the radiation leaks. The IAEA confirmed Kan's claim, saying that radiation contamination in the region had decreased to below legal limits set by the Japanese government. The World Health Organisation went one step ahead and stated that its public health assessment showed there is very little public health risk outside the 30-kilometer (18-mile) evacuation zone![50]

Chief Cabinet Secretary Yukio Edano went to a farmer's market in Tokyo and ate a Fukushima strawberry, and stated: 'Only safe produce is being distributed. Please eat it.' Television talents, sports heroes and popular singers also publicly expressed their support by eating and buying farm produce from Fukushima and outlying prefectures.[51]

On May 21, Japan's Prime Minister Naoto Kan, Chinese Premier Wen Jiabao and South Korean President Lee Myung-Bak travelled to a shelter for Fukushima evacuees 60 kms away from the Fukushima Daiichi nuclear plant, and tasted local cherries, cucumbers and tomatoes in a gesture to show the safety of Fukushima foods![52]

In the United States, President Barack Obama stated on March 17, 2011:

> We do not expect harmful levels of radiation to reach the United States, whether it's the West Coast, Hawaii, Alaska or US territories in the Pacific. Let me repeat that: We do not expect harmful levels of radiation to reach the West Coast, Hawaii, Alaska, or US territories in the Pacific. That is the judgment of our Nuclear Regulatory Commission and many other experts.

A spokeswoman for the US Food and Drug Administration (FDA) assured people that fish from the Pacific was safe to eat: 'So far, there's no reason for concern about Fukushima. The radioactive materials in the water near Fukushima quickly become diluted in the massive volume of the Pacific.'[53] But what if high levels of radiation do reach the United States? Simple. Don't take the readings. The FDA has stopped testing fish for radiation in the Gulf of Alaska. The US Environmental Protection Agency (EPA) has pulled out 8 of its 18 radiation monitors in California, Oregon and Washington—probably because they were giving very high readings. It is also planning to drastically raise the amount of allowable radiation in food, water and the environment![54]

The European Union has acted more speedily. It has secretly increased the permitted amount of radiation in food imported from Japan by up to 20 times previous food standards, without informing the public. Till before the Fukushima accident, a maximum of 600 becquerels of radioactivity per kilogram was allowed. Now, through an Emergency Ordinance issued on March 27, 2011, 12,500 becquerels per kilogram is being permitted![55]

From Prime Minister Kan to President Obama, from the IAEA to FDA, all are claiming that the radiation escaping from Fukushima is not exceeding 'harmful limits' and that contamination levels of food and water are below 'permissible levels' or 'safe limits'. The media, too, has parroted these assurances; in fact, news about Fukushima has virtually disappeared from newspapers and TV channels.

In the *Mahabharata*, there is the story of King Yayati, who took away the youth of one of his sons for enjoying the material pleasures of the present. The present rulers of the world—the Kans, Jiabaos, Obamas and Manmohan Singhs—are a thousand times worse than Yayati; they are willing to pay the price of allowing millions of people to suffer from cancer and other terrible diseases and letting millions of children be born with genetic deformities in our coming generations, just so that the nuclear industry can make more profits in the present, just so that they can continue with their energy-profligate lifestyles.

Part IV: Bigger than Chernobyl

Reality—Level 8 Accident

Japan's Nuclear and Industrial Safety Agency now admits (in its June 6 press release) that the Fukushima nuclear plant, in just the first week after the accident, released 770,000 trillion becquerels of radiation, which is about 40 per cent of the official Soviet estimate of total emissions from Chernobyl.[56] Considering the shamelessness with which it has been hiding facts about the accident, it is obvious that even this high figure is going to be revised much upwards in the coming months and years, like has happened with the Three Mile Island and Chernobyl accidents. Furthermore, this figure does not include the radiation released into the ocean, which is probably as much.[57] And then, of course, the radiation leakages are going to continue for many months; as discussed above, it's going to be many months before the disaster is brought under control. In contrast, at Chernobyl, it took around two weeks to bring the fires and radiation leakages under control.[58]

The Fukushima accident is actually so huge that even the INES scale does not capture its true magnitude—the accident is far bigger than the worst accident imagined by the IAEA. On March 24, 2011, Greenpeace released a report prepared by Dr Helmut Hirsch, a scientific consultant for nuclear safety with 30 years experience. Based on data from the French government's radiation protection agency (IRSN) and the Austrian government's Central Institute for Meteorology and Geodynamics (ZAMG), he calculated that the total radiation releases from the Fukushima plant are so high that they amount to three INES 7 accidents![59]

By the end of May, many nuclear engineers were saying that Fukushima had gone way beyond the scope of the Chernobyl accident and called upon the IAEA to revamp its INES scale and create a new level—Level 8—to categorise it!![60]

Japan Survives, by Luck

According to Arnie Gundersen, the amount of radiation released from the stricken plant during the first few weeks was so much that it could very well have brought the nation of Japan to its knees.[61] Fortunately for Japan, the winds were blowing out towards the sea most of the time during the accident, and so the contamination wound up in the Pacific Ocean as compared to across the nation of Japan. Had the wind been blowing across the island, Japan would have been forced to declare an exclusionary area all the way across the island of Japan. It would have cut Japan in half. This is also admitted to by Dr Saji, a highly respected former member of the Japanese Atomic Nuclear Safety Commission: 'We were very lucky even with a large release from Fukushima 3, due to the most severe hydrogen explosion, that could have induced a heavy land contamination. This resulted from the wind direction towards the sea at the time of the release, although this must have resulted in a wider ocean contamination far from the Fukushima unit.'[62]

Japan has, by sheer luck, survived the enormous amounts of radiation released from the damaged Fukushima plant that would have rendered large parts of Japan uninhabitable for centuries, by dispersing this radiation all over the globe. But for Planet Earth as a whole, as we see below, that has in no way diminished the impact of the Fukushima disaster.

Two Important Scientific Truths

To understand the true devastating implications of the Fukushima disaster, and see through the lies being propagated by the global nuclear industry and its toadies, there are two important scientific facts which need to be emphasised.

i) Any Permitted Radiation is a Permit to Commit 'Murder'

We have discussed this earlier in Chapter 3 also: there is no safe dose of radiation. There is a preponderance of scientific evidence to show that even very low doses of radiation pose a risk of cancer and other health problems and there is no threshold below which exposure can be viewed as harmless. This is admitted to by numerous scientific bodies:[63]

- US National Council on Radiation Protection: 'Every increment of radiation exposure produces an incremental increase in the risk of cancer.'
- US Nuclear Regulatory Commission: 'Any amount of radiation may pose some risk for causing cancer.'
- The US Environmental Protection Agency: '... any exposure to radiation poses some risk, i.e. there is no level below which we can say an exposure poses no risk.'
- Richard R. Monson, chairman of US National Academy of Sciences committee on Biological Effects of Ionising Radiation, and a professor of epidemiology at Harvard's School of Public Health: 'The scientific research base shows that there is no threshold of exposure below which low levels of ionised radiation can be demonstrated to be harmless or beneficial.'

In short, there is no safe dose of radiation. To quote Dr John William Gofman, professor emeritus of Medical Physics at UC Berkeley once again: 'Any permitted radiation is a permit to commit murder.'[64]

ii) Total Number of Cancers is Immutable

Spreading out radiation amongst a very large number of people does not mean that the total numbers of cancers that will be caused by that radiation will be reduced. This very important fact was emphasised by Dr Steven Wing, Associate Professor of Epidemiology at the University of North Carolina Gillings School of Global Public Health, in an interview with Arnie Gundersen:

> It is important for people to know that spreading out a given amount of radiation dose among more people, while it reduces each person's individual risk, does not reduce the total number of cancers that is going to be caused by that amount of radiation. So having millions and millions of people exposed to a particular dose will produce the same number of cancers as having a very small number of people, say a few thousand, exposed to the same dose.[65]

Thus, while the huge amounts of radiation that have escaped from the Fukushima plant will not cause the expected enormous number of cancers in Japan because the winds blew the radiation out to the sea, that does not mean that these cancers have been eliminated; all that has happened is that the cancers have been spread out in a worldwide population (the short-lived radiation might decay before it reaches humans; to that extent, the number of cancers would be less).

Worldwide Impact (Outside Japan)

Radiation from the Fukushima plant has spread to all across the globe. Not only countries near Japan, like South Korea, the Philippines, Vietnam, China and Russia, but also countries far away, across the Pacific Ocean, from Canada to the USA and Mexico, and even Switzerland, Iceland and France, have detected traces of radioactivity from Japan's crippled plant in their soil, air and water.[66]

Thus, China reported that low levels of radiation had been found in the air all across the country, and that spinach had also been found to have been radioactively contaminated in some areas.[67] Radiation from Fukushima has been found in rainwater and seaweed in North

Vancouver in Canada. The Federal Health Office of Switzerland reported that its radiation sensor equipped plane had detected a small amount of iodine-131.[68]

In the US, tests have detected elevated levels of radioactive iodine and cesium in milk and vegetables produced in California; elevated levels of radioactivity have been found in drinking water in numerous municipalities from Los Angeles to Philadelphia; radiation has also been detected in milk in Arizona, Arkansas, Hawaii, Vermont and Washington.[69] Cesium-137 has been found in rainwater samples from Boise, Idaho and Montpelier, Vermont. By 2013, Arnie Gundersen estimates, 'we might see contamination of the water and of the top of the food chain fishes on the (US) West Coast.'[70]

It now also stands revealed that the US EPA's national network of radiation monitoring stations (called RadNet) detected strontium and plutonium from Fukushima in the United States as early as March 18, but hid the fact in its public reports.[71] Americium, which is more toxic than plutonium, has been found in New England (a region in the northeastern corner of America).[72]

Impact on Japan

Pathetic Evacuation Zone

Since the beginning of the nuclear crisis on March 11, 2011, the Japanese government has only grudgingly increased the evacuation zone around the nuclear plant, from 2 miles to 6 miles and then, on March 15, to 12 miles (20 kms). 85,000 people have been evacuated.[73] These people are never going to return to their homes. The contamination levels are huge; deathly radionuclides like plutonium-239 and americium-241 (half-life 433 years) have been found in soil samples near the plant; the area is going to be a 'dead zone' like at Chernobyl.[74]

The Japanese government also initially recommended that people living within a radius of 12 to 18 miles (20 to 30 kms) remain indoors; around 136,000 people live in the region from 12 to 18 miles around the plant.[75]

In the months since the accident, evidence has mounted that there are numerous radiation hot spots beyond the 20-km evacuation zone, at distances up to 200 kilometers and even 300 kms from Fukushima.[76] For instance, a group of independent researchers from Kyoto University and Hiroshima University found radiation levels in soil samples much outside the 30 km zone to be at 4 times the level at which evacuation was ordered for Chernobyl.[77]

Significant amount of radioactive contamination has been detected as far away as Tokyo, Japan's capital and largest city, which is 240 kms from Fukushima. Radioactive cesium of up to 3,200 becquerels per kilogram was found in the soil of Tokyo districts of Koto and Chiyoda, a level comparable to that found in some areas near the Fukushima plant. On March 25, sewage slag at a sewage treatment plant in Tokyo was found to contain cesium and other radioactive materials in very high concentration, 170,000 becquerels per kilogram! Samples at two additional facilities also showed radiation levels of over 100,000 becquerels per kilogram.[78]

In early June, Greenpeace experts found radiation levels to be as high as 45 microsievert per hour in areas outside the 20 km evacuation zone in Japan, a level which equates to an annual dosage of almost 400 millisieverts per year, which is 20 times the annual limit of radiation exposure for nuclear workers and which is so high that it is lethal to 100 per cent of the population within 15 years. Releasing this report, Jen Beranek, a radiology expert from Greenpeace International, recommended that the government widen the evacuation zone to at least 60 or 70 kms from the plant.[79] Professor Chris Busby, one of UK's leading radiation experts who is also the scientific secretary of the Committee on Radiation Risks of the European Parliament, has gone even further. According to him, official data from the Japanese MEXT Ministry shows that contamination levels in the region 100 kms from the damaged reactors are mostly between 6 and 14 microsieverts per hour: this works out to 52-122 millisieverts/year, many times the contamination levels in the Chernobyl exclusion zone![80] (In Chernobyl, areas with annual radiation in excess of 5 mSv were declared mandatory evacuation zones, and in areas registering in excess of 1 mSv/year, residents were given the right to relocate.)[81] Busby has called for the Japanese government to expand its evacuation zone to a minimum of 100 kms from Fukushima, which would mean 3 million people would have to be evacuated.[82] Even the very pro-nuclear and conservative IAEA has recommended that Japan expand its current 20 kms evacuation zone.[83]

In response, instead of expanding the evacuation zone, all that the Japanese government has done is to recommend that people living in such areas evacuate. It has stated that evacuations will not be mandated in the affected areas. In some of these areas where the radiation levels are particularly high, the government has recommended that people evacuate within a month. In other areas, it has said that the households in the radiation hotspots would be contacted individually by the respective local governments, and those wanting to leave will be issued documents certifying them as disaster victims and given government support to evacuate, while those who wish to remain will be able to continue to do so.[84]

It's obvious that Japan's evacuation zone is pathetic. While the Japanese government fiddles, Australia, South Korea and the United States have advised their citizens to stay at least 80 kms away from the plant. Britain has asked its citizens staying north of Tokyo to leave. And the Singapore government has asked its citizens to evacuate from areas 100 kms away from the plant.[85]

Contamination All Over Japan

The evacuation zone may be the more glaring symbol of the horror of a nuclear accident. However, in practical terms, the far bigger tragedy is the heavily contaminated region outside the evacuation zone. Thus, following the Chernobyl accident, while an area 30 kms from the plant was evacuated, an area of 1 lakh square miles was heavily contaminated, and will continue to remain so for thousands of years. Obviously, such a huge area cannot be evacuated. More than 5 million people continue to live in this contaminated region. They live with the knowledge that they and their coming generations are going to suffer from cancer and genetic defects and all kinds of unknown diseases. It is a tragedy of unspeakable proportions!

The Fukushima accident is much bigger than Chernobyl. Three months after the accident, the plant continues to leak radiation into the soil, air and water; as discussed earlier, it is going to take many months, it may even take years, before the disaster is brought fully under control. While during the initial weeks after the accident, Japan was lucky in that much of the radiation was blown out to the sea, now the winds have turned and the radiation is blowing across Japan.[86] The total land area of Japan is only 145,883 sq miles, roughly one and a half times the area contaminated by the Chernobyl accident. It is obvious that by the time the accident is brought under control, large parts of Japan are going to be heavily contaminated. In human terms, the impact of the accident is going to be far more devastating than Chernobyl, as Japan is much more densely populated than Belarus, the country most affected by the Chernobyl accident: Belarus has a population density of 40 persons per square kilometer; Japan in contrast has an average of 800 persons per square kilometer.[87]

Japan is sitting on the edge of an abyss. In a recent interview to *Independent Australia*, Dr Helen Caldicott spoke of the possibility of a grim scenario developing in the near future: 'If there is a very big aftershock, as there very well could be, Reactor 4 will probably collapse along with other buildings. This would create a Chernobyl type catastrophe which, combined with a change in the wind—so its blowing the radiation to the South instead of out to sea as it is at the moment— could make almost all of Japan, including Tokyo, uninhabitable forever.'[88]

Groundwater Contamination

Equally serious is the groundwater pollution, especially at the plant site. The millions of litres of highly radioactive water filling up the trenches below the plant are bound to leak into the groundwater; and if that happens, this contamination is going to remain for a very long period of time. There is evidence that this is happening. TEPCO has admitted that iodine-131 has been found in the groundwater at nearly 50 feet below the reactors, in a concentration 10,000 times higher than the government standard![89] Towns near the Fukushima plant are reporting radioactive sewage sludge, which could be due to radioactive groundwater.[90] In June, TEPCO acknowledged that strontium has also been detected in groundwater samples near the damaged plant.[91] This would indicate that the radioactive cesium and plutonium have also contaminated the water table as well.

In all probability, groundwater contamination in Fukushima is going to be the worst in nuclear history.

Food Contamination

As the hazardous radionuclides escaping from the damaged Fukushima plant get dispersed all over Japan with the winds and come down with rain, to contaminate the soil and groundwater, vegetables, fruits, rice and other crops all over Japan are going to be radioactively contaminated. Cows eat grass, and so the milk is going to get contaminated. The contamination is going to get worse with time as it is many months yet before the radiation leakages stop from the plant.

And, for emphasis, we repeat: there is nothing like safe levels of contamination; the claims by Japanese officials of contamination levels being below safe levels are all bunkum.

Cesium and radioactive iodine has been found in spinach and other green leafy vegetables in many prefectures. In Ibaraki and Fukushima prefectures, farmers are pouring out their milk on the farms as it has been found to be contaminated.[92] High levels of cesium have been found in green tea leaves harvested from farms that are 280 kms from the crippled plant.[93] Cesium has a half-life of 30 years, meaning it is going to take between 300-600 years for its radioactivity to become insignificant. It tends to concentrate in soft tissues, especially muscle tissues, to cause cancer.

Japan's science ministry has admitted that small amounts of strontium have been detected in soil samples and plants at different locations between 40 and 80 kms away from Fukushima plant. Strontium-90 has a half-life of 29 years, that is, it will remain radioactive for between 290-580 years; it tends to accumulate in bones to cause bone cancer and leukaemia.[94]

High concentration of plutonium has been detected in a rice field 50 kms away from the stricken Fukushima reactor.[95] Plutonium has a half-life of 24,000 years, implying it is going to be radioactive for a minimum of 2 lakh years. It is one of the deadliest substances on the planet; one molecule in the body is enough to guarantee the development of cancer, according to radiation medicine experts.[96]

Contamination of the Sea

Probably the worst impact of the Fukushima accident on life on Planet Earth is going to be its contamination of the oceans. Millions of litres of highly radioactive water from the crippled Fukushima plant has leaked or been deliberately released into the Pacific Ocean; scientists have discovered that its radioactive impact far outstrips the Chernobyl disaster. Data released by Japanese scientists show cesium-137 concentrations in the waters immediately adjacent to the reactors at levels more than 1 million times higher than previously existed, and 10 to 100 times higher in the waters off Japan than values measured in the Black Sea after Chernobyl. According to Ken Buesseler, Marine Radiochemist and Senior Scientist at the Woods Hole Oceanographic Institution, a very reputed scientific organisation: 'For the oceans, this is the largest accidental release of radiation we have ever seen.'[97]

Apart from these releases into the ocean, millions of litres of highly radioactive water has also accumulated at the plant site, and its volume continues to increase by hundreds of tons every day—the amount of water TEPCO needs to pour in every day to cool the three reactors. No one knows what to do with it, and much of it may eventually find its way to the Pacific too. By the time the accident is brought under control, what is going to be the level of contamination of the Pacific, God alone knows. Already small fish of the order of 4 to 5 inches, as far away as 50 miles from the coast, have been found to contain cesium levels 10 to 50 times more than 'allowable'.[98] These smaller fish are going to be eaten by the bigger fish, and so the toxins are going to bioconcentrate. Eventually, this cesium and other radionuclides are going to make their way into the tuna and salmon and other fishes that are a very important part of the Japanese diet. To quote Arnie Gundersen from his interview that has now become world-famous: 'In Japan we are saying avoid fish caught in the Pacific, unless you know they are caught a long way away from Fukushima. I am saying 100 miles off Fukushima, don't even consider it.'[99]

Impact on Children

The most devastating impact of Fukushima is going to be on children, as they are the most vulnerable to radiation. Experts consider children to be 10 to 20 times more susceptible to contracting cancer from exposure to radiation than adults.[100] This is because radiation has the greatest effect on cells that are actively dividing; children are still growing and maturing, so a greater proportion of their cells are in that state.[101]

This is borne out by studies on the impact of radioactive contamination on children in the areas around Chernobyl. They have found that children living in contaminated regions in a radius of 250-300 kms from Chernobyl show an increase in mutations.[102] In Ukraine, from the years 1987 to 2004, the incidence of brain tumours in children up to 3 years of age doubled and in infants it increased 7.5-fold, according to the National Ukrainian Medical Academy in Kiev.[103]

The general morbidity of children has drastically increased in the contaminated regions. According to data from the Belarussian Ministry of Public Health, just before the catastrophe (in 1985), 90 per cent of children were considered 'practically healthy.' By 2000, fewer than 20 per cent were considered so and, in the most contaminated Gomel province, fewer than 10 per cent of children were well.[104] Similarly, in Kiev, Ukraine, where before the meltdown up to 90 per cent of children were considered healthy, the figure in 2008 was just 20 per cent. In the heavily contaminated areas, it is difficult to find one healthy child.[105]

There has also been a sharp rise in previously rare multiple congenital malformations (CMs) and severe CMs such as polydactyl, deformed internal organs, absent or deformed limbs, and retarded growth. In the Ukraine, occurrence of officially registered CMs increased 5.7-fold during the first 12 years after the catastrophe,[106] while in Gomel province of Belarus, it was sixfold higher in 1994.[107]

According to Professor Christopher Busby, a UK government and European Parliament Low Level Radiation Expert, a region of at least 100 kms around the Fukushima plant is heavily contaminated, and another 100 kms is significantly contaminated. Three million people live in the region within 100 kms, and another 7 million in the region 100 to 200 kms from the plant.[108] This is what they are going to suffer in the coming decades.

Bioconcentration of Radiation

The actual threat from the radiation being released from Fukushima is much more than what these contamination levels suggest. That is because what is being emitted from the Fukushima plant are radioactive particles—iodine, cesium, strontium, and dozens of other 'hot particles', as they are called. As these toxins enter and move up the food chain—like from soil to grass to cow's milk and meat to humans, or algae to crustaceans to small fish to bigger fish to humans—their concentration increases, through both bioaccumulation and biomagnification. Finally, when these elements—called internal emitters[109]—enter the human body, they migrate to specific organs such as the thyroid, liver, bone and brain, and irradiate small volumes of cells with high doses of alpha, beta and/or gamma radiation. Over many years, this can induce uncontrolled cell replication—that is, cancer. Further, many of the nuclides remain radioactive in the environment for generations, and so will continue to cause cancer and genetic diseases for hundreds and thousands of years.

It may be some time before the hot particles from Fukushima which are contaminating the food, milk and water all over the northern hemisphere find their way into the human body. The hot particles in the air are already being breathed in by people across the world, from Japan to the USA. In an audio talk by Arnie Gundersen on June 12, he stated that (transcript ours):

> Independent scientists are discovering using air filters in Japan that the average person in Tokyo breathed in 10 of these hot particles every day all the way through the month of April. They also found that people in Fukushima were probably breathing in 30-40 times as much radiation as in Tokyo, again in the form of hot particles. But the most surprising finding was that air filters in Seattle (USA) indicated that people there were absorbing 5 hot particles every day for the month of April. What does that mean? It means that that hot particle gets absorbed in your lung or winds up in your intestines or your muscle or your bone. It constantly bombards a very narrow piece of tissue, causing localised damage. The body fights it, and most of the time it wins. Sometimes, however, it can cause a cancer.[110]

What most people don't realise is that the presence of these hot particles cannot be detected by radiation measuring instruments. Arnie Gundersen says: 'You can't run a Geiger counter over someone's lungs on the outside to determine if there is a hot particle there, as those rays don't travel outside the body. They do damage to the local tissue.'[111]

'Some day, we may not be able to live in Japan'

Fukushima is clearly the greatest public health hazard the world has ever seen, apart from the threat of nuclear war. As Helen Caldicott has put it: 'Japan is by orders of magnitude many times worse than Chernobyl. Never in my life did I think that six nuclear reactors would be at risk.'[112]

Basing himself on data provided by the Japanese MEXT Ministry and IAEA bulletins on radiation releases from the stricken Fukushima plant, and applying the radiation risk model of the European Committee on Radiation Risk, British scientist and anti-nuclear activist, Christopher Busby, conservatively estimates that, of the 10 million people living within a 200 kms radius from the stricken Fukushima plant, if they remain in the area for one year, about 400,000 people will develop cancer in the next 50 years, with 200,000 people predicted to develop cancer in the next 10 years due to the radioactive fallout from Fukushima.[113] Since many of the radionuclides will continue to emit radiation for hundreds of years, these effects will continue to be seen for a very long time!

In an unusually frank interview to the *Wall Street Journal* on May 26, 2011, a senior Japanese politician, Ichiro Ozawa, admitted that due to radioactive contamination, areas of Japan are becoming uninhabitable, 'even Tokyo could become off limits', and expressed his anxiety that 'radiation is going to be flowing out for a long period of time', 'some day we may not be able to live in Japan'![114] Even if things don't become so bad for Japan, what can definitely be said is that, by choosing nuclear energy as an energy option, the Japanese political leadership (including Ichiro Ozawa) have condemned the people of Japan and their coming generations to suffer epidemics of cancer, leukaemia and genetic disease for the rest of time.

Part V: Nuclear Policy Responses to Fukushima

Global Response

The global nuclear industry had gone into a tailspin for nearly two decades after the Three Mile Island and Chernobyl disasters: the US alone cancelled 124 reactor orders, and almost 90 per cent of the projected plants globally were never built.[115] Faced with extinction, the global nuclear industry during the past decade launched a massive propaganda drive to revive its fortunes. It led to some countries which had banned or halted nuclear construction to rethink their policies. However, the Japanese nuclear emergency has made many countries put a pause or even reverse their nuclear power revival plans.

The most drastic of these turnarounds has taken place in Germany where, just last year (2010), the Angela Merkel government had got the German lower house to approve plans to extend the working life of Germany's nuclear plants by an average of 12 years. The Fukushima accident led to massive protests in Germany; on March 26, in the largest ever anti-nuclear demonstration in Germany, some 250,000 people demonstrated under the slogan 'heed Fukushima—shut off all nuclear plants.' The government initially shut down 7 of Germany's 17 reactors but, as the protests continued to grow, eventually, on May 30, it announced plans to shut down all of 17 reactors within the next 11 years, by 2022; it also declared that the seven oldest reactors which had been taken off grid immediately after the Fukushima accident would remain offline permanently. The government also announced plans to double the share of renewable energy in the country's power mix, from the present 17 per cent to 35 per cent by 2022.[116]

In Switzerland, too, after huge anti-nuclear protests in the wake of the Fukushima disaster, the Cabinet initially put on hold plans to build new nuclear plants and then, on June 8, 2011, the Swiss parliament approved a government plan to phase out the use of nuclear power and shut down the country's five nuclear power reactors in the medium term.[117] On June 12 and 13, a majority of Italians, 54 per cent, turned out to vote in a nuclear referendum and 94 per cent of them voted in favour of putting a lid on nuclear power indefinitely.

This puts the seal on a moratorium imposed by the Italian Cabinet on nuclear power soon after the Fukushima crisis. Prime Minister Berlusconi had earlier set a target of producing 30 per cent of Italy's electricity needs from nuclear energy by 2030.[118]

The Taiwan government has also suspended plans to build new reactors; it has also frozen the review of its state-run nuclear power utility Taipower's application to extend the license of its No. 1 plant, which has been operating for 33 years.[119] Malaysia, Thailand and Venezuela have also announced a freeze on plans to build their first nuclear power plants.[120]

Even China has announced a slowdown in its nuclear new build program, the most ambitious in the world—60 per cent of all new nuclear plant construction worldwide is taking place in China. The State Council, or Cabinet, has announced suspension of approvals to build new nuclear power plants. It has also announced a cut in its nuclear power targets for 2020 and a greater emphasis on solar power.[121]

On May 10, 2011, Japan's Prime Minister announced a major shift in Japan's energy policy. Speaking at a press conference in Tokyo, he declared that Japan's energy policy must now 'start from scratch', with a sharp turn to green technologies. He also announced that plans to build 14 new nuclear reactors would now be abandoned. If completed, those reactors would have raised nuclear power's share of Japan's electricity generation to about 50 per cent.[122]

Two months later, the Japanese PM made an even bolder declaration. In a television address to the country on July 13, he stated that Japan should learn from the disaster and called for a complete phase-out of nuclear power. 'We should seek a society that does not rely on nuclear energy,' Kan said. 'We should gradually and systematically reduce reliance on nuclear power and eventually aim at a society where people can live without nuclear power plants.' Taking a stand against the government's long-peddled slogan about the safety of nuclear power—the 'safety myth' that allowed for the construction of 54 reactors over four decades, he stated: 'Through my experience of the March 11 accident, I came to realise the risk of nuclear energy is too high. It involves technology that cannot be controlled according to our conventional concept of safety.'[123]

India's Nuclear Program: Full Steam Ahead

The huge scale of the tragedy in Fukushima has, however, left the Indian government unfazed. In a statement to the Lok Sabha two days after the earthquake and tsunami devastated Japan, Prime Minister Manmohan Singh assured the house that India's nuclear plants are safe and that India attaches 'the highest importance to nuclear safety'.[124] Cocking a snook at global concerns about nuclear safety after the Fukushima accident, the Prime Minister chose the 25th anniversary of the Chernobyl disaster (April 26, 2011) to call a meeting and announce his government's resolve to go ahead with the Jaitapur atomic power project![125]

To assuage public concerns after the Japanese accident, the government announced a safety review of India's nuclear installations. On March 19, the AERB announced the formation of a committee to carry out this review.[126] However, as we have seen earlier, the AERB is only a lapdog of India's non-transparent and autocratic nuclear establishment, and a safety audit done by it has no meaning. That the safety review was always going to be a farce is obvious from the statement given to the press by top officials and scientists of India's atomic energy establishment on March 15, that events in Japan would not affect India's nuclear program in any way.[127] True to form, the AERB committee conducted a hasty and technologically superficial exercise, and declared all installations perfectly safe.[128]

On learning that the Prime Minister had ordered a safety audit of India's nuclear installations after the Fukushima accident, Dr Gopalakrishnan, a former chief of the AERB, wrote an article about three previous nuclear safety audits undertaken by the AERB/DAE that he was aware of, to illustrate how serious the DAE is about these audits. Gopalakrishnan writes that the first such audit was ordered in 1979 by the then Prime Minister Morarji Desai after the Three Mile Island accident in the USA. The audit report identified serious deficiencies needing immediate correction. However, the DAE management classified this report as 'Top Secret' and shelved it. No action was taken on the committee's findings. The second safety audit was ordered by Prime Minister Rajiv Gandhi immediately after the Chernobyl accident in April 1986. This report, too, was marked by the DAE as 'Top Secret' and, again, no action was taken on it. Gopalakrishnan writes,

> After I took over as AERB chairman in June 1993, officials told me about the earlier safety audit reports. I insisted and got these reports from the DAE. Upon reviewing them, I was appalled at the clearly dangerous lack of safety in the various hazardous nuclear installations at that time due to unattended safety problems accumulated over the previous 15 or so years, while the DAE continued to operate these installations at extremely high risk to the public.

As head of the AERB, Gopalakrishnan decided in July 1995 that the AERB must carry out an overall safety assessment of all DAE facilities. He writes that this safety audit 'was discussed and approved by the AERB Board at its 46th meeting on November 7, 1995, and then submitted to the Atomic Energy Commission.' However, the DAE 'promptly classified the report as "Top Secret". To date, no details are known about concrete corrective actions taken, if any, on each of these recommendations.'[129]

Circus Clowns or Mercenaries

The kind of statements being made by India's political leadership and our top establishment scientists after the Fukushima accident to justify India's nuclear program should have made us all double with laughter, but are actually a cause of deep concern because a nuclear accident can have calamitous consequences. Here are two gems:

- Soon after the Fukushima accident, Mr S.K. Jain, Chairman of the NPCIL, India's nuclear operator, stated, 'There is no nuclear accident or incident in Japan's Fukushima plants. It is a well planned emergency preparedness program which the nuclear operators of the Tokyo Electric Power Company are carrying out to contain the residual heat after the plants had an automatic shut-down following a major earthquake.'[130]
- Not to be left behind, his boss, Dr Srikumar Banerjee D.Sc. (Honouris Clausa, Dhanbad School of Mines), the Chairman of India's Atomic Energy Commission, decried media reports

about the 'emergency drill' in Japan as being a nuclear catastrophe: 'It was purely a chemical reaction and not a nuclear emergency as described by some sections of media.'[131]

Lies Unlimited

The absolute nonchalance of India's nuclear establishment towards the terrible safety situation at India's nuclear power plants is shocking, to say the least. Despite the fact that there have been hundreds of accidents at India's nuclear installations, and on several occasions a Chernobyl-like accident has only narrowly been avoided, India's nuclear authorities and political leadership continue to be in public denial about this dangerous state of affairs. Even after the Fukushima catastrophe, rather than be concerned about the possibility of a major accident at India's nuclear reactors, a high-level meeting of Cabinet ministers, government officials, nuclear scientists and disaster management authorities at Prime Minister Manmohan Singh's residence 'noted with satisfaction that there was no accident in any nuclear facility in the past in the country'![132]

Speaking to the media a few days after the Fukushima accident, India's top bureaucrat-scientists came up with the most outrageous lies regarding the safety situation at India's reactors. Mr S.K. Jain (CMD, NPCIL), stated, "India was uniquely placed as it had a centralised emergency operating centre with well drawn procedures scrutinised by regulators."[133] This when India is the only country in the world whose nuclear regulator (the AERB) is not an independent body but a part of the Department of Atomic Energy (DAE)—it does not even have a separate building to house itself. This violates all international norms. India is unique indeed! To end all confusion about the uniqueness of India's nuclear reactors, Jain further added: 'Our plants also have multiple level of heat removal system'.[134] A misleading statement. This is a common feature in reactors all over the world. What is unique about India's reactors is that many of them have unsafe or untested Emergency Core Cooling Systems (ECCS)—'No pressurized heavy water reactor anywhere in the world currently operates with such obsolete and unsafe ECCS', in the words of Dr Gopalakrishnan, a former Chairperson of the AERB.[135]

Jain and Banerjee (Chairman, AEC) also claimed that "station blackout" was the "root cause" of the Japanese accident and "such a thing will not happen in the existing as well as future Indian reactors".[136] They are lying through their hats. In the accident at the Narora Atomic Power Station on the night of March 31, 1993, a fire in the turbine room led to complete loss of station power for a period of 17 hours—it is only the good luck of the people of Bundelkhand that this accident did not snowball into a Chernobyl-like disaster.[137]

Countering concerns expressed by anti-nuclear activists regarding the EPR reactors planned to be imported from France for the Jaitapur plant, these two consummate liars stated that the design of the EPR was based on the design experience of 58 reactors running in Europe, and when the Indian EPR will come up, it would have seen the experience of five such similar plants in Finland, France, China and UK.[138] Their dodgy statement gives the impression that many EPR reactors are in operation around the world, whereas the reality is that not a single EPR reactor has yet been built anywhere in the world. The two EPR reactors being constructed in France and Finland are years over schedule. Recent developments have in fact called into question the very future of the EPR design itself. The design has been refused approval by US and UK nuclear regulators;[139] and the president of the French Nuclear Safety Authority has stated that he cannot rule out a moratorium on the EPR reactor under construction in Flamanville in France.[140] But what about the 58 reactors running in Europe whose experience is supposed to reassure us? Again our top nuclear bureaucrats are lying. They are all of different design![141]

Another Prizefighter Joins the Line-up

Another 'atomic expert', Dr Anil Kakodkar, the former chief of India's Atomic Energy Commission, has become a prizefighter for the French nuclear corporation, Areva. He is going around the country giving lectures on the viability and safety of the Jaitapur nuclear plant, especially after the Fukushima accident. (The irony is, his self-belief in his arguments is so low that he refuses to take any questions after his lecture!) Defending the siting of the Jaitapur Nuclear Plant in an earthquake prone zone, Kakodkar has been saying that the Jaitapur plant is located in a less seismically active zone as compared to the Fukushima plant, and so is inherently safer.[142] Speaking at the fifth convocation ceremony of the Defence Institute of Advanced Technology (DIAT) in Pune, Kakodkar explained that the worst earthquake in the region will be taken as the reference point and the infrastructure will be such that .it can withstand a bigger calamity than those recorded in the past.[143] This is a rather imbecilic argument. Obviously, the Japanese had planned their reactor designs to withstand the largest possible earthquakes they could visualise, and yet an earthquake bigger than the maximum they planned for did take place. This implies that the Jaitapur plant can always be hit by an earthquake of bigger intensity than that for which it is designed![144]

Even assuming that a big earthquake does not occur, there can be some other big natural calamity which can lead to a Fukushima-type accident. David Lochbaum, one of the most eminent nuclear engineers in the USA, in an interview on the US National Public Radio after the Fukushima accident, said a Fukushima-kind of accident can potentially occur in the United States—though of course the 'exact scenario might not be the same, the earthquake and tsunami'. Thus: 'On the East Coast and the northeast you have ice storms and northeasters. In the southeast you have hurricanes that hit Florida and the Gulf Coast. So we might not find it following the exact same script, but we could end up with the exact same ending.'[145] Such a possibility exists for India's reactors too, including the new ones we are building.

The DAE is so casual about designing its reactors to withstand the impact of large natural disasters that, till before 2004, its reactors were not designed to withstand tsunamis. It had assumed that such a calamity will never strike India's coast! This, even though many tsunamis have struck India's coasts in the past, including one on November 27, 1945 which hit Mumbai and caused much loss of life. The DAE had in fact been warned about this possibility, but chose to ignore it. In 2004, a tsunami hit India's east coast, damaging the Madras Atomic Power Station at Kalpakkam, devastating DAE's quarters, and killing several employees. An unknown number of contract workers working at the construction site of the Prototype Fast Breeder Reactor also lost their lives.[146]

Fortunately, it was a small tsunami. However, what if a massive tsunami, like the one that hit Japan on March 11, 2011, hits MAPS, or worse, the Kudankulam plant? Then, there is also the possibility of a killer cyclone striking one of DAE's reactors, like the one that hit Diviseema (in Andhra Pradesh) in 1977, killing more than 10,000 people and 10 lakh heads of cattle, or the one which struck Orissa in 1999, killing more than 15,000 people. Will the Jaitapur or Kudankulam reactors be able to withstand the impact of such a huge natural calamity? It is a scary thought.

Part VI: Nuclear Accidents are Inevitable

Till before Fukushima happened, in the intervening 25 years after the Chernobyl accident, the global nuclear industry and its apologists were arguing that lessons had been learnt from Chernobyl, the necessary design modifications had been made in nuclear reactors and no major nuclear accident will occur in the future. Now, after Fukushima, they are arguing that this was a one-in-a-million chance occurrence, as the accident was caused by a huge earthquake followed by a massive tsunami. Such a double natural calamity will not occur again, so there is no need to worry. (Now, of course, this argument stands discredited, as it is now well established that the meltdown in Fukushima's Reactor 1 had begun before the tsunami stuck, that is, it was caused by the earthquake). Other official scientists are putting the blame for the accident on the Japanese, that they shouldn't have built the reactor in a place prone to such massive earthquakes (and so Indian nucleocrats are claiming that our reactors are safe as they are built in less earthquake-prone regions). Still others are arguing that the reactor was of an old design and should have been scrapped long ago. (This argument actually rebounds on these nucleophiles as there are 23 operating reactors in the US with the same General Electric Mark I design as the Fukushima reactors,[147] while the Tarapur reactors are of an even older vintage, pre-Mark I design.[148]) On the whole, the general argument of the nuclear establishments of the pro-nuclear countries from the US and France to China and India is that the Fukushima accident occurred due to some reasons particular to Japan, and that their nuclear reactors are safe.

The inherent assumption in the arguments of all these 'nuclear energy lovers' is that nuclear technology is inherently safe, and that if an accident has occurred, its reasons can be identified, lessons drawn and design modifications made to make the technology even safer for the future.

Whereas the reality is the exact opposite. M.V. Ramana, a noted nuclear safety expert, writes: 'It is a complex technology, involving large quantities of radioactive materials, and relatively high temperatures and pressures... it is in the very nature of such systems that serious accidents are inevitable. In other words, that accidents are a "normal" part of the operation of nuclear reactors, and no amount of safety devices can prevent them.'[149]

The argument being given by nuclear authorities in the US/ France/India that a Fukushima cannot occur in their countries as their safety systems are more advanced is inherently fallacious. M.V. Ramana explains:

> Accidents are inevitable ... no two major accidents are alike. Historically, severe accidents at nuclear plants have had varied origins, progressions, and impacts. These have occurred in multiple reactor designs in different countries. This means, unfortunately, that while it may be possible to guard against an exact repeat of the Fukushima disaster, the next nuclear accident will probably be caused by a different combination of initiating factors and failures. There are no reliable tools to predict what that combination will be, and therefore one cannot be confident of being protected against such an accident ... The lesson from the Fukushima, Chernobyl, and Three Mile Island accidents is simply that nuclear power comes with the inevitability of catastrophic accidents.

To sum up, in Ramana's own words: 'Catastrophic nuclear accidents are inevitable, because designers and risk modellers cannot envision all possible ways in which complex systems can fail.'[150]

Numerous eminent nuclear scientists from around the world have come to the same conclusion. Following a near-miss in the Forsmark nuclear reactor in 2006, some of the world's most distinguished nuclear scientists examined the safety records of nuclear plants in several countries to find out if there had been any other near-misses after the Chernobyl accident. Their report, presented to the European Parliament in 2007, concluded: 'Many nuclear safety related events occur year after year, all over the world, in all types of nuclear plants and in all reactor designs ... the widespread belief that lessons learnt from the past have enhanced nuclear safety turns out ill-conceived.'[151]

Mycle Schneider, a well-known nuclear consultant and coordinator of this study, writes, 'In the course of the last twenty years, the world has lived with the illusion that it is possible to make nuclear reactors safe. In reality, every day, countless incidents occur in nuclear reactors, and, since Chernobyl, catastrophe has, on several occasions, only narrowly been avoided.'[152]

In other words, sooner or later, a catastrophic nuclear accident was bound to happen, in one or the other reactor, in some or the other country around the world. Dr Helen Caldicott, the pioneering Australian anti-nuclear activist, had prophetically warned in 2006: 'Statistically speaking, an accidental meltdown is almost a certainty sooner or later in one of the 438 nuclear power plants located in thirty-three countries around the world.'[153]

It happened in Fukushima. An accident needs a reason. The earthquake happened to be it.

After Fukushima, if we still don't learn the lesson and do not shut down each and every nuclear reactor all over the world, sooner or later, another catastrophic accident is bound to happen again, in one of the world's 442 operating reactors.[154]

Part VI : Unite, to Save India from Inevitable Destruction

Even for a technologically advanced and rich country like Japan, it is going to take years before it is able to bring the Fukushima disaster under control. Providing medical relief to the lakhs of people who will be affected by radiation induced illnesses in the coming years is going to be another gargantuan task. And then, of course, there is the huge task of rehabilitating the tens of thousands who have been permanently evacuated.

The public health care system in India is virtually non-existent. Our relief and rehabilitation systems are so abysmally inefficient and corrupt that even 26 years after the Bhopal gas tragedy, we have not been able to provide succour to the victims. Forget medical and economic rehabilitation, we have not been able to provide them even safe drinking water (the groundwater is poisoned)! A nuclear accident will be hundreds of times bigger than the Bhopal gas tragedy. If a nuclear accident even one-fourth the size of Fukushima takes place in a poor and technologically backward country like India, it will have apocalyptical consequences.

If the government of India continues with its diabolical nuclear program, sooner or later, a major nuclear accident is bound to take place in one of our nuclear reactors. It will destroy India. We cannot allow it to happen. We must join the countrywide anti-nuclear struggle and demand of the government of India:

1. Scrap the Jaitapur and Kudankulam nuclear power projects! Scrap all new nuclear power plants!!
2. Shut down Tarapur-1 and 2 and Rawatbhata-1 reactors immediately.
3. Phase out all other operating nuclear power plants as early as possible.
4. Invest massively in energy saving and development of renewable technologies!

Postscript

While we were giving the finishing touches to this book came the fantastic news that the West Bengal government, bowing to the resolute struggle of the people of Bengal, has decided to scrap the Haripur Nuclear Park. Not only that, the West Bengal Power Minister announced on the floor of the state Assembly that the government was opposed to setting up of a nuclear plant anywhere in the state, as there is 'no fool-proof technology to prevent accidents.'[155]

The people/united/will never be defeated!

They Did Not Register Us

Lyubov Sirota

(To all past and future victims of Chernobyl)

They did not register us
and our deaths
were not linked to the accident.
No processions laid wreaths,
no brass bands melted with grief.
They wrote us off as
lingering stress,
cunning genetic disorders . . .
But we—we are the payment for rapid progress,
mere victim of someone else's sated afternoons.
It wouldn't have been so annoying for us to die
had we known
our death would help
to avoid more "fatal mistakes"
and halt replication of "reckless deeds"!
But thousands of "competent" functionaries
count our "souls" in percentages,
their own honesty, souls, long gone—
so we suffocate with despair.
They wrote us off.
They keep trying to write off
our ailing truths
with their sanctimonious lies.
But nothing will silence us!
Even after death,
from our graves
we will appeal to your Conscience

not to transform the Earth
into a sarcophagus!

* * *

Peace unto your remains,
unknown fellow-villager!
We'll all end up there sooner or later.
Like everyone, you wanted to live.
As it turned out,
you could not survive ...
Your torment is done.
Our turn will come:
prepare us a roomier place over there.
Oh, if only our "mass departure"
could be a burning lump of truth
in duplicity's throat!
May God not let anyone else
know our anguish!
May we be extinction's limit.
For this, you died.
Peace unto your remains,
my fellow-villager
from abandoned hamlets.

Translated from Russian by Leonid Levin and Elisavietta Ritchie

(Lyubov Sirota, Ukrainian poet and playwright, is a victim of the Chernobyl nuclear accident. This is from her collection, *Burden.*)

NOTES

Introduction

1. 'India To Achieve 20,000 MW Nuclear Power Production Target By 2020', Feb 19, 2009, http://www.india-server.com
2. *Integrated Energy Policy*, 2006, p. 22, http://planningcommission.nic.in; Devirupa Mitra, 'India's Nuclear Power Corp targets 63,000 MW by 2032', Oct 8, 2009, http://www.thaindian.com
3. M.V. Ramana, *The Future of Nuclear Power in India*, April 1, 2010, http://princeton.academia.edu
4. Jeremy Page, 'India promises 12,000% boost in nuclear capacity by 2050', *The Sunday Times*, Sept 30, 2009, http://www.allvoices.com
5. 'Installed Capacity to Rise to 40,000 MWe by 2020', *Asian Nuclear Energy*, Aug 18, 2009, http://www.asiannuclearenergy.com
6. CEA, http://www.cea.nic.in
7. Yogesh Pawar, 'Tarapur an example of clean, economic and safe energy: Manmohan Singh', Jan 8, 2011, http://www.dnaindia.com
8. 'Statement by the Prime Minister of India Dr Manmohan Singh at the Nuclear Security Summit', Press Release, Embassy of India, Washington, DC, April 13, 2010, http://www.indianembassy.org
9. 'Tarapur Atomic Power Station 3&4 Dedicated to the Nation', *Nuclear India*, Vol. 41, Sept.-Oct. 2007, http://www.dae.gov.in/ni/nisep07/ni.pdf
10. 'Statement by Prime Minister Dr Manmohan Singh in Parliament on his Visit to the United States', Press Release, Embassy of India, Washington, DC, July 29, 2005, http://www.indianembassy.org
11. Darryl D'Monte, 'Politics of power plants: Is Jaitapur a safe game?', Jan 28, 2011, http://southasia.oneworld.net
12. 'Government, experts back Jaitapur project', Jan 19, 2011, http://www.thehindu.com
13. Darryl D'Monte, 'Politics of power plants: Is Jaitapur a safe game?', op. cit.; 'Government, experts back Jaitapur project', ibid.
14. 'Jaitapur N-project: Chavan allays fears over safety', Jan 18, 2011, http://www.zeenews.com/news681318.html
15. Chittaranjan Tembhekar, 'Jaitapur locals boycott meet with Chavan, scientists', *TNN*, Jan 19, 2011, http://timesofindia.indiatimes.com

16. Darryl D'Monte, 'Politics of power plants: Is Jaitapur a safe game?', op. cit.; 'Government, experts back Jaitapur project', op. cit.
17. Chittaranjan Tembhekar, 'Jaitapur locals boycott meet with Chavan, scientists', op. cit.
18. For some information about these struggles, see back issues of *Anumukti*, especially Vol. 1, No. 1; Vol. 2, No. 6; Vol. 3, Nos. 1, 6; Vol. 5, Nos. 1, 2: published by Dr Surendra Gadekar, Sampoorna Kranti Vidyalaya, Vedchhi, India-394641; available online at http://www.nonuclear.in/anumukti
19. Lison Joseph, 'Protests hold up uranium mining projects in Andhra, Meghalaya', June 20, 2008, http://www.livemint.com
20. 'Kakodkar to hold talks on uranium mining in Meghalaya', August 22, 2008, http://www.rediff.com
21. See *Anumukti,* Vol. 2, No. 6; Vol. 6, Nos. 1, 3: op. cit.
22. For more information on Kudankulam struggle: Praful Bidwai and M.V. Ramana, 'Indian Mega Nuclear Plant Protested: Government short-circuits environmental hearings', June 14 and 23, 2007, http://www.japanfocus.org/-MV-Ramana/2449; Patibandla Srikant, *Koodankulam Anti-Nuclear Movement: A Struggle for Alternative Development*?, The Institute for Social and Economic Change, Bangalore, http://www.isec.ac.in/WP%20232%20-%20P%20Srikant.pdf
23. For more on Haripur struggle: Rina Mukherji, 'No to Nuclear Plant', Mar 2007 Edition, http://www.civilsocietyonline.com; Achintyarup Ray, 'A Nuclear Flashpoint', *TNN*, Oct 26, 2009, http://www.cndpindia.org; 'National Meet Against Nuclear Power', http://www.doccentre.net/Announcements/National-Meet-Against-Nuclear-Power.php
24. 'Popular Resistance stops site preparations for NPP in India', *Nuclear Monitor*, June 18, 2010, http://www.nirs.org/mononline/nm712.pdf
25. See Chapter 3, Part V, Section 5.
26. See Epilogue for details about the Fukushima accident.
27. See the Epilogue, Part V for these facts and their references, and also more on the response of the Indian atomic energy establishment and the government to the Fukushima accident.

Chapter 1: Nuclear Energy: From Slowdown to 'Renaissance'

1. 'Atoms for Peace', http://www.u-s-history.com
2. Facts taken from: 'Nuclear power', *Wikipedia*, http://en.wikipedia.org, accessed on Dec 1, 2010.

3. Ibid.
4. Antony Froggatt and Mycle Schneider, 'The global nuclear decline', *China Dialogue*, Jan 7, 2008, http://www.chinadialogue.net
5. I. Semendeferi, 'Legitimating a Nuclear Critic: John Gofman, Nuclear Safety, and Cancer Risks', 2008, http://www.deepdyve.com; 'Anti-nuclear movement in the United States', *Wikipedia*, http://en.wikipedia.org, accessed on Dec 1, 2010.
6. 'Environment: The Struggle over Nuclear Power', *TIME*, March 8, 1976, http://www.time.com; 'Anti-nuclear movement in the United States', *Wikipedia*, ibid.
7. *Anumukti*, Vol. 9, No. 1: published by Dr Surendra Gadekar, Sampoorna Kranti Vidyalaya, Vedchhi, India-394641; available online at http://www.nonuclear.in/anumukti
8. Mycle Schneider, et al., *The World Nuclear Industry Status Report 2009*, Commissioned by German Federal Ministry of Environment, Nature Conservation and Reactor Safety, August 2009, pp. 91-2, http://gala.gre.ac.uk
9. 'Nuclear power', *Wikipedia*, op. cit.
10. Praful Bidwai, 'No global "nuclear renaissance", only limited nuclear power expansion in India', March 2010, http://www.prafulbidwai.org
11. Cited in: *Nuclear Energy no Solution to Climate Change: A background paper*, http://archive.greenpeace.org
12. Praful Bidwai, 'No global "nuclear renaissance", only limited nuclear power expansion in India', op. cit.; The Rocky Mountain Institute is an independent non-profit organisation in the United States dedicated to fostering efficient and restorative use of resources.
13. 'Nuclear power', *Wikipedia*, op.cit.
14. Tyson Slocum, 'Nuclear's Power Play: Give Us Subsidies or Give Us Death', *Multinational Monitor*, Sep/Oct 2008, http://www.multinationalmonitor.org
15. This figure obtained by adding up the the figures given in the following two articles: Tyson Slocum, 'Nuclear's Power Play: Give Us Subsidies or Give Us Death', ibid. and Judy Pasternak, 'Nuclear energy lobby working hard to win support', Jan 24, 2010, http://investigativereportingworkshop.org
16. Mike Stuckey, 'New nuclear power 'wave' – or just a ripple?', Jan 23, 2007, http://www.msnbc.msn.com
17. Judy Pasternak, 'Nuclear energy lobby working hard to win support', op. cit.
18. Mike Stuckey, 'New nuclear power 'wave' – or just a ripple?', op. cit.

19. Praful Bidwai, 'No global "nuclear renaissance", only limited nuclear power expansion in India', op. cit.
20. 'Nuclear lobby buoyant as Europe warms up to atomic energy', *Deutsche Welle*, May 19, 2009, http://www.dw-world.de
21. Helen Caldicott, *Nuclear Power is not the answer to Global Warming or anything else*, Melbourne University Press, 2006, p. xvii.
22. 'The Secret Pro-Nuclear Push In British Schools', Dec 15, 2008, http://www.sourcewatch.org
23. See for example: 'MEP-turned-lobbyist shuns voluntary register', May 20, 2009, http://brusselsbubble.blogspot.com
24. 'G8 Statement Supports Nuclear', Sept 2009, http://www.nei.org; 'G8 Summit Reaffirms Support for IAEA's Work', July 10, 2009, http://www.iaea.org
25. *Energy, Electricity and Nuclear Power Estimates for the Period up to 2030*, IAEA, 2010 edition, http://www-pub.iaea.org
26. Roland M. Frye, Jr, 'The Current Nuclear Renaissance in the United States, Its Underlying Reasons, and Its Potential Pitfalls', *Energy Law Journal*, 2008, http://www.felj.org

Chapter 2: What is Nuclear Energy?

1. Strictly speaking, plutonium does occur naturally, in very low concentrations: see 'Do transuranic elements such as plutonium ever occur naturally?' Mar 23, 1998, http://www.scientificamerican.com
2. *Uranium Resources and Nuclear Energy*, Background paper by Energy Watch Group, Germany, Dec 2006, http://www.lbst.de/ressources/docs2006
3. Low-level waste theoretically means waste containing short-lived radioactivity.
4. See Helen Caldicott, *Nuclear Power is not the answer to Global Warming or anything else*, Melbourne University Press, 2006, pp. 12-3; 'Decommissioning The Nuclear Power Industry: Rubble, Rubble, Toil And Trouble', http://www.nirs.org/mononline/decommiss.htm
5. Helen Caldicott, ibid., p. 14.
6. 'Nuclear power', *Wikipedia*, http://en.wikipedia.org, accessed on Jan 10, 2011.
7. This is so even in the US. See for example: Robert Alvarez, 'Advice for the Blue Ribbon Commission', March 24, 2010, http://www.thebulletin.org; also: 'Nuclear power', *Wikipedia*, http://en.wikipedia.org, accessed on Jan 10, 2011.
8. Intermediate-level waste contains higher amounts of radioactivity than low-level waste, and may require shielding.

9. *Nuclear Power Reactors in the World*, IAEA Reference Data Series No. 2, 2010 Edition, p. 12, http://www-pub.iaea.org
10. *Anumukti*, Vol. 1, No. 1, p. 13: published by Dr Surendra Gadekar, Sampoorna Kranti Vidyalaya, Vedchhi, India-394641; available online at http://www.nonuclear.in/anumukti

Chapter 3: Is Nuclear Energy Safe?

1. 'President Bush's Radio Address Focuses on Energy Issues', Feb 18, 2006, http://www.energy.gov/news/3222.htm
2. 'Nuclear energy safe and clean: US envoy', June 6, 2008, http://www.thaindian.com/newsportal
3. 'Manmohan Singh wants public to develop faith in nuclear safety', *ANI*, Sept 30, 2009, http://trak.in/news; 'Nuclear energy essential for India: Manmohan Singh', March 24, 2008, http://www.thaindian.com/newsportal
4. See for instance: 'N-power the new green fuel, says PM's scientific adviser', Aug 13, 2009, http://www.thaindian.com/newsportal
5. Cited in Stephen Lendman, 'Global Warming Beats Global Poisoning', Aug 8, 2006, http://www.amazon.com/gp/pdp/profile
6. Z is the atomic number, that is, the number of protons in the nucleus (defined in Chapter 2).
7. A is the mass number, that is, the sum of protons and neutrons in the nucleus (defined in Chapter 2).
8. For long, Bismuth (Z=83) was regarded as stable, till in 2003 it was demonstrated that it was radioactive with a half-life of 19 x 1018 years, that is, over a billion times longer than the current estimated age of the universe. Due to its extraordinarily long half-life, for nearly all applications, Bismuth can be treated as if it is stable and non-radioactive.
9. Information for this section is taken from Helen Caldicott, *Nuclear Power is not the answer to Global Warming or anything else*, Melbourne University Press, 2006, Chapter 3, and also from a wide variety of sources easily available on the Internet; also see S. Shawky, 'Depleted uranium: an overview of its properties and health effects', *Eastern Mediterranean Health Journal*, March 2002, http://www.emro.who.int/Publications.
10. Rosalie Bertell, 'Nuclear Radiation and its Biological Effects: Permissible Levels of Exposure', http://www.ratical.org; Isaac Asimov cited in 'An Evolutionist's Paradise', http://www.pathlights.com/ce_encyclopedia

11. Burton Goldberg, 'The Dangers of Mammograms: What You Need To Know,(Part 2)', http://alternativecancerresources.com. Information about Dr Gofman is available on many websites.
12. John W. Gofman in: *Nuclear Witnesses, Insiders Speak Out*, 1982, http://www.ratical.org
13. Cited in: 'Radiation - No safe dose', The Alliance for Clean Environment, http://www.acereport.org
14. 'US Radiation Panel Recognizes No Safe Radiation Dose', Washington, DC, July 2005, http://www.nirs.org/radiation
15. 'Next Time Someone Talks About "Safe Levels" of Radiation', Missourians for Safe Energy, March 19, 2011, http://www.facebook.com
16. Helen Caldicott, *Nuclear Power is not the answer to Global Warming or anything else*, op. cit., pp. 44-5.
17. R. Deolalikar, 'Safety in nuclear power plants in India', 2008, http://www.ijoem.com
18. Peter Eyre, 'Part 2: Uranium Weapons - Does anyone care about our planet?', http://www.paltelegraph.com/opinions
19. Even Dr APJ Abdul Kalam gives this argument: 'Address at the Launch of Paryavaran Mitra Programme of Centre for Environment Education', Pune, Dec 26 2010, http://www.abdulkalam.com
20. Helen Caldicott, *Nuclear Power is not the answer to Global Warming or anything else*, op. cit., p. 44.
21. Ibid., p. 46.
22. We have made only very cursory modifications in the main book after we finished writing it towards the end of February 2011. This section is one exception, it has been added after the Fukushima accident.
23. 'Low levels of radiation found in U.S. Milk', March 31, 2011, http://www.msnbc.msn.com
24. 'UCB Milk Sampling Results', April 13, 2011, http://www.nuc.berkeley.edu/node/2174
25. ' "Prescription for Survival": A Debate on the Future of Nuclear Energy Between Anti-Coal Advocate George Monbiot and Anti-Nuclear Activist Dr Helen Caldicott', *Democracy Now*, March 30, 2011, http://www.democracynow.org
26. Helen Caldicott, 'How nuclear apologists mislead the world over radiation', April 11, 2011, http://www.guardian.co.uk
27. Cited in: 'Fukushima Radiation Update', *Sovereigns' Health Freedom Network Newsletter,* May 3, 2011, http://www.healthfreedom.info
28. Helen Caldicott, *Nuclear Power is not the answer to Global Warming or*

anything else, op. cit., pp. 48-51; Chip Ward, 'Big Bad Boom: Radioactive Déjà Vu in the American West', June 19, 2008, *TomDispatch.com*; Peter Diehl, 'Uranium Mining and Milling Wastes: An Introduction', http://www.wise-uranium.org/uwai.html

29. Helen Caldicott, ibid., p. 48.
30. Helen Caldicott, ibid., pp. 48-51; Chip Ward, 'Big Bad Boom: Radioactive Déjà Vu in the American West', op. cit.; Peter Diehl, 'Uranium Mining and Milling Wastes: An Introduction', op. cit.
31. Material for this section, unless otherwise mentioned, taken from Peter Diehl, 'Uranium Mining and Milling Wastes: An Introduction', ibid.; Helen Caldicott, ibid., p. 50.
32. 'Navajos: recalling disaster, forging a green future', *MAC: Mines and Communities*, July 27, 2009, http://www.minesandcommunities.org
33. Peter Diehl, 'Uranium Mining in Europe: The Impacts on Man and Environment', Chapter 3, http://www10.antenna.nl/wise
34. 'Uranium Waste Pile Makes Colorado River Most Endangered', Washington, DC, April 14, 2004, http://www.ens-newswire.com/ens/apr2004; 'Decommissioning of Moab, Utah, Uranium Mill Tailings', http://www.wise-uranium.org/udmoa.html
35. Helen Caldicott, *Nuclear Power is not the answer to Global Warming or anything else*, op. cit., pp. 51-3.
36. 'Nuclear Regulatory Commission Ignores Depleted Uranium Risks, Votes to Ignore Sound Science, Its Own Prior Analysis, and Radiological Safety', Press Release, IEER, March 18, 2009, http://www.ieer.org; Helen Caldicott, ibid., p. 51.
37. 'Cogéma's Depleted Uranium Storage Project (France)', http://www.wise-uranium.org/ediss.html
38. Thomas D. Williams, 'The Depleted Uranium Threat', Aug 13, 2008, www.truthout.org; Peter Eyre, 'Israeli Use of Depleted Uranium on Gaza Residents', Aug 15, 2009, http://ccun.org
39. Arun Shrivastava, 'Depleted Uranium is "blowing in the wind" ', *Global Research*, March 2, 2006, http://www.globalresearch.ca
40. Ali al-Fadhily and Dahr Jamail, 'Iraq: "Special Weapons" Have a Fallout on Babies', *Global Research*, June 13, 2008, http://www.globalresearch.ca
41. Helen Caldicott, *Nuclear Power is not the answer to Global Warming or anything else*, op. cit., p. 54.
42. Annie Makhijani and Arjun Makhijani, 'Radioactive Rivers and Rain: Routine Releases of Tritiated Water From Nuclear Power Plants', *Science for Democratic Action*, August 2009, IEER publication, http://www.ieer.org

43. Helen Caldicott, *Nuclear Power is not the answer to Global Warming or anything else*, op. cit., pp. 54, 59.
44. Ibid., pp. 54, 59.
45. Ibid., pp. 54, 55.
46. Ibid., p. 55.
47. Ibid., p. 56.
48. Ibid., pp. 62-4; 'What are the Health Risks of Living Near a Nuclear Power Plant?', SEED Coalition, http://www.nukefreetexas.org
49. 'Radioactive Waste and Pollution', Riverkeeper, http://www.riverkeeper.org
50. Victoria Cho, 'How Safe Is Our Seafood?', Aug 31, 2010, http://ecohearth.com
51. Helen Caldicott, *Nuclear Power is not the answer to Global Warming or anything else*, op. cit., pp. 56-7; Annie Makhijani and Arjun Makhijani, 'Radioactive Rivers and Rain: Routine Releases of Tritiated Water From Nuclear Power Plants', op. cit.
52. Helen Caldicott, ibid., pp. 56-7.
53. Annie Makhijani and Arjun Makhijani, 'Radioactive Rivers and Rain: Routine Releases of Tritiated Water From Nuclear Power Plants', op. cit.; 'Nuclear Energy is Dirty Energy', NIRS Briefing Paper, Jan 25, 2011, http://www.nirs.org
54. 'Seven Ontario Hydro CANDU Reactors to Shut Down', Nuclear Awareness Project (Canada), http://www.ccnr.org
55. Helen Caldicott, *Nuclear Power is not the answer to Global Warming or anything else*, op. cit., pp. 58-9.
56. Ibid., p. 59.
57. Victoria Cho, 'How Safe Is Our Seafood?', op. cit.
58. Baker PJ, Hoel D., 'Meta-analysis of standardized incidence and mortality rates of childhood leukaemia in proximity to nuclear facilities', *European Journal of Cancer Care*, July 2007, pp. 355-363, cited in: M.V. Ramana, 'Nuclear Power: Economic, Safety, Health, and Environmental Issues of Near-Term Technologies', *Annual Review of Environment and Resources*, 2009; also cited in 'Children and young people show elevated leukaemia rates near nuclear facilities', http://www.eurekalert.org
59. Spix C., Schmiedel S., Kaatsch P., Schulze-Rath R., Blettner M., 'Case-control study on childhood cancer in the vicinity of nuclear power plants in Germany 1980–2003', *European Journal of Cancer Care*, pp. 275–84, 2008, cited in: M.V. Ramana, ibid.; paper also available online at http://www.ncbi.nlm.nih.gov

60. *Anumukti*, Vol. 5, No. 2: published by Dr Surendra Gadekar, Sampoorna Kranti Vidyalaya, Vedchhi, India-394641; available online at http://www.nonuclear.in/anumukti
61. Helen Caldicott, *Nuclear Power is not the answer to Global Warming or anything else*, op. cit., p. 60.
62. Ibid.; Poornima Gupta, 'U.S. needs fresh look at nuclear waste issue: Chu', Mar 5, 2010, http://www.reuters.com
63. *Reprocessing*, June 27, 2006, http://www.greenpeace.org
64. Helen Caldicott, *Nuclear Power is not the answer to Global Warming or anything else*, op. cit., pp. 61-2.
65. Ibid., p. 109.
66. David Biello 'Spent Nuclear Fuel: A Trash Heap Deadly for 250,000 Years or a Renewable Energy Source?', *Scientific American*, January 28, 2009, http://www.scientificamerican.com
67. Helen Caldicott, *Nuclear Power is not the answer to Global Warming or anything else*, op. cit., pp. 107-114; Steve Tetreault, 'Federal panel to examine nuclear waste storage', *Las Vegas Review Journal*, Jan 30, 2010, http://www.lvrj.com; Channing Turner, 'Liability for Scrapping Yucca Mountain Could Run In the Billions', July 27, 2010, http://www.mainjustice.com
68. 'German Nuclear Storage Facility Hit by Safety Scandal', Sept 4, 2008, http://www.dw-world.de; James Owen, 'Photos: Leaking Nuclear Waste Fills Former Salt Mine', July 8, 2010, http://news.nationalgeographic.com; 'Germany must rethink its permanent nuclear storage', Sept 3, 2008, http://www.thelocal.de
69. Frank von Hippel, Editor, *The Uncertain Future of Nuclear Energy*, Research report of: International Panel on Fissile Materials, Sept 2010, p. 53, www.fissilematerials.org
70. *Reprocessing and Nuclear Waste*, Union of Concerned Scientists, July 2009, http://www.ucsusa.org
71. Cited in Frank von Hippel, Editor, *The Uncertain Future of Nuclear Energy*, op. cit., p. 19.
72. http://en.wikipedia.org/wiki/Nuclear_reprocessing#Economics
73. *Reprocessing*, Greenpeace, June 27, 2006, http://www.greenpeace.org
74. Ibid.
75. *The Possible Toxic Effects from the Nuclear Reprocessing Plants at Sellafield and Cap de La Hague*, Report commissioned by the Scientific and Technological Options Assessment Panel of the European Parliament, and prepared by WISE-Paris, published on Nov 21, 2001, http://www.n-base.org.uk

76. 'Radioactive Flow', June 30, 1997, http://www.downtoearth.org
77. *The Possible Toxic Effects from the Nuclear Reprocessing Plants at Sellafield and Cap de La Hague*, op. cit.; 'Reprocessing', Greenpeace, op. cit.
78. *Reprocessing*, Greenpeace, ibid.; 'Nuclear Power and Children's Health', http://www.helencaldicott.com/childrenshealth_proc.pdf
79. Linda Gunter, 'The French Nuclear Industry Is Bad Enough in France; Let's Not Expand It to the U.S.', Mar 23, 2009, http://www.alternet.org
80. 'Areva trying to block TV documentary on residual contamination left around former uranium mine sites in France', Decommissioning Projects – France, updated Oct 2010, http://www.wise-uranium.org
81. Linda Gunter, 'The French Nuclear Industry Is Bad Enough in France; Let's Not Expand It to the U.S.', op. cit.
82. Ibid.; 'Radioactive spill at AREVA Uranium mine in Niger', Dec 18, 2010, http://www.greenpeace.org
83. Henry Sokolski, Editor, *Nuclear Power's Global Expansion: Weighing its costs and risks*, Dec 2010, p. 250, http://www.dtic.mil; *Residual Risk: An Account Of Events In Nuclear Power Plants Since The 1986 Chernobyl Accident*, WISE, May 18, 2005, http://www.klimaatkeuze.nl
84. 'Nuclear power in France', http://en.wikipedia.org, accessed on Jan 15, 2010; Linda Gunter, 'The French Nuclear Industry Is Bad Enough in France; Let's Not Expand It to the U.S.', op. cit.
85. Greg Keller, 'Nuclear accidents in France raise concern', July 19, 2008, http://seattletimes.nwsource.com
86. *The nuclear waste crisis in France*, Greenpeace document, May 30, 2006, http://www.greenpeace.org
87. Linda Gunter, 'The French Nuclear Industry Is Bad Enough in France; Let's Not Expand It to the U.S.', op. cit.
88. Annie Makhijani, Linda Gunter, Arjun Makhijani, 'Cogema: Above the Law?', May 7, 2002, http://www.ieer.org
89. Linda Gunter, 'The French Nuclear Industry Is Bad Enough in France; Let's Not Expand It to the U.S.', op. cit.; 'Mudding the Waters: Cogema's Hidden Environmental Crimes', http://www.nirs.org
90. Annie Makhijani, Linda Gunter, Arjun Makhijani, 'Cogema: Above the Law?', op. cit.; 'Mudding the Waters: Cogema's Hidden Environmental Crimes', http://www.nirs.org; 'France Acknowledges Massive Radioactive Pollution at La Hague', Aug 5, 1997, http://naturalscience.com
91. Full report and executive summary available on the website of NIRS: http://www.nirs.org/reactorwatch. NIRS was founded in 1978 as a national information and networking center for citizens and

environmental activists concerned about nuclear power, radioactive waste, radiation and sustainable energy issues. It is headquartered in Maryland, USA.

92. Roger Witherspoon, 'Ravishing the Waterways: DEP vs. the Power Plants', Dec 13, 2010, http://spoonsenergymatters.wordpress.com
93. Ibid.; *Licensed To Kill Executive Summary* available on the website of NIRS: http://www.nirs.org/reactorwatch
94. Helen Caldicott, *Nuclear Power is not the answer to Global Warming or anything else*, op. cit., pp. 67-71.
95. Sue Sturgis, 'Investigation: Revelations about Three Mile Island disaster raise doubts over nuclear plant safety', *Facing South*, April 2, 2009, http://www.southernstudies.org
96. Helen Caldicott, *Nuclear Power is not the answer to Global Warming or anything else*, op. cit., pp. 67-8.
97. Sue Sturgis, 'Investigation: Revelations about Three Mile Island disaster raise doubts over nuclear plant safety', op. cit.
98. Ibid.
99. 'Three Mile Island Had Lasting Consequences', Three Mile Island Alert, http://www.tmia.com; Helen Caldicott, *Nuclear Power is not the answer to Global Warming or anything else*, op. cit., pp. 71-4.
100. 'Caldicott: Japan may spell end of nuclear industry worldwide', *Global Research*, Mar 16, 2011, http://www.globalresearch.ca
101. Helen Caldicott, *Nuclear Power is not the answer to Global Warming or anything else*, op. cit., p. 75.
102. '20 years after Chernobyl - The ongoing health effects', International Physicians for the Prevention of Nuclear War (IPPNW) powerpoint presentation, Mar 23, 2006, http://www.ippnw-students.org
103. Ibid.; Helen Caldicott, *Nuclear Power is not the answer to Global Warming or anything else*, op. cit., pp. 75-80.
104. 'Chernobyl Radiation Killed Nearly One Million People: New Book', April 26, 2010, http://www.ens-newswire.com; Alexey V. Yablokov and Vassily B. Nesterenko and Alexey V. Nesterenko, 'Chernobyl: The Consequences of the Catastrophe for People and the Environment', *Global Research*, http://www.globalresearch.ca/index.php?context=va&aid=23745
105. We have derived the figure of 9000 reactor years from the figure given in Dr Georgui Kastchiev, et al., *Residual Risk: An Account of Events in Nuclear Power Plants Since the Chernobyl Accident in 1986* (see Endnote 108 on next page). They say that in 21 years, 8000 reactor years experience has accumulated, so from this we have roughly calculated for 24 years.

106. A leading nonprofit science advocacy group based in the United States, whose membership includes many eminent professional scientists.
107. The IAEA maintains an international database of these 'nuclear events', and has created the International Nuclear Event Scale (INES) to categorise these events. On this scale, events are classified on a scale from 1 to 7. INES defines events as 'deviations' (Level 0), 'anomalies' (Level 1), 'incidents' (Level 2), 'serious incidents' or 'near accidents' (Level 3) and 'accidents' (Levels 4 to 7).

    According to the IAEA, a total number of 23 Level 3 (serious incidents) and one Level 4 (accident, Tokai Mura, Japan, 1999) events have occurred in nuclear power facilities worldwide since the introduction of the INES in 1991.

    However, the scientists who did this study found that there are a host of problems with this classification system. For one, it is entirely arbitrary, there is no internationally agreed definition as to what is a serious 'event'. Secondly, the IAEA has left it to the individual member countries to report on the incidents/accidents taking place in their nuclear plants; and the countries often either do not report them, or, under political pressure, under-report the severity of the event. Then, the INES scale rates the severity of a given event only from the point of view of immediate radiological impact and not from the potential risk. It may so happen that the direct material and environmental consequences of an event are insignificant, but the event may have come very close to a serious disaster.
108. Dr Georgui Kastchiev, Prof. Wolfgang Kromp, Dipl.-Ing. Stephan Kurth, David Lochbaum, Dr Ed Lyman, Dipl.-Ing. Michael Sailer, and Mycle Schneider, *Residual Risk: An Account of Events in Nuclear Power Plants Since the Chernobyl Accident in 1986*, May 2007, http://www.greens-efa.org
109. Helen Caldicott, *Nuclear Power is not the answer to Global Warming or anything else*, op. cit., pp. 81-3.
110. http://ecobridge.org/content/n_wdo.htm
111. Philip Bethge and Sebastian Knauer, 'How Close Did Sweden Come to Disaster?' *Spiegel Online International*, Aug 7, 2006, http://www.spiegel.de
112. M.V. Ramana is a physicist who works at the Nuclear Futures Laboratory and the Program on Science and Global Security, both at Princeton University. He is a member of the International Panel on Fissile Materials and the Bulletin of the Atomic Scientists' Science and Security Board.

113. M.V. Ramana, Suchitra J.Y., 'Flaws in the pro-nuclear argument', *Infochange India*, Issue 5, 2006, http://infochangeindia.org
114. http://www.planetark.com/enviro-news/item/53381
115. Christian Parenti, 'Zombie Nuke Plants', *The Nation*, Dec 7, 2009, http://www.britannica.com
116. Karl Grossman, 'The Nuclear Phoenix', http://www.emagazine.com
117. Bob Herbert, 'We're Not Ready', July 19, 2010, http://www.nytimes.com
118. Christian Parenti, 'Zombie Nuke Plants', op. cit.
119. Ibid.
120. Ibid.
121. Stephen Thomas, et al., *The Economics of Nuclear Power*, Report prepared for Greenpeace International, May 2007, p. 19, http://www.greenpeace.org; Mycle Schneider, et al., *The World Nuclear Industry Status Report 2009*, Commissioned by German Federal Ministry of Environment, Nature Conservation and Reactor Safety, August 2009, pp. 42-3, http://gala.gre.ac.uk
122. Helmut Hirsch, Oda Becker, Mycle Schneider, Antony Froggatt, *Nuclear Reactor Hazards: Ongoing Dangers of Operating Nuclear Technology in the 21st Century*, Greenpeace International, April 2005, www.greenpeace.org
123. Ibid.
124. Helen Caldicott, *Nuclear Power is not the answer to Global Warming or anything else*, op. cit., p. 83.
125. Art Levine, 'Helen Caldicott Slams Environmental Groups on Climate Bill, Nuclear Concessions', *Truthout*, Jan 28, 2010, http://www.climateshift.com

Chapter 4: Is Nuclear Energy Cheap?

1. For example, in 1968, Pacific Gas & Electric projected the total cost of building Unit 1 of its Diablo Canyon plant at $445 million. By 1984, the final bill was $3750 million, an 843 per cent increase. Likewise, the Fermi Unit 2 plant in Michigan, the Public Power Supply System Unit 2 reactor in Washington state, the River Bend reactor in Louisiana, and the Clinton plant in Illinois all cost approximately 600 per cent more than originally estimated: *Nuclear power: Too expensive to solve global warming*, The National Environmental Trust, Washington D.C., November 1, 1999, Report available on: http://www.democraticunderground.com; Steve Thomas, *Can nuclear power plants be built in Britain without public subsidies and guarantees?*,

Presentation at a Conference: 'Commercial Nuclear Energy in an Unstable, Carbon Constrained World', Prague, March 17-18, 2008, p. 10, http://gala.gre.ac.uk

2. *An Analysis of Nuclear Power Plant Operating Costs: A 1995 Update*, Energy Information Administration, US Department of Energy, Washington, DC, April 1995, http://tonto.eia.doe.gov
3. Mycle Schneider, et al., *The World Nuclear Industry Status Report 2009*, Commissioned by German Federal Ministry of Environment, Nature Conservation and Reactor Safety, August 2009, p. 40, http://gala.gre.ac.uk; Same point is made by: Steve Thomas, *Can nuclear power plants be built in Britain without public subsidies and guarantees?*, op. cit., p. 7.
4. Mycle Schneider, et al., ibid., p. 40.
5. Steve Thomas, *Can nuclear power plants be built in Britain without public subsidies and guarantees?*, op. cit., p. 14.
6. *The New Economics of Nuclear Power*, Report of the World Nuclear Association, London, p. 7, http://www.world-nuclear.org/reference/pdf/economics.pdf
7. See Chapter 3, Part V, Section 5.
8. Stephen Thomas, et al., *The Economics of Nuclear Power*, Report prepared for Greenpeace International, May 2007, p. 3, http://www.psiru.org/reports/2007-06-E-the-economics-of-nuclear-power.pdf
9. Stephen Thomas has called it a Generation III design, while Mycle Schneider has called it a Generation III+ design. See: Stephen Thomas, et al., *The Economics of Nuclear Power*, ibid., p. 16; and Mycle Schneider, et al., *The World Nuclear Industry Status Report 2009*, op. cit., pp. 42-3.
10. Stephen Thomas, et al., ibid., pp. 16-7.
11. 'The Other Economic Summit' (TOES) was a counter to the annual G-7 summits, and was first held in the UK in 1984 by a diverse group of activists concerned with social and environmental justice. TOES later became an umbrella term and similar conferences were held around the world for the next two decades.
12. Andrew Simms, et al., *Mirage and Oasis: Energy Choices in an Age of Global Warming*, June 29, 2005, http://www.neweconomics.org
13. Cited in: John Murawski, 'Putting a Price on Nuclear Power', *The News & Observer*, Raleigh, N.C., Feb 21, 2006, http://www.redorbit.com/news/science
14. *The Economic Future of Nuclear Power*, University of Chicago, August 2004, http://www.eusustel.be

15. Trevor Findlay, *The Future of Nuclear Energy to 2030 and its Implications for Safety, Security and Nonproliferation*, The Centre for International Governance Innovation (CIGI), Waterloo, Ontario, Canada, February 2010, http://www.cigionline.org
16. The levelised cost is the sum of all the costs incurred during the lifetime of the project divided by the units of energy produced during the lifetime of the project, expressed as $/kWh. Since most of the costs or expenses as well as the sales revenue occur in a future time, one has to take the Present Value of these cash flows in one's calculations.
17. *Update of the MIT 2003 Future of Nuclear Power Study*, Massachusetts Institute of Technology, 2009, http://web.mit.edu/nuclearpower/pdf/nuclearpower-update2009.pdf
18. 'Costs', http://www.downtheyellowcakeroad.org/html/costs.html
19. Ibid.
20. Mycle Schneider, et al., *The World Nuclear Industry Status Report 2007*, Brussels, Nov 2007, Commissioned by the Greens-EFA Group in the European Parliament, p. 11, http://www.energiestiftung.ch
21. Ibid.
22. Cited in: Mycle Schneider, et al., *The World Nuclear Industry Status Report 2009*, op. cit., p. 55.
23. David Schlissel, et al., *Nuclear Loan Guarantees: Another Taxpayer Bailout Ahead?*, Union of Concerned Scientists, March 2009, p. 19, www.ucsusa.org
24. Tyler Hamilton, '$26bn cost killed nuclear bid', July 14, 2009, http://www.thestar.com
25. Anton Caputo, 'Nuclear expansion could cost $18.2 billion', Dec 23, 2009, http://www.mysanantonio.com; Richard W. Caperton, 'Protecting Taxpayers from a Financial Meltdown', March 8, 2010, http://www.americanprogress.org
26. 'Experts: Three latest industry setbacks further dim nuclear "renaissance", taxpayer-backed loan guarantees can't fix fundamental problems with new reactors', Dec 16, 2009, www.psr.org/nuclear-bailout/nuclear4.pdf
27. 'Olkiluoto provision could mean losses for Areva', *The Nuclear N-Former*, June 24, 2010, http://www.nuclearcounterfeit.com; same figure is cited in: 'AREVA covers up extent of massive nuclear reactor cost overrun', June 24, 2010, http://www.greenpeace.org/international/en/news
28. Mycle Schneider, et al., *The World Nuclear Industry Status Report 2009*, op. cit., p. 45.

29. James Kanter, 'In Finland, Nuclear Renaissance Runs Into Trouble', May 28, 2009, http://www.nytimes.com
30. Mycle Schneider, et al., *The World Nuclear Industry Status Report 2009*, op. cit., pp. 45-6; Frank von Hippel, Editor, *The Uncertain Future of Nuclear Energy*, A research report of the International Panel on Fissile Materials, September 2010, p. 41, http://www.fissilematerials.org
31. 'AREVA covers up extent of massive nuclear reactor cost overrun', op. cit.
32. Ibid.
33. Mycle Schneider, et al., *The World Nuclear Industry Status Report 2009*, op. cit., p. 46.
34. Ibid., p. 47.
35. 'Nuclear lightning strikes twice, three times…', July 6, 2010, http://www.greenpeace.org/international/en/news
36. Dan Yurman, 'Areva under pressure', July 31, 2010, http://theenergycollective.com
37. For example, see the statement by French Foreign Minister Bernard Kouchner, cited in Mycle Schneider, 'The reality of France's aggressive nuclear power push', June 3, 2008, http://www.thebulletin.org
38. Mycle Schneider, 'The reality of France's aggressive nuclear power push', ibid.
39. Mark Cooper, 'France's Nuclear "Miracle" is More Fantasy that Fact', Sept 23, 2010, http://thehill.com; *France's Nuclear Failures - The great illusion of nuclear energy*, Greenpeace, Nov 30, 2008, http://www.greenpeace.org
40. Mycle Schneider, 'The reality of France's aggressive nuclear power push', op. cit.
41. *France's Nuclear Failures - The great illusion of nuclear energy*, op. cit.
42. Helen Caldicott, *Nuclear Power is not the answer to Global Warming or anything else*, Melbourne University Press, 2006, p. 32; same figure cited by Steve Thomas, *Can nuclear power plants be built in Britain without public subsidies and guarantees?*, op. cit., p. 12.
43. David H. Martin, *Canadian Nuclear Subsidies: Fifty Years of Futile Funding*, Campaign for Nuclear Phaseout, Jan 2003, www.cnp.ca/resources/nuclear-subsidies-at-50.pdf
44. Mycle Schneider, et al., *The World Nuclear Industry Status Report 2009*, op. cit., p. 88.
45. Ibid., p. 44.
46. David Schlissel, et al., *Nuclear Loan Guarantees: Another Taxpayer Bailout Ahead?*, op. cit., pp. 1, 11.

47. Ibid., pp. 1, 11.
48. Ibid., p. 11.
49. Ibid., pp. 12-3; Mycle Schneider, et al., *The World Nuclear Industry Status Report 2009*, op. cit., p. 77.
50. Mycle Schneider, et al., ibid., pp. 76-7.
51. The levelised cost is the sum of all the costs incurred during the lifetime of the project divided by the units of energy produced during the lifetime of the project, expressed as $/kWh. Since most of the costs or expenses as well as the sales revenue occur in a future time, one has to take the Present Value of these cash flows in the calculations.
52. Mycle Schneider, et al., *The World Nuclear Industry Status Report 2009*, op. cit., p. 73.
53. Steve Thomas, *Can nuclear power plants be built in Britain without public subsidies and guarantees?*, op. cit., p. 4.
54. David Schlissel, et al., *Nuclear Loan Guarantees: Another Taxpayer Bailout Ahead?*, op. cit., p. 2.
55. 'Nuclear Pork – Enough is Enough', May 9, 2008, http://climateprogress.org
56. 'President Barack Obama grants loan guarantee to build new nuclear reactors in Georgia', Feb 16, 2010, http://www.oregonlive.com
57. Cited in: Mycle Schneider, et al., *The World Nuclear Industry Status Report 2009*, op. cit., p. 56.
58. 'Senate Currently Proposing $40 Billion to More Than $140 Billion in Subsidies for Nuclear Industry, New Analysis Finds', Union of Concerned Scientists, July 1, 2010, http://www.ucsusa.org
59. Estimate by Doug Koplow, an economist and founder of Earth Track, based in Cambridge, Massachusetts, which campaigns against environmentally harmful subsidies. Cited in: Diana S. Powers, 'Nuclear Energy Loses Cost Advantage', July 26, 2010, http://www.nytimes.com
60. 'Vogtle's new reactors to benefit not only from Georgia CWIP, but also from federal loan guarantee', Jan 1, 2010, http://www.beyondnuclear.org; Frank von Hippel, Editor, *The Uncertain Future of Nuclear Energy*, op. cit., p. 40.
61. J. J. Dooley, *U.S. Federal Investments in Energy R&D: 1961-2008*, Report prepared for US Department of Energy, Oct 2008, http://www.pnl.gov
62. Cindy Folkers, 'Price-Anderson Act: Unnecessary & Irresponsible', Oct 2001, http://www.nirs.org
63. 'Price-Anderson Nuclear Industries Indemnity Act', *Wikipedia*, http://en.wikipedia.org

64. 'Price-Anderson Act: The Billion Dollar Bailout for Nuclear Power Mishaps,' Sept 2004, http://www.citizen.org
65. Gopal Krishna, 'Nuclear Liability Bill needs scrutiny', Feb 25, 2010, http://news.rediff.com
66. Cindy Folkers, 'Price-Anderson Act: Unnecessary & Irresponsible', op. cit.
67. Silva Herrmann, 'The Price Tag of Nuclear Power', Friends of the Earth Europe, http://www.foeeurope.org; Mycle Schneider, et al., *The World Nuclear Industry Status Report 2009*, op. cit., pp. 68, 75.
68. See Helen Caldicott, *Nuclear Power is not the answer to Global Warming or anything else*, op. cit., pp. 99-105 for a more detailed discussion of this.
69. Mycle Schneider, et al., *The World Nuclear Industry Status Report 2009*, op. cit., p. 44.
70. Kim Cawley, 'The Federal Government's Responsibilities and Liabilities Under the Nuclear Waste Policy Act', Statement before Committee on the Budget, US House of Representatives, http://www.cbo.gov/ftpdocs
71. 'Yucca Mountain cost estimate rises to $96 billion', Aug 6, 2008, http://www.world-nuclear-news.org
72. Steve Tetreault, 'Federal panel to examine nuclear waste storage', *Las Vegas Review Journal*, Jan 30, 2010, http://www.lvrj.com
73. 'Yucca Mountain cost estimate rises to $96 billion', op. cit.
74. 'Decommissioning Risks', http://www.banktrack.org; 'Everything you want to know about Nuclear Power', Nuclear Power Education, http://nuclearinfo.net
75. Steve Thomas, cited in: Mycle Schneider, et al., *The World Nuclear Industry Status Report 2009*, op. cit., p. 44.
76. 'Economics of new nuclear power plants', *Wikipedia*, http://en.wikipedia.org
77. Bruce Biewald and David White, 'Stranded Nuclear Waste: Implications of Electric Industry Deregulation for Nuclear Plant Retirements and Funding Decommissioning and Spent Fuel', Jan 15, 1999, http://www.synapse-energy.com
78. 'Experts: Three latest industry setbacks further dim nuclear "renaissance", taxpayer-backed loan guarantees can't fix fundamental problems with new reactors', op. cit.
79. Steve Thomas, *The Economics of Nuclear Power*, Heinrich Boll Stiftung, Dec 2005, p. 29, http://www.nirs.org
80. Ibid., pp. 11, 18.

81. Mycle Schneider, et al., *The World Nuclear Industry Status Report 2009*, op. cit., p. 46.
82. Ibid., p. 58.
83. Ibid., p. 58; Steve Thomas, *Can nuclear power plants be built in Britain without public subsidies and guarantees?*, op. cit., p. 16.
84. Mycle Schneider, et al., ibid., p. 56; Steve Thomas, *The Economics of Nuclear Power*, Heinrich Boll Stiftung, op. cit., p. 11.

Chapter 5: Is Nuclear Energy Green?

1. 'Nuclear energy essential for India: Manmohan Singh', March 24, 2008, http://www.thaindian.com/newsportal
2. 'N-power the new green fuel, says PM's scientific adviser', August 13, 2009, http://www.siliconindia.com/hownews
3. 'Canadian Nuclear Association clings to inaccurate, unsupported claims', Sept 30, 2010, http://www.northumberlandview.ca
4. Helen Caldicott, *Nuclear Power is not the answer to Global Warming or anything else*, Melbourne University Press, 2006, pp. 8-9.
5. Ibid., p. 9.
6. Helen Caldicott, 'Nuclear Power is the Problem, Not a Solution', *Australian*, April 15, 2005, http://www.helencaldicott.com
7. Jan Willem Storm van Leeuwen, 'Nuclear power – the energy balance', http://www.stormsm :th.nl; Helen Caldicott, *Nuclear Power is not the answer to Global Warming or anything else*, op. cit., p. 6.
8. Surendra Gadekar, 'India's nuclear fuel shortage', *Bulletin of the Atomic Scientists*, Aug 6, 2008, http://www.thebulletin.org
9. Ibid.
10. Statistics taken from the flowchart: *World Greenhouse Gas Emissions 2005*, World Resources Institute, http://www.wri.org
11. Ibid.
12. The IEA is an intergovernmental organisation which acts as energy policy advisor to 28 member countries.
13. *Energy Technology Perspectives 2008*, IEA/OECD, June 2008, cited in: *Nuclear power: a dangerous waste of time*, Greenpeace, Jan 2009, http://www.greenpeace.org; also cited in Praful Bidwai, 'No global "nuclear renaissance", only limited nuclear power expansion in India', March 25, 2010, http://www.prafulbidwai.org
14. Praful Bidwai, ibid.
15. Estimate made in: *Getting Serious about Nuclear Power*, Greenpeace, www.greenpeace.org; and *Nuclear power: a dangerous waste of time*, Greenpeace, Jan 2009, http://www.greenpeace.org

16. Bill Dougherty, senior scientist, Stockholm Environmental Institute, cited in: *Nuclear Power and Children's Health*, symposium Proceedings, Chicago, Illinois, Oct 15-16, 2004, http://www.helencaldicott.com/childrenshealth_proc.pdf
17. *Nuclear power: a dangerous waste of time*, op. cit., p. 11.

Chapter 6: Global Nuclear Energy Scenario: Reviewing the Renaissance

1. *IAEA-PRIS: IAEA-Power Reactor Information System*, January 9, 2010, http://www.iaea.org
2. *Energy, Electricity and Nuclear Power Estimates for the Period up to 2030*, IAEA, 2010 edition, http://www-pub.iaea.org
3. Statistics taken from 2009, 2008 and 2007 editions of IAEA reports: *Energy, Electricity and Nuclear Power Estimates for the Period up to 2030*, IAEA, http://www-pub.iaea.org
4. http://www-pub.iaea.org
5. Statistics taken from 2010, 2009, 2008 and 2007 editions of IAEA reports: *Energy, Electricity and Nuclear Power Estimates for the Period up to 2030*, IAEA, http://www-pub.iaea.org
6. Mycle Schneider, et al., *The World Nuclear Industry Status Report 2009*, Commissioned by German Federal Ministry of Environment, Nature Conservation and Reactor Safety, Aug 2009, p. 9, http://gala.gre.ac.uk
7. *IAEA-PRIS: IAEA-Power Reactor Information System*, January 9, 2010, http://www.iaea.org
8. *Energy, Electricity and Nuclear Power Estimates for the Period up to 2030*, IAEA, 2010 edition, op. cit.
9. Mycle Schneider, et al., *The World Nuclear Industry Status Report 2009*, op. cit., pp. 9, 10; Trevor Findlay, *The Future of Nuclear Energy to 2030 and its Implications for Safety, Security and Nonproliferation – Overview*, The Centre for International Governance Innovation (CIGI), Waterloo, Ontario, Canada, February 2010, p. 10, http://www.cigionline.org
10. Mycle Schneider, et al., ibid., pp. 10, 11; *Energy, Electricity and Nuclear Power Estimates for the Period up to 2030*, IAEA, 2010 edition, op. cit.
11. Statistics taken from 2010, 2009, 2008 and 2007 editions of IAEA reports: *Energy, Electricity and Nuclear Power Estimates for the Period up to 2030*, IAEA, http://www-pub.iaea.org
12. *IAEA, International Status and Prospects of Nuclear Power, 2008*, cited in: Mycle Schneider, et al., *The World Nuclear Industry Status Report 2009*, op. cit., p. 15.

13. Mycle Schneider, et al., ibid., p. 14.
14. Ibid., p. 11.
15. Ibid., pp. 14-7, 115-6.
16. Ibid., pp. 17-8.
17. Fact admitted to by Trevor Findlay, *The Future of Nuclear Energy to 2030 and its implications for safety, security and nonproliferation – Overview*', op. cit., p. 9.
18. Saibal Dasgupta, 'No hand in NKorea N-facilities: China', *The Times of India*, April 20, 2009, http://timesofindia.indiatimes.com
19. Hideo Kubota, 'China's Nuclear Industry at a Turning Point', *EJAM*, Vol. 1, No. 3 (Nov 2009), http://www.jsm.or.jp
20. Dinah Deckstein, et al., 'Is the nuclear industry in meltdown?', Oct 21, 2009, http://www.presseurop.eu
21. Compiled from: Gordon G. Chang, 'The Best And Worst Of Times For China's Environment', July 9, 2010, http://www.forbes.com; Andreas Lorenz and Wieland Wagner, 'The downside of the boom: China's Poison for the Planet', Feb 1, 2007, http://www.global-sisterhood-network.org; Jonathan Watts, 'China's "cancer villages" reveal dark side of economic boom', June 7, 2010, http://www.guardian.co.uk
22. *Energy, Electricity and Nuclear Power Estimates for the Period up to 2030*, IAEA, 2010 edition, op. cit.
23. *Anumukti*, Vol. 8, No. 3, p. 13 and Vol. 12, No. 4, p. 10: published by Dr Surendra Gadekar, Sampoorna Kranti Vidyalaya, Vedchhi, India-394641; available online at http://www.nonuclear.in/anumukti
24. Compiled from: *Mayak*, Greenpeace, May 2002, http://www.greenpeace.org; *Greenpeace highlights 50 years of nuclear disaster in Mayak, Russia*, Sept 28, 2007, http://www.greenpeace.org; *Mayak: A 50-Year Tragedy*, Greenpeace, Sept 2007, http://www.greenpeace.org; 'Ural Mountains Radiation Pollution', www1.american.edu; *Anumukti*, Vol. 12, No. 4, p. 10 and Vol. 2, No. 5, pp. 14-5, ibid.
25. Testimony submitted by David Lochbaum, Director, Nuclear Safety Project, to the Select Committee on Energy Independence and Global Warming, United States House of Representatives, Mar 12, 2008, http://globalwarming.house.gov
26. Mycle Schneider, et al., *The World Nuclear Industry Status Report 2009*, op. cit., pp. 91-2.
27. Ibid.
28. 'Obama Administration Announces $8.33B in Loan Guarantees for New Nuclear Power Reactors in Georgia; First New US Nuclear Power

Plant Project in Nearly 3 Decades', Feb 17, 2010, http://www.greencarcongress.com

29. Mycle Schneider, et al., *The World Nuclear Industry Status Report 2009*, op. cit., p. 92.
30. Ibid., p. 92.
31. Tyson Slocum, 'Nuclear's Power Play: Give Us Subsidies or Give Us Death', *Multinational Monitor*, Sep/Oct 2008, http://www.multinationalmonitor.org
32. Mycle Schneider, et al., *The World Nuclear Industry Status Report 2009*, op. cit., p. 92; 'Federal Loan Guarantees for Nuclear Reactors', Jan 2010, http://www.nonukesyall.org
33. Jackson Browne, 'It's a Nuclear Retreat, not Renaissance', Aug 27, 2009, http://www.jacksonbrowne.com; Caroline Stetler, 'Nuclear lobbying turns to states', Jan 24, 2010, http://investigativereportingworkshop.org
34. 'Vermont Senate vote shows that Obama's nuclear renaissance is dead on arrival', Feb 25, 2010, http://members.greenpeace.org/blog
35. 'Experts: Three latest industry setbacks further dim nuclear "renaissance", taxpayer-backed loan guarantees can't fix fundamental problems with new reactors', Dec 16, 2009, www.psr.org/nuclear-bailout/nuclear4.pdf
36. Mycle Schneider, et al., *The World Nuclear Industry Status Report 2009*, op. cit., p. 92.
37. 'Experts: Energy Department Should "Immediately Halt" Plans to Issue Taxpayer-Backed Loan Guarantees in Wake of Major NRC Safety Objection to Westinghouse Reactor Design', October 22, 2009, http://www.prnewswire.com/news-releases
38. Sue Sturgis, 'New safety issues documented with nuclear reactors planned for Southeast', April 22, 2010, http://www.southernstudies.org; 'Groups Urge Feds to Suspend Nuclear Licensing; Westinghouse Reactor Defect Was Missed By Regulators', April 21, 2010, http://gawand.org
39. Mycle Schneider, et al., *The World Nuclear Industry Status Report 2009*, op. cit., p. 92.
40. 'Aborted nuclear loan guarantee for Calvert Cliffs 3 causes global ambitions for EPR to "cool" ', *Beyond Nuclear*, Oct 26, 2010, http://www.beyondnuclear.org; Peggy Hollinger, Stephanie Kirchgaessner, 'Constellation pulls out of EDF project', *Financial Times*, Oct 9, 2010, http://www.ft.com
41. 'Exelon's Rowe: Low gas prices and no carbon price push back nuclear

renaissance a "decade, maybe two" ', *Climate Progress*, Oct 12, 2010, http://climateprogress.org

42. Richard W. Caperton, 'Protecting Taxpayers from a Financial Meltdown', March 8, 2010, http://www.americanprogress.org
43. Robert K. Temple, 'Saving Japan's Nuclear Reactor Industry', August 4, 2010, http://www.japaninc.com; Katarzyna Klimasinska, 'NRG Energy May Stop Texas Nuclear-Plant Expansion (Update3)', Jan 29, 2010, http://www.bloomberg.com
44. Mycle Schneider, et al., *The World Nuclear Industry Status Report 2009*, op. cit., pp. 90-1.
45. David H. Martin, 'The CANDU Syndrome: Canada's Bid to Export Nuclear Reactors to Turkey', Sept 1997, http://www.ccnr.org
46. 'Nuclear Power in Canada', Nov 30, 2010, http://www.world-nuclear.org
47. Mycle Schneider, et al., *The World Nuclear Industry Status Report 2009*, op. cit., pp. 90-1; Trevor Findlay, *The Future of Nuclear Energy to 2030 and its Implications for Safety, Security and Nonproliferation – Overview*, op. cit., p. 12.
48. For our definition of Western Europe, we have included the 15 countries which constituted the European Union before its expansion in 2004, and added to it the 3 countries whose names are also there in list of West European countries as defined by National Geographic Society. For this latter list, see 'Western Europe', http://en.wikipedia.org. Other definitions of Western Europe include tiny countries like Andorra, Liechtenstein, Malta, Monaco and San Marino, but that makes no difference to our discussion as all these are non-nuclear countries.
49. Mycle Schneider, et al., *The World Nuclear Industry Status Report 2009*, op. cit., pp. 98-9; we have added figures for Switzerland to the figures given by Mycle Schneider.
50. Ibid., pp. 98-9.
51. 'Nuclear energy policy', *Wikipedia*, http://en.wikipedia.org, accessed on Dec 6, 2010; *Coming Clean: How clean is nuclear energy*, published by Stichting GroenLinks in the European Union, October 2000, http://www10.antenna.nl/wise
52. 'Anti-nuclear Movement in Austria', July 10, 2010, http://www.greenkids.de
53. 'Nuclear Energy in Denmark', May 2010, http://www.world-nuclear.org
54. 'Greece announces nuclear moratorium', Feb 11, 2009, http://

www.edie.net

55. 'Nuclear Power in Italy', May 2010, http://www.world-nuclear.org
56. 'Nuclear Power in the Netherlands', Nov 2010, http://www.world-nuclear.org
57. Mycle Schneider, et al., *The World Nuclear Industry Status Report 2009*, op. cit., pp. 104-5.
58. Ibid., p. 99.
59. Ibid., p. 103; 'Anti-nuclear movement in Germany', *Wikipedia*, http://en.wikipedia.org, accessed on Dec 6, 2010.
60. 'Nuclear Power in Spain', *Wikipedia*, http://en.wikipedia.org, accessed on Dec 6, 2010; Mycle Schneider, et al., ibid., p. 104.
61. 'Nuclear Power in the Netherlands', op. cit.
62. 'Netherlands Looking into Second New Nuclear Power Plant', Sept 8, 2010, http://nuclearstreet.com
63. Sonal Patel, 'Belgium, Germany Edge Toward Nuclear Future', Dec 1, 2009, http://www.powermag.com
64. 'The postponement of the nuclear phase-out postponed', May 18, 2010, http://belgianenergylaw.blogspot.com; 'Belgium asked the outgoing government to manage the country during the crisis', Oct 6, 2010, http://www.pisqa.com; *Nuclear power and coalition government*, Greenpeace, May 12, 2010, http://www.greenpeace.org
65. Mycle Schneider, et al., *The World Nuclear Industry Status Report 2009*, op. cit., pp. 104-5; Terry Macalister, 'Sweden lifts ban on nuclear power', Feb 5, 2009, guardian.co.uk
66. 'Swedish parliament narrowly reverses ban on new nuclear power plants', June 17, 2010, http://www.dw-world.de
67. 'PM defends minority government formation', Oct 10, 2010, http://www.stockholmnews.com
68. 'Swedish parliament narrowly reverses ban on new nuclear power plants', op. cit.; Mycle Schneider, et al., *The World Nuclear Industry Status Report 2009*, op. cit., pp. 104-5.
69. Elisabeth Rosenthal, 'Italy Embraces Nuclear Power', *The New York Times*, May 23, 2008, http://www.nytimes.com/2008; Mycle Schneider, et al., ibid., p. 25.
70. Israel Rafalovich, 'Italy's Nuclear Plans at Risk', Feb 4, 2010, http://www.energytribune.com; Guy Dinmore, 'Berlusconi crisis hits nuclear revival', Nov 22, 2010, http://guydinmore.wordpress.com
71. Tony Czuczka, 'Merkel's Nuclear-Power Plans Are Passed as Greens Cry "Putsch" ', Oct 28, 2010, http://www.businessweek.com
72. Kate Connolly, 'Germany agrees to extend life of nuclear power

stations', Sept 6, 2010, http://www.guardian.co.uk

73. Mycle Schneider, et al., *The World Nuclear Industry Status Report 2009*, op. cit., p. 104; Cristina Mateo-Yanguas, 'Nuclear energy might not be on its way out in Spain after all', Oct 26, 2009, http://www.globalpost.com; 'Nuclear Power in Spain', *Wikipedia*, http://en.wikipedia.org, accessed Nov 19, 2010.
74. Mycle Schneider, et al., ibid., pp. 105-7.
75. Carsten Volkery, 'Britain's Nuclear Renaissance in Doubt under New Government', *Spiegel Online*, July 21, 2010, http://www.spiegel.de/international/europe
76. Mycle Schneider, et al., *The World Nuclear Industry Status Report 2009*, op. cit., pp. 107-8.
77. 'Finland approves two new nuclear reactors', July 1, 2010, http://www.swedishwire.com
78. Mycle Schneider, et al., *The World Nuclear Industry Status Report 2009*, op. cit., pp. 101-3.
79. *Framatome EPR Brochure*, cited in *EPR: The French Reactor*, Greenpeace, Nov 2008, http://www.greenpeace.org
80. *Factsheet: Olkiluoto 3*, Greenpeace, Apr 23, 2008, http://www.greenpeace.org
81. Ibid.; 'French, UK, Finnish Regulators: Have Raised Areva EPR Issues', Nov 6, 2009, http://www.chinanuclear.cn
82. *Factsheet: Olkiluoto 3*, ibid.; H. Hirsch and W. Neumann, *Progress and Quality Assurance Regime at the EPR Construction at Olkiluoto*, Report prepared for Greenpeace Nordic, Hanover, May 14, 2007, http://www.greenpeace.org
83. H. Hirsch and W. Neumann, ibid.; see also *Factsheet: Olkiluoto 3*, ibid.
84. H. Hirsch and W. Neumann, ibid.; see also *Factsheet: Olkiluoto 3*, ibid.
85. *Factsheet: Olkiluoto 3*, ibid.
86. Dr Helmut Hirsch has about 30 years of experience as nuclear expert. He has worked for the Austrian Federal Government as well as for German State Government and municipal administrations. He is also a member of OECD's Nuclear Energy Agency expert group.
87. H. Hirsch and W. Neumann, *Progress and Quality Assurance Regime at the EPR Construction at Olkiluoto*, op. cit.
88. *EPR: The French Reactor*, Greenpeace International, Nov 2008, http://www.greenpeace.org; A. Gopalakrishnan, 'Reject French reactors for Jaitapur', Dec 2, 2010, http://www.cndpindia.org
89. Discussed in Chapter 4.

90. Stephen Thomas, 'Really, Mr Huhne, you should brush up on your French', Sept 1, 2010, http://www.parliamentarybrief.com
91. 'Flamanville: spinning the cost increases', Dec 5, 2008, http://weblog.greenpeace.org
92. Mycle Schneider, et al., *The World Nuclear Industry Status Report 2009*, op. cit., p. 102.
93. 'France hit by anti-nuclear protests', *Evening Echo News*, March 17, 2007, http://www.eveningecho.ie; 'Greenpeace assault on EPR', *Nuclear Engineering International*, May 1, 2007, http://www.neimagazine.com
94. *EPR: The French Reactor*, op. cit.; 'Flamanville: spinning the cost increases', op. cit.
95. STUK, Press Release, July 12, 2006, cited in: Mycle Schneider, et al., *The World Nuclear Industry Status Report 2009*, op. cit., p. 100.
96. Mycle Schneider, et al., ibid., p. 100; *Factsheet, Olkiluoto-3*, Greenpeace, op. cit.
97. Stephen Thomas, 'Really, Mr Huhne, you should brush up on your French', op. cit.
98. Ibid.; *Factsheet, Olkiluoto-3*, Greenpeace, op. cit.; 'French, UK, Finnish Regulators: Have Raised Areva EPR Issues', op. cit.
99. Stephen Thomas, ibid.
100. Robin Pagnamenta, 'UK regulator raises French nuclear concerns', July 1, 2009, http://business.timesonline.co.uk
101. 'Nuclear Renaissance in Disarray', August 5, 2010, http://www.nirs.org/home/080510nirsconstellationnewsrelease.pdf; Stephen Thomas, 'Really, Mr Huhne, you should brush up on your French', op. cit.
102. 'Nuclear Renaissance in Disarray', ibid.; Stephen Thomas, ibid.
103. Stephen Thomas, ibid.
104. Ibid.
105. 'Prognos: Nuclear power losing in importance world-wide', Jan 23, 2010, http://www.lovisamovement.eu; 'Nuclear power losing in importance world-wide', Feb 11, 2010, http://cndpindia.org
106. Trevor Findlay, *The Future of Nuclear Energy to 2030 and its Implications for Safety, Security and Nonproliferation – Overview*, op. cit.; *The Future of Nuclear Energy to 2030: Final Report of the Nuclear Energy Futures Project*, Feb 4, 2010, http://www.cigionline.org

Chapter 7: India's Nuclear Energy Program

1. M.V. Ramana, *Nuclear Power in India: Failed Past, Dubious Future*, March 2007, http://www.isn.ethz.ch

2. Dhirendra Sharma, *India's Nuclear Estate*, Lancers Publishers, New Delhi, 1983, pp. 28-29 (available on internet at www.psaindia.org); Yash Thomas Mannully & V.N. Haridas, 'Civil Liability For Nuclear Damage Bill, 2010: An Ideological Twin Of Indian Atomic Energy Establishment', *Countercurrents.org*, Aug 25, 2010, http://www.countercurrents.org
3. Book Reviews: C.V. Sundaram, L.V. Krishnan and T.S. Iyengar, 'Atomic Energy in India - 50 years', DAE, Government of India, 1998, *Current Science*, Vol. 77, No. 11, Dec. 10, 1999, http://www.ias.ac.in/currsci/dec101999/BOOKREVIEWS.PDF
4. M.V. Ramana, 'Nehru, Science and Secrecy', http://www.reocities.com/m_v_ramana/nucleararticles/Nehru.pdf
5. Dhirendra Sharma, *India's Nuclear Estate*, op. cit., pp. 20, 31; Yash Thomas Mannully & V.N. Haridas, 'Civil Liability For Nuclear Damage Bill, 2010: An Ideological Twin Of Indian Atomic Energy Establishment', op. cit.
6. M.V. Ramana, *Nuclear Power in India: Failed Past, Dubious Future*, op. cit.
7. Dhirendra Sharma, *India's Nuclear Estate*, op. cit., p. 32.
8. Ibid., pp. 36-7, 149.
9. Ibid., pp. 16-7, 31-2, 36-7; *Anumukti*, Vol. 11, Nos. 5 & 6, pp. 4-5: published by Dr Surendra Gadekar, Sampoorna Kranti Vidyalaya, Vedchhi, India-394641; available online at http://www.nonuclear.in/anumukti
10. Praful Bidwai, 'Playing with Nuclear Fire: Lessons from the Kursk Catastrophe', *Peacework magazine*, Oct 2000.
11. M.V. Ramana, *Nuclear Power in India: Failed Past, Dubious Future*, op. cit.
12. Ibid.
13. Ibid.; Dhirendra Sharma, *India's Nuclear Estate*, op. cit., pp. 51-2.
14. Dhirendra Sharma, ibid., pp. 51-2.
15. M.V. Ramana, 'The Future of Nuclear Power in India', April 1, 2010, http://princeton.academia.edu
16. M.V. Ramana, *Nuclear Power in India: Failed Past, Dubious Future*, op. cit.
17. Shankar Sharma, 'Shadow Integrated Energy Policy (IEP)', p. 27, National Alliance of Movements Against Coal, www.indiaenvironmentportal.org.in
18. CEA, http://www.cea.nic.in
19. Shankar Sharma, 'Shadow Integrated Energy Policy (IEP)', op. cit., p. 27.

20. CEA, http://www.cea.nic.in
21. M.V. Ramana, *Nuclear Power in India: Failed Past, Dubious Future,* op. cit.
22. Rahul Goswami, 'India's "stage three" fantasy', *Himal South Asian,* Oct 2009, www.indiaenvironmentportal.org.in
23. CEA, http://www.cea.nic.in
24. 'Uranium imports boost Indian reactor output', Oct 12, 2010, http://www.world-nuclear-news.org
25. 'Nuclear Power in India', updated Nov 30, 2010, http://www.world-nuclear.org; 'Mohuldih Uranium Project', http://www.ucil.gov.in
26. 'Tummalapalle mine to start producing uranium by year end', Aug 10, 2010, http://www.zeenews.com
27. Srinivas Sirnoorkar, 'India's third uranium mining unit at Gogi', May 13, 2010, http://www.deccanherald.com
28. U.A. Shimrary, M.V. Ramana, 'Uranium mining in Meghalaya: a simmering problem', Dec 29, 2007, http://www.minesandcommunities.org; 'Uranium mining project in the West Khasi Hills of Meghalaya, India', March 19, 2010, http://coalgeology.com
29. 'Nuclear Power in India', op. cit.
30. Nuclear Power Corporation of India website: http://www.npcil.nic.in, accessed on Jan 6, 2011.
31. 'RAPP-6 attains criticality', Jan 24, 2010, http://news.webindia123.com
32. Nuclear Power Corporation of India website: http://www.npcil.nic.in, accessed on Jan 6, 2011.
33. 'Nuclear Power in India', op. cit.; M.V. Ramana, *The Indian Nuclear Industry: Status and Prospects,* The Centre for International Governance Innovation Nuclear Energy Futures Paper No. 9, December 2009, http://www2.carleton.ca/cctc/ccms/wp-content/ccms-files/Nuclear_Energy_WP9.pdf
34. M.V. Ramana, ibid.; 'India - Country Nuclear Waste Profile - NEWMDB – IAEA', http://newmdb.iaea.org/profiles.aspx?ByCountry=IN
35. M.V. Ramana, ibid.
36. 'Nuclear Non-Proliferation Treaty', *Wikipedia,* http://en.wikipedia.org
37. More detailed discussion of this is beyond the scope of this book. For more details, see: Neeraj Jain, *Globalisation or Recolonisation,* January 2007, and Neeraj Jain, *India Becoming a Colony Again,* Oct 2010, both published by Lokayat, Pune; available on internet at 'Lokayat', http://www.lokayat.org.in

38. For the full statement, see: *Joint Statement Between President George W. Bush and Prime Minister Manmohan Singh*, http://georgewbush-whitehouse.archives.gov
39. Brahma Chellaney, 'Revelations unravel hype and spin', *The Hindu*, Sept 5, 2008, http://www.hindu.com
40. 'Nuke deals' first gains for India, power from imported uranium', Sept 1, 2009, http://economictimes.indiatimes.com
41. 'Uranium imports boost Indian reactor output', Oct 12, 2010, http://www.world-nuclear-news.org
42. 'Nuclear Power in India', op. cit.
43. Nuclear Power Corporation of India website: http://www.npcil.nic.in, accessed on January 6, 2011.
44. Parnab Dhal Samanta, 'India, France ink nuclear deal, first after NSG waiver', Sept 30, 2008, http://www.indianexpress.com
45. Anil Sasi, '10-fold n-power capacity scale-up on', *The Hindu Business Line*, Jan 31, 2010, http://www.thehindubusinessline.in
46. T. S. Subramanian, 'Kudankulam power early next year', *Frontline*, Aug 1, 2009, http://www.hinduonnet.com; 'Russia, India sign defence, nuclear deals', Dec 21, 2010, http://economictimes.indiatimes.com
47. 'Fission statement', *The Telegraph*, Dec 5, 2010, http://www.telegraphindia.com
48. 'Impetus to Indo-French ties, N-plant to come up in Maha', Dec 6, 2010, http://www.theindiapost.com
49. 'Haripur Energy Park Project On Track', Jan 13, 2010, http://www.kolkatascoop.com; 'Revenue officials begin land survey for Kovvada power plant', Nov 5, 2010, http://timesofindia.indiatimes.com
50. 'Nuclear Power in India', op. cit.
51. Srikumar Banerjee, 'India harnesses nuclear energy for development', August 7, 2010, http://www.commodityonline.com; Preeti Parashar, 'Haryana's first N-power plant to come up at Fatehabad village', Mar 9, 2010, http://www.financialexpress.com

Chapter 8: India: The True Economic Costs of Nuclear Electricity

1. 'Statement by Prime Minister Dr Manmohan Singh in Parliament on his Visit to the United States', Press Release, Embassy of India, Washington, DC, July 29, 2005, http://www.indianembassy.org; see the 'Introduction' of this book for the full quote.
2. 'Tarapur Atomic Power Station-3&4 Dedicated to the Nation', *Nuclear*

India, Vol. 41, Sept.-Oct. 2007, http://www.dae.gov.in/ni/nisep07/ni.pdf; see 'Introduction' of this book for the full quote.

3. Brahma Chellaney, 'Nuclear deal: Elusive benefits, tangible costs', *The Hindu*, Aug 19, 2010, http://www.hindu.com. A recent press release of the NPCIL says the average per unit cost of electricity from its nuclear power plants has been lower, in the range of Rs.2.28 to Rs.2.34 per unit for last four years: 'Jaitapur N-plant to produce power at competitive rate: NPCIL', August 31, 2009, http://in.news.yahoo.com. However, as we argue ahead, all these official nuclear electricity tariffs have no relation to reality.
4. Surendra Gadekar, 'Revival of the Nuclear Dream', http://www.members.tripod.com
5. M.V. Ramana, 'Heavy Subsidies in Heavy Water: Economics of Nuclear Power in India', *Economic and Political Weekly*, Mumbai, Aug 25, 2007, p. 3489, http://www.cised.org
6. Ibid., p. 3486.
7. Located at Kaiga Atomic Power Station, near Karwar, Karnataka.
8. Figure taken from: M.V. Ramana, Antonette D'Sa, Amulya K.N. Reddy, 'Economics of Nuclear Power from Heavy Water Reactors', E*conomic and Political Weekly*, Mumbai, April 23, 2005, p. 1765, http://www.laka.org
9. Jason Morgan, 'Nuclear Waste Disposal and Plant Decommissioning Costs', Mar 22, 2010, http://nuclearfissionary.com
10. M.V. Ramana, J.Y. Suchitra, *Economic and Environmental Costs of Nuclear Power*, paper presented at the Ninth Biennial Conference of the International Society of Ecological Economics, New Delhi.
11. Ibid.
12. M.V. Ramana, Antonette D'Sa, Amulya K.N. Reddy, 'Economics of Nuclear Power from Heavy Water Reactors', op. cit., p. 1765.
13. Located at Rawatbhata Atomic Power Station, near Kota, Rajasthan.
14. M.V. Ramana, Antonette D'Sa, Amulya K.N. Reddy, 'Economics of Nuclear Power from Heavy Water Reactors', op. cit., p. 1765.
15. Brahma Chellaney, 'Nuclear deal: Elusive benefits, tangible costs', op. cit.
16. Jayanth Jacob, 'France signs pact on Jaitapur N-plant', Dec 7, 2010, http://www.hindustantimes.com
17. See, for example: 'India, France sign nuclear reactor sale deals', Dec 6, 2010, http://www.khaleejtimes.com
18. See Chapter 4.
19. 'We are partners over the long haul', interview with Anne Lauvergeon,

CEO of Areva, Nov 25, 2010, http://www.thehindu.com

20. 'Jaitapur: Repeating Enron Once Again – Prabir Purkayastha', Jan 11, 2011, http://indiacurrentaffairs.org
21. Submission made by Dr Vivek Monteiro in response to *Environmental Impact Assessment (EIA) Report of the Jaitapur-Madban nuclear power plant* prepared by NEERI, at the statutory public hearing on the EIA report held on May 16, 2010 in Madban village (unpublished).
22. Praful Bidwai, 'Spurring nuclear Bhopals?', *Frontline*, Aug 1-14, 2009, http://www.hinduonnet.com; also available at: http://groups.yahoo.com/group/SAAN_/message/1286
23. Brahma Chellaney, 'Nuclear deal: Elusive benefits, tangible costs', op. cit.
24. Ibid.; Vinay Kumar and J. Balaji, ' "Intent" dropped; Lok Sabha adopts nuclear liability Bill', *The Hindu*, Aug 26, 2010, http://www.hindu.com; 'Lok Sabha passes Nuclear Liability Bill', Aug 25, 2010, http://timesofindia.indiatimes.com
25. Vinay Kumar and J. Balaji, ' "Intent" dropped; Lok Sabha adopts nuclear liability Bill', ibid.
26. The Bill as passed by the Parliament allows the government to increase the maximum liability by notification. Copy of bill as passed by Lok Sabha available at: http://www.prsindia.org
27. Praful Bidwai, 'Spurring nuclear Bhopals?', op. cit.
28. 'Nuclear liability bill needs to go through, says Ficci chief', Mar 21, 2010, http://economictimes.indiatimes.com; KS Narayanan, 'FICCI wants Clause 17 off NL Bill', Aug 25, 2010, http://expressbuzz.com
29. Praful Bidwai, 'Passing the Nuclear Buck', *Tehelka*, Vol 7, Issue 16, April 24, 2010, http://www.tehelka.com
30. Praful Bidwai, 'Wrong call on nuclear liability', July 20, 2010, http://www.prafulbidwai.org
31. Brahma Chellaney, 'Revisions in N-liability bill a must', Apr 6, 2010, http://epaper.timesofindia.com
32. Brahma Chellaney, 'Kill the nuclear liability Bill', March 10, 2010, http://www.livemint.com
33. Siddharth Varadarajan, 'India-Russia nuclear talks hit liability snag', July 30, 2010, http://www.thehindu.com; 'US fears Indian nuclear liability law may deter foreign suppliers', Sept 7, 2010, http://economictimes.indiatimes.com
34. Arun Kumar, 'US presses India to further water down nuclear liability law', Sept 25, 2010, http://www.wsws.org
35. See: Neeraj Jain, *India Becoming a Colony Again*, October 2010, Lokayat Publication, Pune

Chapter 9: India's Nuclear Installations: Safety and Other Issues

1. T.S. Gopi Rethinaraj, 'In The Comfort Of Secrecy', *The Bulletin of Atomic Scientists*, November/December 1999, Vol. 55, No. 6, pp. 52-7, available online at: *South Asians Against Nukes Dispatch 2* (3 Nov, 99), http://www.insaf.net
2. A.S. Panneerselvan, 'Close To A Critical Mess', *Outlook*, Nov 8, 1999, http://www.outlookindia.com
3. T.S. Gopi Rethinaraj, 'In The Comfort Of Secrecy', op. cit.
4. Dhirendra Sharma, *India's Nuclear Estate*, Lancers Publishers, New Delhi, 1983, pp. 121-2; available on internet at *www.psaindia.org*
5. T.S. Gopi Rethinaraj, 'In the Comfort of Secrecy', op. cit.
6. Sunita Dubey, 'India's Uranium Nightmare', *Siliconeer, News Feature*, Mar 19, 2007, http://news.newamericamedia.org
7. 'Frequently asked Questions to UCIL', http://www.ucil.gov.in/web/faq.html
8. Praful Bidwai, 'Kaiga: Question mark over nuclear safety', *Rediff News*, Dec 4, 2009, http://news.rediff.com
9. Praful Bidwai, 'Lessons from Kaiga', *Frontline*, Dec. 19, 2009, http://www.hinduonnet.com
10. Sugata Srinivasaraju, 'Atoms and Leaks: The Kaiga N-plant leak leaves too many questions unanswered', *Outlook*, Dec 14, 2009, http://www.outlookindia.com; we have converted microcuries into megabecquerels.
11. Ibid.
12. 'Sabotage in Kaiga: Tritium added to drinking water', *Economic Times*, Nov 30, 2009, http://economictimes.indiatimes.com; Praful Bidwai, 'Kaiga: Question mark over nuclear safety', op. cit.
13. Praful Bidwai, 'Lessons from Kaiga', op. cit.; M.V. Ramana and Ashwin Kumar, 'Safety First? Kaiga and Other Nuclear Stories', *Economic and Political Weekly*, Mumbai, Feb 13, 2010, available online at http://princeton.academia.edu
14. Praful Bidwai, 'Kaiga: Question mark over nuclear safety', op. cit.
15. Ibid.
16. *Anumukti*, Vol. 2, No. 6, p. 3: published by Dr Surendra Gadekar, Sampoorna Kranti Vidyalaya, Vedchhi, India-394641; available online at http://www.nonuclear.in/anumukti
17. T.S. Gopi Rethinaraj, 'In the Comfort of Secrecy', op. cit.; Buddhi Kota Subbarao, 'India's Nuclear Prowess: False Claims and Tragic Truths', *Manushi*, Issue 109, Nov-Dec 1998, http://www.indiatogether.org/manushi; M.A. Rane, 'Dangers from nuclear

power projects', *PUCL Bulletin*, Nov 2006, http://www.pucl.org
18. 'India: Jadugoda Case and Supreme Court Judgment', May 12, 2004, http://groups.yahoo.com
19. Moushumi Basu, 'Uranium Corporation of India Limited: Wasting Away Tribal Lands', *CorpWatch,* Oct 7, 2009, http://www.corpwatch.org; 'Frequently asked Questions to UCIL', http://www.ucil.gov.in/web/faq.html
20. 'Frequently asked Questions to UCIL', ibid.
21. Tarun Kanti Bose, 'Adivasis live under nuclear terror in Jadugoda, Jharkhand', *Jharkhand Forum*, http://jadugoda.jharkhand.org.in
22. Sunita Dubey, 'India's Uranium Nightmare', op. cit.; Suhrid Sankar Chattopadhyay, T. S. Subramanian, 'Villages and woes', *Frontline*, Aug 28 - Sep 10, 1999, http://www.hinduonnet.com
23. M. Channa Basavaiah, 'No More Jadugoda : People of Nalgonda, A.P. Oppose Uranium Mining', *Peace Now*, Special bulletin, Dec 2004, http://cndpindia.org
24. Hiroaki Koide, *Radioactive contamination around Jadugoda uranium mine in India*, Research Reactor Institute, Kyoto University, April 27, 2004, http://www.rri.kyoto-u.ac.jp
25. Aparna Pallavi, 'Uranium mine waste imperils villages in Jaduguda', *Down to Earth*, Mar 15, 2008, http://www.downtoearth.org.in; Moushumi Basu, 'Uranium Corporation of India Limited: Wasting Away Tribal Lands', op. cit.
26. *Tehelka* is an independent national newsmagazine published from Delhi which has acquired reputation for its investigative reporting.
27. G. Vishnu, 'High Grade Energy, Low Grade Safety', *Tehelka*, Sept 25, 2010, http://www.tehelka.com
28. 'Issues at Jaduguda Uranium Mine, Jharkhand, India', http://www.wise-uranium.org/umopjdg.html
29. BIRSA is a research and documentation centre started by intellectuals and activists related to various people's movements of Jharkhand.
30. Tarun Kanti Bose, 'Adivasis live under nuclear terror in Jadugoda, Jharkhand', op. cit.
31. Richard Mahapatra, 'Eyewitness: Radioactivity doesn't stop at the mines in Jaduguda', *Down to Earth*, April 30, 2004, http://www.downtoearth.org.in; Lina Krishnan, 'Jadugoda: Four decades of nuclear exposure', *Infochange News and Features*, June 2008, http://www.scribd.com
32. Aparna Pallavi, 'Uranium mine waste imperils villages in Jaduguda', op. cit.; Moushumi Basu, 'Uranium Corporation of India Limited:

Wasting Away Tribal Lands', op. cit.

33. http://www10.antenna.nl/wise/index.html?http://www10.antenna.nl/wise/362-3/region.html
34. M. Channa Basavaiah, 'No More Jadugoda : People of Nalgonda, A.P. Oppose Uranium Mining', op. cit.
35. Bengt G. Karlsson, 'Nuclear Lives: Uranium Mining, Indigenous Peoples, and Development in India', *Economic and Political Weekly*, Mumbai, Aug 22, 2009; available online at: http://www.global-sisterhood-network.org/content/view/2345/59/
36. Buddhi Kota Subbarao, 'India's Nuclear Prowess: False Claims and Tragic Truths', op. cit.; T.S. Gopi Rethinaraj, 'In The Comfort Of Secrecy', op. cit.; The Hanford Works, located in the US state of Washington, was home to a number of nuclear reactors and plutonium processing complexes during the 1950s-60s, and produced plutonium for the nuclear weapons in the US military arsenal. While all the reactors have since been decommissioned, the manufacturing processes have left behind so many millions of litres of high-level radioactive waste that the site is today the most contaminated nuclear site in the United States.
37. The *Nuclear Engineering International* magazine is an independent magazine being published since 1956 and is renowned for providing independent technical and business analysis for the nuclear power industry.
38. Cited in: T.S. Gopi Rethinaraj, 'In The Comfort Of Secrecy', op. cit.
39. Helen Caldicott, *Nuclear Power is not the answer to Global Warming or anything else*, Melbourne University Press, 2006, p. 152.
40. The Safe Energy Communication Council is a non-profit council in Washington, DC of ten environmental and public interest media groups that promote environmentally safe energy.
41. Cited in: Adnan Gill, 'Indian nuclear program: disasters in making', July 10, 2006, http://www.bhopal.net/opinions
42. Nicholas Lenssen, Christopher Flavin, 'Meltdown', *World Watch Magazine*, May-June, 1996, http://findarticles.com
43. Lekha Rattanani, 'How safe are our N-reactors?', *Outlook*, July 10, 1996, http://www.outlookindia.com
44. Ibid.; M.V. Ramana, *Nuclear Power in India: Failed Past, Dubious Future*, March 2007, http://www.isn.ethz.ch
45. Buddhi Kota Subbarao, 'India's Nuclear Prowess: False Claims and Tragic Truths', op. cit.
46. A. Gopalakrishan, 'Issues of nuclear safety', *Frontline*, Mar. 13 - 26,

1999, http://www.hinduonnet.com

47. Buddhi Kota Subbarao, 'India's Nuclear Prowess: False Claims and Tragic Truths', op. cit.
48. Lekha Rattanani, 'How safe are our N-reactors?', op. cit.
49. Manika Gupta, 'Centre to set up two nuclear power plants', *Economic Times*, Mar 6, 2007, http://articles.economictimes.indiatimes.com; also cited in: V.K. Shashikumar, 'Leaks at India's nuclear-power plants: cause for concern?' *The Christian Science Monitor*, October 11, 2002, http://www.csmonitor.com; Prof. T. Shivaji Rao, 'Myths of Nuclear Safety Dose and Economics', http://sites.google.com/site/profshivajirao/nuclearsafety-4
50. Benjamin K. Sovacool, 'A Critical Evaluation of Nuclear Power and Renewable Electricity in Asia', *Journal of Contemporary Asia*, Vol. 40, No. 3, August 2010, pp. 393-400: cited in 'Nuclear Power in India', *Wikipedia*, http://en.wikipedia.org
51. Frederick Wolf, cited in M.V. Ramana and Ashwin Kumar, 'Safety First? Kaiga and Other Nuclear Stories', op. cit.
52. *Anumukti*, Vol. 7, No. 5, p. 3, op. cit.
53. Ibid., pp. 3-6; A.S. Panneerselvan, 'Close To A Critical Mess', op. cit.; T.S. Gopi Rethinaraj, 'In The Comfort Of Secrecy', op. cit.; A. Gopalakrishnan, 'Issues of nuclear safety', op. cit.
54. We are discussing Tarapur 3&4 only in brief, as we do not know much about the state of these reactors.
55. Ram Parmar, Yogesh Sadhwani, 'Radioactive leak averted at Tarapur Atomic Plant', July 7, 2010, http://www.mumbaimirror.com
56. *Anumukti*, Vol. 5, No. 2, pp. 12-3 & Vol. 5, No. 6, p. 7, op. cit.
57. We are not discussing RAPS-3, 4, 5 & 6, as we do not know anything about the state of these reactors.
58. Buddhi Kota Subbarao, 'India's Nuclear Prowess: False Claims and Tragic Truths', op. cit.; 'Nuclear Power in India', updated Nov 30, 2010, http://www.world-nuclear.org; Adnan Gill, 'Indian nuclear program: Disasters-in-Making', July 10, 2006, http://www.paktribune.com; T.S. Gopi Rethinaraj, 'In the Comfort of Secrecy', op. cit.; *National Report to The Convention on Nuclear Safety*, Government of India, September 2007, http://www.dae.gov.in
59. T.S. Gopi Rethinaraj, 'In the Comfort of Secrecy', ibid.; A.S. Panneerselvan, 'Close To A Critical Mess', op. cit.
60. *National Report to The Convention on Nuclear Safety*, op. cit.
61. A.S. Panneerselvan, 'Close To A Critical Mess', op. cit.; A. Gopalakrishnan, 'Issues of nuclear safety', op. cit.

62. A.S. Panneerselvan, ibid.
63. Buddhi Kota Subbarao, 'India's Nuclear Prowess: False Claims and Tragic Truths', op. cit.
64. J. Sri Raman, 'Tsunamis and a Nuclear Threat in the South of India', Jan 2, 2005, www.truthout.org
65. A. Gopalakrishnan, 'Issues of nuclear safety', op. cit.
66. Buddhi Kota Subbarao, 'India's Nuclear Prowess: False Claims and Tragic Truths', op. cit.; T.S. Gopi Rethinaraj, 'In the Comfort of Secrecy', op. cit.; M.V. Ramana, *The Indian Nuclear Industry: Status and Prospects*, The Centre for International Governance Innovation Nuclear Energy Futures Paper No. 9, December 2009, http://www2.carleton.ca/cctc/ccms/wp-content/ccms-files/Nuclear_Energy_WP9.pdf; *Anumukti*, Vol. 6, No. 6, p. 5, op. cit.
67. Buddhi Kota Subbarao, ibid.; *Anumukti*, Vol. 7, No. 6, pp. 4-5 and Vol. 8, No. 4, pp. 8-9, ibid.
68. *Anumukti*, Vol. 5, No. 2, p. 13, ibid.
69. *Anumukti*, Vol. 7, No. 6, pp. 1-2 and Vol. 8, No. 3, p. 3, ibid.
70. 'Kakrapar nuclear reactor shut down', *The Hindu*, April 23, 2004, http://hindu.com
71. A. Gopalakrishnan, 'Issues of nuclear safety', op. cit.
72. T.S. Gopi Rethinaraj, 'In The Comfort Of Secrecy', op. cit.; Buddhi Kota Subbarao, 'India's Nuclear Prowess: False Claims and Tragic Truths', op. cit.
73. Buddhi Kota Subbarao, ibid.; *Anumukti*, Vol. 6, No. 1, pp. 3-4, op. cit.
74. Kunal Majumdar, 'Accident Sites - radiation, cancer, blindness, tardiness, cover-ups: The lessons from the Kalpakkam nuclear facility', *Tehelka*, Sept 11, 2010, http://www.tehelka.com
75. C. Shivakumar, 'Kalpakkam's forgotten people', Oct 20, 2008, http://www.sacw.net
76. A. Gopalakrishnan, 'Undermining nuclear safety', *Frontline*, June 24-July 07, 2000, http://www.hinduonnet.com
77. M.V. Ramana and Ashwin Kumar, 'Safety First? Kaiga and Other Nuclear Stories', op. cit.
78. M.V. Ramana, *The Indian Nuclear Industry: Status and Prospects*, op. cit.
79. *Anumukti*, Vol. 7, No. 6, pp. 3-4 and Vol. 8, No. 4, pp. 8-9, op. cit.
80. *Anumukti*, Vol. 7, No. 6, pp. 1-2 and Vol. 8, No. 3, p. 5-7, ibid.
81. *Anumukti*, Vol. 5, No. 6, p. 3, Vol. 6, No. 1, pp. 1, 14, and Vol. 6, No. 2, p. 4, ibid.
82. M.V. Ramana and Ashwin Kumar, 'Safety First? Kaiga and Other

Nuclear Stories', op. cit.; M.V. Ramana, *The Indian Nuclear Industry: Status and Prospects*, op. cit.

83. 'Nuclear brass allay fears over Jaitapur plant', Jan 7, 2010, http://www.newkerala.com
84. 'Indian nuclear plants are safe: Scientists', *The Hindu*, March 15, 2011, http://www.hindu.com
85. *Anumukti*, Vol. 1, No. 5, p. 5 and Vol. 7, No. 5, pp. 9-10, op. cit.; 'When will we ever learn: Surendra Gadekar', http://www.anumukti.in
86. 'List of Boiling Water Reactors', *Wikipedia*, http://en.wikipedia.org, accessed on July 20, 2011.
87. 'Doubts raised on Tarapur nuclear plants safety', *The Siasat Daily*, April 13, 2011, http://www.siasat.com
88. 'Fukushima - Radioactive cloud over nuclear renaissance - Prabir Purkayastha', March 27, 2011, http://indiacurrentaffairs.org
89. 'DAE claims Tarapur spent fuel rods not a health hazard', *The Asian Age*, March 5, 2011, http://www.asianage.com
90. Discussed earlier, in Chapter 3, Part III, Section 6.
91. Matin Zuberi, *The Nuclear Deal: India cannot be coerced*, ORF Issue Brief No. 4, Nov 2005, http://www.observerindia.com
92. Robert Alvarez is a senior scholar at Institute for Policy Studies and served as senior policy advisor to the Energy Department's secretary from 1993 to 1999. Cited in Helen Caldicott, *Nuclear Power is not the answer to Global Warming or anything else*, op. cit., pp. 99-105.
93. Helen Caldicott, ibid.
94. 'Nuclear power in India', http://en.wikipedia.org
95. Praful Bidwai, 'Lessons from Kaiga', op. cit.
96. M.V. Ramana and Ashwin Kumar, 'Safety First? Kaiga and Other Nuclear Stories', op. cit.; T.S. Subramanian, 'The Kalpakkam 'incident'', *Frontline*, August 16, 2003, http://www.hinduonnet.com; M.V. Ramana, *The Indian Nuclear Industry: Status and Prospects*, op. cit.
97. Kunal Majumdar, 'Accident Sites—radiation, cancer, blindness, tardiness, cover-ups: The lessons from the Kalpakkam nuclear facility', op. cit.
98. Ibid.
99. Praful Bidwai, 'Lessons from Kaiga', op. cit.; M.V. Ramana and Ashwin Kumar, 'Safety First? Kaiga and Other Nuclear Stories', op. cit.; for a detailed discussion of the impact of tritium on the human body, see Chapter 3, Part III, section 4.
100. See for instance: *Anumukti*, Vol. 7, No. 4, p. 8, op. cit.; Elsa Fayner,

'Subcontractors: sacrificed for nuclear energy', Aug 15, 2011, http://owni.eu; Hiroko Tabuchi, 'Temporary nuclear laborers paid little for taking big risk', April 10, 2011, http://articles.boston.com; ''Nuclear nomads' in German nuclear power plants', http://economicsnewspaper.com; John Chan, 'Untrained labourers working in Japan's nuclear industry', April 14, 2011, http://www.wsws.org

101. Both examples cited in: M.V. Ramana and Ashwin Kumar, 'Safety First? Kaiga and Other Nuclear Stories', op. cit.
102. *Anumukti*, Vol. 8, No. 3, p. 3, op. cit.
103. Buddhi Kota Subbarao, 'India's Nuclear Prowess: False Claims and Tragic Truths', op. cit.
104. V.K. Shashikumar, 'Leaks at India's nuclear-power plants: cause for concern?', *The Christian Science Monitor*, October 11, 2002, http://www.csmonitor.com
105. 'Activists to protest against proposed nuke power plant', April 22, 2007, http://news.webindia123.com; M.V. Ramana, J.Y. Suchitra, *Economic and Environmental Costs of Nuclear Power*, paper was originally presented at the Ninth Biennial Conference of the International Society of Ecological Economics, New Delhi; Dr Surendra Gadekar and Dr Sanghamitra Gadekar, 'Health Survey Around an Indian Nuclear Power Plant', *Science for Democratic Action*, IEER publication, November 2002, http://www.friendsofbruce.ca/Print%20Press%20Files/IndianCanduHealthSurvey.pdf
106. Darryl D'Monte, 'Politics of power plants: Is Jaitapur a safe game?', Jan 28, 2011, http://southasia.oneworld.net
107. http://aratichokshy.blogspot.com
108. The study found the average prevalence of goiter among women above the age of 14 to be 11.5% in the local area within a distance of 6 kms around the MAPS, 3.8% at a distance of 40 kms from the plant, and 1.4% 400 kms away from the plant; the average prevalence of AITD was 2% in the local area, 0.9% 40 kms away, and 0.2% in the region 400 kms away from the plant.
109. V. Pugazhenthi, et al., *Prevalence of goiter and autoimmune thyroid disorder in the local area of Madras Atomic Power Station, India – Results of a cross sectional epidemiological study*, unpublished study: for copies – V. T. Padmanabhan, P. O. Pathayakkunnu, Tellicherry, Kerala, India, Pin – 670 691; some results from this survey published in: C. Shivakumar, 'Kalpakkam's forgotten people have a tale to tell', Oct 20, 2008, http://www.sacw.net
110. C. Shivakumar, ibid.

111. Cited in: Buddhi Kota Subbarao, 'India's Nuclear Prowess: False Claims and Tragic Truths', op. cit.; C. Shivakumar, ibid.
112. Thomas B. Cochran, et al., 'It's Time to Give Up on Breeder Reactors', *Bulletin of the Atomic Scientists*, May-June 2010, available on internet at www.princeton.edu/sgs
113. M.V. Ramana, *Nuclear Power in India: Failed Past, Dubious Future*, op. cit.
114. Thomas B. Cochran, et al., 'It's Time to Give Up on Breeder Reactors', op. cit.
115. Ibid.; M.V. Ramana, *Nuclear Power in India: Failed Past, Dubious Future*, op. cit.
116. M.V. Ramana, ibid.; 'IEER-CNIC release re: Rokkasho', Sept 27, 2001, http://www.ieer.org
117. Thomas B. Cochran, et al., 'It's Time to Give Up on Breeder Reactors', op. cit.
118. Ibid.; Thomas B. Cochran, et al., *Fast Breeder Reactor Programs: History and Status*, Research report of: International Panel on Fissile Materials, Feb 2010, pp. 10, 18-9, www.fissilematerials.org; Frank von Hippel, Editor, *The Uncertain Future of Nuclear Energy*, Research report of: International Panel on Fissile Materials, September 2010, p. 34, www.fissilematerials.org
119. Thomas B. Cochran, et al., 'It's Time to Give Up on Breeder Reactors', ibid.; See also statements by scientists Pavel Podvig and Thomas B. Cochran in: 'History and status of fast breeder reactor programs worldwide', Feb 17, 2010, http://www.fissilematerials.org; Allison Macfarlane, 'Russia's Nukes – Canning Plutonium: Faster and Cheaper', *Bulletin of the Atomic Scientists*, May / June 1999, http://belfercenter.ksg.harvard.edu
120. M.V. Ramana, 'India and Fast Breeder Reactors', 2009, www.princeton.edu; M.V. Ramana, *The Indian Nuclear Industry: Status and Prospects,* op. cit.
121. M.V. Ramana, 'India and Fast Breeder Reactors', ibid.; M.V. Ramana, *Nuclear Power in India: Failed Past, Dubious Future*, op. cit; Praful Bidwai, 'No global "nuclear renaissance", only limited nuclear power expansion in India', Mar 25, 2010, http://www.prafulbidwai.org
122. M.V. Ramana, *Nuclear Power in India: Failed Past, Dubious Future*, ibid.
123. Ibid.
124. Rahul Goswami, 'India's "stage three" fantasy', *Himal South Asia*, Oct 2009, http://himalmag.com

125. J.Y. Suchitra and M.V. Ramana, *The Costs of Power: Plutonium and the Economics of India's Prototype Fast Breeder Reactor*, submitted for publication, cited in: Frank von Hippel, Editor, *The Uncertain Future of Nuclear Energy*, op. cit., p. 32.
126. Most of our reactors are 220 MW, and the latest and biggest ones are 540 MW, which are therefore one-eighth to one-third the size of the 1650 MW EPR proposed to be built in Jaitapur.
127. Cited in: Linda Gunter, 'The French Nuclear Industry Is Bad Enough in France; Let's Not Expand It to the U.S.', Mar 23, 2009, http://www.alternet.org
128. Vineeta Pandey, 'Deal done, Jaitapur nuke power plant gets a leg up', Dec 7, 2010, http://www.dnaindia.com
129. http://aratichokshy.blogspot.com
130. Ibid.
131. Praful Bidwai and M.V. Ramana, 'Home, Next to N-Reactor', *Tehelka*, June 23, 2007 and *Japan Focus*, June 18, 2007, http://www.tehelka.com
132. N. Myers, R.A. Mittermeier, C.G. Mittermeier, G.A.B. Da Fonseca, and J. Kent, 'Biodiversity Hotspots for Conservation Priorities', *Nature* 403:853–858, Feb 24, 2000, cited in 'Western Ghats', *Wikipedia*, http://en.wikipedia.org, accessed on Jan 30, 2010; Anil Singh, 'Ecology expert Prof Madhav Gadgil visits Western Ghats', Oct 11, 2010, http://timesofindia.indiatimes.com
133. Praful Bidwai, 'People vs Nuclear Power in Jaitapur, Maharashtra', *Economic and Political Weekly*, Mumbai, Feb 19, 2011, http://www.tni.org
134. Cited in: *Submission for the Public Hearing to be held on June 2, 2007 regarding the Koodankulam Nuclear Power Project*, June 1, 2007, made by CISED, ISEC Campus, Nagarabhavi, Bangalore-72
135. Dhaval Kulkarni, 'Jaitapur project safe, say nuclear experts', Jan 8, 2011, http://www.indianexpress.com; Vinaya Deshpande, 'It's paradoxical that environmentalists are against nuclear energy: Jairam Ramesh', Nov 29, 2010, http://www.hindu.com
136. Figures given in: *Comprehensive EIA for Additional Units 3 to 6 of Nuclear Power Plant at Kudankulam, Dist. Tirunelveli, Tamil Nadu*, prepared by National Environmental Engineering Research Institute, Nagpur, 2004, p. 3.26, and cited in *Submission for the Public Hearing to be held on June 2, 2007 regarding the Koodankulam Nuclear Power Project*, June 1, 2007, made by CISED, ISEC Campus, Nagarabhavi, Bangalore-72; also cited in: Praful Bidwai and M.V. Ramana, 'Home, Next to N-Reactor', op. cit.

137. T.N. Raghunatha, 'Pawar, Chavan's nod to increasing fish export irks Maha fishermen', Jan 11, 2011, http://www.dailypioneer.com; same figures also cited in: Praful Bidwai, 'People vs Nuclear Power in Jaitapur, Maharashtra', op. cit.
138. Praful Bidwai and M.V. Ramana, 'Home, Next to N-Reactor', op. cit.
139. These and more details about safety problems with VVER-1000 reactors are given in a letter written by Friends of Earth, Sydney, Australia to the Indian Prime Minister questioning the decision of the Indian government to build two VVER-1000 reactors in Kudankulam. Letter available at: 'Nuking the public interest?' April 1999, http://www.indiatogether.org
140. Praful Bidwai and M.V. Ramana, 'Indian Mega Nuclear Plant Protested: Government short-circuits environmental hearings', *Japan Focus*, June 14, 2007, http://www.japanfocus.org/-MV-Ramana/2449; Praful Bidwai and M.V. Ramana, 'Home, Next to N-Reactor', op. cit.
141. Cited in: Buddhi Kota Subbarao, 'India's Nuclear Prowess: False Claims and Tragic Truths', op. cit.
142. Ibid.; Donella Meadows, 'A Nuclear Power Plant out of the Old World Order', 2004, http://www.sustainer.org
143. Fuel burnup is a measure of how much actual energy is extracted from the nuclear fuel. It is measured as the actual energy released per mass of initial fuel, in gigawatts-days/ton of heavy metal, or similar units.
144. A. Gopalakrishnan, 'Reject French reactors for Jaitapur', Dec 2, 2010, http://www.cndpindia.org
145. Praful Bidwai and M.V. Ramana, 'Indian Mega Nuclear Plant Protested: Government short-circuits environmental hearings', op. cit.
146. For some more details on this, see the 'Introduction' to this book.
147. *Courting Nuclear Disaster in Maharashtra*, CNDP report, Jan 2011, p. 12, report available at http://www.cndpindia.org; Nikhil Ghanekar, 'The nuclear park at Jaitapur will be huge. So will the human cost', *Tehelka*, Sept 18, 2010, http://www.tehelka.com
148. *Courting Nuclear Disaster in Maharashtra*, ibid., p. 12.
149. M.V. Ramana, *Preliminary Observations on the Environmental Impact Assessment of the proposed reactors for Jaitapur, Maharashtra carried out by National Environmental Engineering Research Institute (NEERI), Nagpur*, unpublished, submitted to the MoEF at the public hearing organised in Madban on the EIA prepared by NEERI, in May 2010.
150. M.V. Ramana, ibid.; see also: Meena Menon, 'Critical mass', *Frontline*, June 5, 2010, http://www.frontlineonnet.com
151. M.V. Ramana, ibid.

152. Murty, C.V.R., 'Learning earthquake design and construction: 4. Where are the seismic zones in India?', 2004, www.ias.ac.in/resonance/Sept2004
153. Viju B., '20 years, 92 quakes: Ground trembles beneath Jaitapur's feet', *The Times of India*, March 16, 2011, http://articles.timesofindia.indiatimes.com
154. 'Japan Earthquake Rattles World's Largest Nuclear Power Plant', *Environment News Service*, July 16, 2007, http://www.ens-newswire.com; 'Japan nuke plant leak bigger than thought', July 18, 2007, http://www.msnbc.msn.com; *2007 Niigata Chuetsu-Oki Japan Earthquake Reconnaissance Report*, p. 1, http://www.grmcat.com
155. Ashwin Kumar, M.V. Ramana, 'Nuclear safety lessons from Japan's summer earthquake', Dec 4, 2007, http://www.thebulletin.org; Gavan McCormack, 'Japan as a Nuclear State', Aug 1, 2007, http://www.japanfocus.org
156. *Environmental Impact Assessment for Proposed Nuclear Power Park at Jaitapur, District Ratnagiri, Maharashtra*, Vol. 1, p. 369, NEERI, Nagpur, February 2010.
157. Ibid., Vol. 1, p. 383; Gopal Krishna, 'Why the Jaitapur nuclear plant must be opposed', Dec 29, 2010, http://www.rediff.com
158. 'Environmental Clearance for JNPP by MOEF', Nov 26, 2010, www.npcil.co.in
159. T.N. Raghunatha, 'Jaitapur nuclear power project gets conditional clearance', Nov 29, 2010, http://www.dailypioneer.com
160. M.A. Rane, 'Dangers from nuclear power projects', op. cit.

Chapter 10: The Sustainable Alternative to Nuclear Energy

1. *Integrated Energy Policy*, 2006, http://planningcommission.nic.in, pp. xiii, 20, 32.
2. Ibid., p. 22.
3. 'Nuclear energy essential for India: Manmohan Singh', *ANI* report, March 24, 2008, http://www.andhranews.net/India/2008/March/24-Nuclear-energy-essential-38627.asp
4. See for example: Bharat Jhunjhunwala, 'Will nuclear power really help?', *The Hindu Business Line*, Mar 26, 2008, http://www.thehindubusinessline.in
5. *Integrated Energy Policy*, op. cit., p. 20.
6. Neeraj Jain, *India becoming a colony again*, Oct 2010, Lokayat publication, Pune, available online at 'Lokayat', http://www.lokayat.org.in

7. Sagar Dhara, 'Is thermal power going to be the biggest risk to Indian biodiversity this century?', *NBSAP: Thermal power and biodiversity review*, Jan 2003.
8. Even the *IEP 2006* admits that while officially 85 per cent of villages are considered electrified, around 57 per cent of the rural households and 12 per cent of the urban households, i.e. 84 million households (over 44.2% of total) in the country did not have electricity in 2000. Source: *Integrated Energy Policy*, op. cit., p. 2.
9. Source: CEA, http://www.cea.nic.in
10. *Still Waiting: A Report on Energy Injustice*, Report produced by Greenpeace India Society, Oct 2009, available on: http://www.scribd.com/doc/26256876/Still-Waiting
11. *Integrated Energy Policy*, op. cit., p. 22.
12. Ibid., p. 46.
13. Hydropower capacity was 36 GW in 2010, so projected increase in installed capacity by 2032 is 150 – 36 = 114 or 110 GW approx; installed coal-based power capacity was 84 GW in 2010, so that means an additional capacity increase of 270-84 = 186 or 190 GW upto 2032.
14. *National Electricity Plan, Vol I*, Central Electricity Authority, Ministry of Power, Government of India, cited in: *Strategies to Reduce GHG Emissions from India's Coal-Based Power Generation*, The Energy and Resources Institute, Project Report No. 2008IE14, May 19, 2009, p. 20, http://assets.panda.org
15. *Integrated Energy Policy*, op. cit., pp. 46-7.
16. *Green Energy*, WISE, May-June 2009, pp. 16-7, http://www.wise-cleanenergy.info/green-energy/GE-May-June09.pdf
17. See for example: Philippe Dozolme, 'Common Mining Accidents', http://mining.about.com; Simon Elegant / Zhangjiachang, 'Where The Coal Is Stained With Blood', *Time Magazine*, Mar 2, 2007, http://www.time.com; 'Coal Mining', http://www.perryopolis.com
18. http://www.psr.org/assets/pdfs/coals-assault-executive.pdf
19. Sagar Dhara, 'Is thermal power going to be the biggest risk to Indian biodiversity this century?', op. cit.
20. *The True Cost of Coal*, http://www.greenpeace.org
21. Sagar Dhara, 'Is thermal power going to be the biggest risk to Indian biodiversity this century?', op. cit.
22. 'Coal Ash Pollution Contaminates Groundwater, Increases Cancer Risks', Sept 4, 2007, http://www.earthjustice.org
23. Mara Hvistendahl, 'Coal Ash Is More Radioactive than Nuclear Waste',

December 13, 2007, http://www.scientificamerican.com; 'Environmental impact of coal mining and burning', *Wikipedia*, http://en.wikipedia.org

24. Joseph V. Spadaro, et al., *Greenhouse Gas Emissions of Electricity Generation Chains: Assessing the Difference*, IAEA bulletin, http://www.iaea.org/Publications/Magazines/Bulletin/Bull422/article4.pdf
25. http://www.pewclimate.org/technology/overview/electricity
26. http://moef.nic.in/downloads/public-information/Report_INCCA.pdf, p. i.
27. http://www.narmada.org/gcg/gcg.html
28. 'A Trojan Horse for Large Dams', http://www.thecornerhouse.org.uk
29. 'Greenhouse Gas Emissions from Dams FAQ', May 1, 2007, http://www.internationalrivers.org
30. 'Lesson for India: Large dams lead to global warming too', May 28, 2007, http://www.rediff.com/news/2007
31. See, for example: 'A Trojan Horse for Large Dams', op. cit.; Himanshu Thakkar, 'Future Water Solutions for India', http://indiawaterportal.org
32. http://www.sandrp.in/hydropower/Hydro_Tariff_CERC_Proposal_CWP_Submission_March08.pdf
33. A discussion on this can be found in: 'A Trojan Horse for Large Dams', op. cit.
34. Cited in: Balam Vincent and Patrick McCully, 'Supreme Contradictions', http://www.indiatogether.org/opinions/sc-ssp.htm
35. As on March 2010; source: CEA, http://www.cea.nic.in
36. A. Shyam, 'Energy Efficiency options for thermal power plants in India', Jan 21, 2010, http://www.energycentral.com
37. Source: *Integrated Energy Policy*, op. cit., pp. 81-82; and *Still Waiting: A Report on Energy Injustice*', op. cit.; we have modified the figure for 'end-use efficiency in other areas (domestic, street lighting, et cetera)': Greenpeace gives it as 20-30%, but other studies, like the one by Prof. Saifur Rahman mentioned in endnote 45 below, puts the energy saving potential in end-uses (lighting, cooling, ventilation, refrigeration) at between 20-50%. International best practice is 80% for this category. Therefore, end-use efficiency in India for this category: 80 minus 20-50 = 30-60.
38. CEA, *Monthly Report, March 2010*, http://www.cea.nic.in
39. 'India's Electricity Transmission and Distribution Losses', July 26, 2008, http://cleantechindia.wordpress.com
40. 'South Korea to help India cut power T&D losses', Jun 3, 2010, http://economictimes.indiatimes.com

41. Shankar Sharma, 'Power sector reforms: a pilot study on Karnataka', http://www.indiaenvironmentportal.org.in
42. CEA, *Monthly Report, March 2010,* http://www.cea.nic.in
43. Same point has been made by Puneet Tayal, 'Power Generation: Power saved is power generated', http://www.waterenergynexus.com
44. Jayant A. Sathaye, et al., *Implementing End-use Efficiency Improvements in India: Drawing from Experience in the US and Other Countries,* Ernest Orlando Lawrence Berkeley National Laboratory, May 8, 2006, http://ies.lbl.gov
45. Professor Saifur Rahman, *Energy Use and National Energy Efficiency Strategies in China, India the USA and the EU,* International Conference on Industrial & Commercial Use of Energy, South Africa, May 22-23, 2006, http://www.ceage.vt.edu
46. Girish Sant, et al., *Energy saving potential in Indian households from improved appliance efficiency,* March 2010, http://www.prayaspune.org
47. Assuming T&D losses of 15% and average availability of 90% for new capacity: Girish Sant, et al., ibid.
48. Shankar Sharma, 'Power sector reforms: a pilot study on Karnataka', op. cit.
49. 'Bachat Lamp Yojana', *Wikipedia,* http://en.wikipedia.org; 'Kerala, Karnataka roll out CFL scheme', July 6, 2011, http://www.thehindubusinessline.com
50. CEA, *Annual Report 2007-08*, pp. 54-5, http://www.cea.nic.in
51. Shankar Sharma, a well-known consultant to the electrical industry, has made these calculations for the State of Karnataka and shown that the power supply deficit in the State can be met just by adopting energy efficiency measures. The gross power availability in the state was 8954 MW as on Oct 21, 2008. Therefore, net power availability —after accounting for auxiliary consumption and unplanned outages —should have been about 8046 MW. The peak power requirement for the state was only 6583 MW for the year 2007-08. This indicates that had the state electricity sector functioned at the international T&D loss level of less than 10%, the electricity generation capacity in the state was more than enough to meet not just the peak demand but also the annual energy requirement of 40,320 MU. In fact, there would have been a surplus generating capacity, and there would have been no need to set up new power plants in the state for the next few years. Instead of this, due to the inefficiency prevailing in the power sector, there was a power supply deficit! The maximum demand the state met during that year was only 5,567 MW, and the annual energy supply

was just 39,230 MU, resulting in a deficit of 15.7% and 2.7% respectively. (Reference: Shankar Sharma, 'Power sector reforms: a pilot study on Karnataka', op. cit.)

52. Shankar Sharma, 'Power sector reforms: a pilot study on Karnataka', ibid.
53. A.K.N. Reddy, 'Energy Strategies for a Sustainable Development in India', http://amulya-reddy.org.in
54. *Integrated Energy Policy*, op. cit., pp. 81-2.
55. John Bellamy Foster, Brett Clark and Richard York, 'Capitalism and the Curse of Energy Efficiency: The Return of the Jevons Paradox', *Monthly Review*, November 2010, Volume 62, Issue 06, http://monthlyreview.org
56. Helen Caldicott, *Nuclear Power is not the answer to Global Warming or anything else*, Melbourne University Press, 2006, pp. 161, 168; 'South Dakota school wants to build wind farm; other news', *CropChoice news*, June 29, 2005, http://www.cropchoice.com
57. http://en.wikipedia.org/wiki/Wind_power; *How wind energy works*, Union of Concerned Scientists, http://www.ucsusa.org; http://www.wwindea.org/home/index.php
58. 'Wind Power', http://en.wikipedia.org, accessed on Feb 25, 2011; *World Wind Energy Report 2009, Executive Summary*, Mar 10, 2010, World Wind Energy Association, http://www.wwindea.org/home/index.php?option=com_content&task=view&id=266 &Itemid=43
59. *How wind energy works*, Union of Concerned Scientists, http://www.ucsusa.org
60. *Global wind energy markets will continue to boom*, Global Wind Energy Council, Sept 4, 2009, http://www.gwec.net
61. G.M. Pillai, 'Need for a new energy policy', *Green Energy*, May-June 2009, p. 17: publication of WISE, www.wisein.org; available online at: http://www.wise-cleanenergy.info/green-energy/GE-May-June09.pdf
62. *How wind energy works*, Union of Concerned Scientists, op. cit.
63. *How solar energy works*, Union of Concerned Scientists, http://www.ucsusa.org
64. Ned Haluzan, 'Renewable energy general info', Nov 23, 2010, http://www.renewables-info.com; 'Renewable Energy Services', http://greensideelectric.ca/renewableenergy-services.html; 'Alternative Energy', http://www.activeonearth.com/alternativeenergysources.html
65. Y.B. Ramakrishna (Chairman of Taskforce on Bio-fuels, government of Karnataka), Keynote address at workshop on: 'Renewable Energy:

Employment & Entrepreneurial opportunities', June 6-7, 2009, Mangalore.

66. Lester Brown, 'On Rooftops Worldwide, A Solar Water Heating Revolution', http://www.celsias.com; *How solar energy works*, Union of Concerned Scientists, op. cit.
67. *How solar energy works*, Union of Concerned Scientists, ibid.
68. *Concentrated Solar Power, Factsheet*, August 2009, http://www.eesi.org/files/csp_factsheet_083109.pdf
69. http://www.erec.org/renewableenergysources/csp-solar-power.html
70. Keith Bradsher, 'China Tries a New Tack to Go Solar', *New York Times*, Jan 9, 2010, http://www.nytimes.com
71. *Concentrated Solar Power, Factsheet*, op. cit.
72. 'CSP and photovoltaic solar power', Aug 23, 2009, http://www.reuters.com
73. 'Emerging Energy Research' is a leading provider of market intelligence on the global energy industry.
74. G.M. Pillai, 'Need for a new energy policy', op. cit., p. 17.
75. http://en.wikipedia.org/wiki/Concentrating_solar_power
76. http://en.wikipedia.org/wiki/Photovoltaics
77. US cost figures taken from: 'Solar Energy Costs/Prices', http://www.solarbuzz.com/statsCosts.htm
78. http://en.wikipedia.org/wiki/Photovoltaics
79. *How solar energy works*, Union of Concerned Scientists, op. cit.
80. http://en.wikipedia.org/wiki/Photovoltaics
81. *Renewables Global Status Report 2009 Update*, http://www.ren21.net
82. Ibid.
83. 'EU renewable energy policy', Aug 2, 2007, http://www.euractiv.com/en/energy
84. G.M. Pillai, 'Need for a new energy policy', op. cit., pp. 15, 17; '50% of Electricity Must Come From Renewables by 2050 to Avert Global Warming: IEA', http://www.treehugger.com
85. Report actually talks of renewables, and not new renewables. We have excluded figures for hydro while giving the statistics here for new renewables—but therefore the potential of small hydropower is not included in these statistics.
86. *Energy [R]Evolution: A Sustainable World Energy Outlook*, Greenpeace–European Renewable Energy Council Report, 2007, p. 94: we have excluded the figures for hydropower to arrive at the figures for new renewables. Source: http://www.energyblueprint.info
87. Mycle Schneider, et al., *World Nuclear Industry Status Report,*

2010-2011, Commissioned by Worldwatch Institute, Washington, D.C., April 2011, p. 7.

88. Ministry of New and Renewable Energy, Official Website: http://mnre.gov.in
89. Wp or Watt-peak is a measure of nominal power of a photovoltaic solar energy device under laboratory illumination conditions.
90. http://mnre.gov.in/achievements.htm
91. Ministry of New and Renewable Energy, *Annual Report 2009-10*, http://mnre.gov.in
92. 'The Nuclear Lies of Expensive Solar power', July 22, 2011, http://www.dianuke.org; M. Ramesh, 'Solar mission: Firms tie up Rs.4,500 cr for projects of over 500 MW capacity', July 15, 2011, http://www.thehindubusinessline.com
93. http://mnre.gov.in/annualreport/2009-10EN; http://mnre.gov.in/achievements.htm; 'Wind Energy', Indian Wind Energy Association, http://www.inwea.org
94. 'Indian Wind Energy Outlook 2009 – Executive Summary', www.businesswireindia.com
95. Ministry of New and Renewable Energy, *Annual Report 2009-10*, http://mnre.gov.in
96. Ministry of New and Renewable Energy, *Annual Report 2009-10* and *Annual Report 2008-09*, http://mnre.gov.in
97. Ministry of New and Renewable Energy, *Annual Report 2009-10*, http://mnre.gov.in
98. Gopal Krishna, 'Waste-to-energy or waste-to-pollution?', *InfoChange News & Features*, June 2006, http://infochangeindia.org; 'Angry Rally Against Incinerator-based Waste to Energy Projects', March 23, 2011, http://toxicswatch.blogspot.com
99. 'Ocean Energy in India', *Energy Alternatives India*, 2009, http://www.eai.in/ref/ae/oce/oce.html; 'Ocean Energy', Our Energy, http://www.our-energy.com/ocean_energy.html; 'Energy Story, Chapter 14: Ocean Energy', http://www.energyquest.ca.gov/story/chapter14.html
100. *How Geothermal Energy Works,* Union of Concerned Scientists, http://www.ucsusa.org; 'Geothermal electricity', *Wikipedia*, http://en.wikipedia.org, accessed in May, 2011.
101. 'Geothermal Energy–India Energy Portal', http://www.indiaenergyportal.org; 'Geothermal Energy Gathers Steam in India', Clean Technica, December 31, 2010, http://cleantechnica.com
102. G.M. Pillai, 'Need for a new energy policy', op. cit., p. 14.

103. Figure of 50,000 MW is quoted from: http://direc2010.gov.in/solar.html
104. G.M. Pillai, 'Need for a new energy policy', op. cit., p. 18.

Chapter 11: Unite, to Fight this Madness!

1. J. Sri Raman, 'The U.S.-India nuclear deal – one year later', *Bulletin of the Atomic Scientists*, Oct 1, 2009, http://www.thebulletin.org; 'India demands lift of ban on dual use items; $270 Billion dangles', *IANS*, March 24, 2009, http://www.siliconindia.com
2. J. Sri Raman, ibid.
3. Ibid.
4. 'Commercial Motivations Add Impetus to Indo-US Nuclear Agreement', *WMD Insights*, May 2007 issue, http://www.wmdinsights.com
5. 'Nuclear Power in India', updated Nov 30, 2010, http://www.world-nuclear.org
6. 'Commercial Motivations Add Impetus to Indo-US Nuclear Agreement', op. cit.
7. Ibid.
8. 'End to India nuclear isolation opens huge market', Sept 9, 2008, http://www.nuclearpowerdaily.com
9. 'India-France nuclear energy cooperation agreement enters into force', Jan 14, 2010, http://www.india.carbon-outlook.com
10. The Olkiluoto-3 reactor is already costing more than $7 billion (see Chapter 4), and construction is only at the halfway stage; so even this figure of $14 bn for Areva's first two reactors at Jaitapur is actually an underestimate.
11. Charmaine Noronha, 'Canada close to nuclear deal with India', Nov 29, 2009, http://www.prdailysun.com
12. 'End to India nuclear isolation opens huge market', op. cit.
13. Anil Sasi, Richa Mishra, 'ONGC enters the fray for n-power; looks for tie-up with NPCIL', Aug 6, 2010, http://www.thehindubusinessline.in
14. 'Nuclear Power in India', op. cit.; J. Sri Raman, 'The U.S.-India nuclear deal – one year later', op. cit.; 'L&T, Rolls-Royce sign nuclear equipment pact', *Financial Express*, April 2, 2010, http://www.financialexpress.com
15. Ravi Teja Sharma, 'India's "nuclear renaissance" ', Dec 12, 2010, http://economictimes.indiatimes.com
16. Vineeta Pandey, 'Deal done, Jaitapur nuke power plant gets a leg up',

Dec 7, 2010, http://www.dnaindia.com

17. J. Sri Raman, 'The U.S.-India nuclear deal – one year later', op. cit.
18. 'Privatisation of Indian nuclear power plants soon: Ansari', Aug 26, 2009, http://www.thaindian.com
19. Siddharth Srivastava, 'India on power trip as nuke deal advances', *Asia Times*, Jun 29, 2006, http://www.atimes.com; Ravi Teja Sharma, 'India's "nuclear renaissance" ', op. cit.
20. For a more detailed discussion of why India began globalisation, see Neeraj Jain, *Globalisation or Recolonisation*, published by Lokayat, Pune, Jan 2007; available online at 'Lokayat', http://www.lokayat.org.in
21. Sarah Anderson and John Cavanagh, 'Top 200: The Rise of Corporate Global Power', Institute for Policy Studies, Washington, DC, Dec 2000, http://www.ips-dc.org
22. Ibid.
23. This entire section is summarised from Neeraj Jain, *India becoming a Colony Again*, Lokayat publication, Pune, October 2010; for references and the book itself, see 'Lokayat', http://www.lokayat.org.in
24. 'India boasts record 69 billionaires: Forbes', Sept 30, 2010, http://www.channelnewsasia.com
25. 'Asia beats Europe in wealth race', *The Deccan Chronicle*, Sept 29, 2010, http://www.deccanchronicle.com
26. Utsa Patnaik, 'Neoliberal roots', *Frontline*, Chennai, Mar 15-28, 2008, http://www.thehindu.com/fline
27. 'Nearly 80 pct of India lives on half dollar a day', Aug 10, 2007, http://www.reuters.com
28. Neelam Raaj, 'Malnutrition ails Indian Kids', *Times of India*, Feb 12, 2007; A.K. Shiva Kumar, 'Why are child malnutrition levels not improving?' *The Hindu*, June 22, 2007.
29. Utsa Patnaik, 'Neoliberal roots', op. cit.; see also Neeraj Jain, *India becoming a Colony Again*, op. cit.
30. 'India's external debt rises to $273.1 billion', Oct 13, 2010, http://timesofindia.indiatimes.com
31. *India's External Debt: A Status Report*, Government of India, Ministry of Finance, Department of Economic Affairs, p. 1, http://finmin.nic.in
32. 'Monthly Market Report', June, 2010, http://www.indiafirstlife.com
33. 'Economic Survey', cited in *Economic and Political Weekly*, Mumbai, No. 11, Mar 11, 2000, p. 859.
34. 'Rising current account deficit a worry, says FM', Oct 14, 2010, http://www.business-standard.com
35. We have deliberately avoided mentioning the growing crisis of global

warming, as the developed countries are more responsible for it. But it is also true that India's contribution is rapidly increasing.

Epilogue: The Fukushima Catastrophe in Japan

1. 'Fukushima Daiichi Nuclear Power Plant', *Wikipedia*, http://en.wikipedia.org, accessed on May 30, 2011.
2. Yuji Okada, Tsuyoshi Inajima and Shunichi Ozasa, 'Fukushima Station May Have Leaked Radiation Before Tsunami', *Bloomberg Businessweek*, May 19, 2011, http://www.businessweek.com; 'Nuclear Crisis in Japan', Update, May 20 and May 16, 2011, http://www.nirs.org/fukushima/crisis.htm
3. Most of this summary of the Fukushima accident taken from: Surendra Gadekar, 'The Depth in the "Defence in Depth", Proves Insufficient', May 2, 2011, http://www.nonuclear.in
4. Arnie Gundersen is a former nuclear industry senior vice president. He earned his Bachelor's and Master's Degrees in nuclear engineering, holds a nuclear safety patent, and was a licensed reactor operator. During his nuclear industry career, Arnie managed and coordinated projects at 70 nuclear power plants around the country.
5. William Whitlow, 'Fukushima radiation levels rise to highest levels yet', May 2, 2011, http://www.wsws.org; 'Fukushima reactor 3 explosion scattered spent fuel rods for miles', June 18, 2011, http://mgx.com
6. Mitsuru Obe, 'Cores Damaged at Three Reactors', *Wall Street Journal*, May 16, 2011, http://online.wsj.com
7. 'Fukushima: It's much worse than you think', June 16, 2011, http://english.aljazeera.net
8. 'Exclusive Arnie Gundersen interview: The dangers of Fukushima are worse and longer-lived than we think', June 3, 2011, http://www.chrismartenson.com
9. Mitsuru Obe, 'Cores Damaged at Three Reactors', op. cit.
10. '3 nuclear reactors melted down after quake, Japan confirms', *CNN*, June 6, 2011, http://edition.cnn.com; Alexander Higgins, 'Japan Admits "Worse Than Meltdown" – Nuclear "Melt-Through" In 3 Reactors', June 7, 2011, http://blog.alexanderhiggins.com
11. 'Exclusive Arnie Gundersen interview: The dangers of Fukushima are worse and longer-lived than we think', op. cit.; 'Containment vessels also damaged', *The Yomiuri Shimbun*, May 26, 2011, http://www.yomiuri.co.jp; Ami Sedghi, 'Fukushima nuclear power plant update: get all the data', June 7, 2011, http://www.guardian.co.uk

12. Mike Adams, Editor, 'TEPCO now confirms nuclear meltdown in Fukushima reactor No. 1', *Natural News*, May 12, 2011, http://www.naturalnews.com; David Wright, 'Fukushima: 7-Week Update', Union of Concerned Scientists, May 3, 2011, http://allthingsnuclear.org
13. Ethan A. Huff, 'MOX plutonium fuel used in Fukushima's Unit 3 reactor two million times more deadly than enriched uranium', *Natural News*, March 17, 2011, http://www.naturalnews.com
14. Yuji Okada, Tsuyoshi Inajima and Shunichi Ozasa, 'Fukushima Station May Have Leaked Radiation Before Tsunami', op. cit.
15. David Wright, 'Fukushima: 7-Week Update', op. cit.
16. David Ferguson, 'Fukushima Reactor No. 1 more radioactive than ever', June 4, 2011, http://www.rawstory.com
17. 'Fukushima: It's much worse than you think', June 16, 2011, http://english.aljazeera.net
18. 'Japan Doubles Nuclear Radiation Leak Estimate', *Discovery News*, June 7, 2011, http://news.discovery.com; 'Fukushima Water has more Radiation than Released into Air', *Bloomberg Businessweek*, June 8, 2011, http://www.businessweek.com
19. 'Exclusive Arnie Gundersen interview: The dangers of Fukushima are worse and longer-lived than we think', op. cit.
20. 'Radioactive debris hampers efforts to cool reactor at Japan quake-hit nuke plant', *RIA Novosti*, May 22, 2011, http://en.rian.ru
21. 'Exclusive Arnie Gundersen interview: The dangers of Fukushima are worse and longer-lived than we think', op. cit.
22. Ibid.
23. 'Japan's nightmare-in-progress overshadowed by media mush,' Special to *Vancouver Courier*, May 20, 2011, http://www.vancourier.com/news
24. Kurt Nimmo, 'Media Silent about Plutonium Contamination of Japanese Rice', *Infowars.com*, May 16, 2011, http://www.infowars.com
25. 'Japan's nightmare-in-progress overshadowed by media mush', Special to *Vancouver Courier*, op. cit.; Harvey Wasserman, 'Fukushima's Apocalyptic Threat', *Counterpunch*, May 20-22, http://www.counterpunch.org
26. Eben Harrell, 'Was Fukushima a China Syndrome?', *Ecocentric Feed*, May 16, 2011, http://ecocentric.blogs.time.com
27. Bryan Hood, 'TEPCO Announces Shutdown Plan for Fukushima', *The Atlantic Wire*, Apr 17, 2011, http://www.theatlanticwire.com
28. 'Fukushima: It's much worse than you think', June 16, 2011, http://english.aljazeera.net

29. Ibid.
30. See for instance the views of Professor Hiroaki Koide of Kyoto University Research Reactor Institute and Dr Michio Kaku, one of the world's best known theoretical physicists who teaches at City University of New York: 'Hiroaki Koide of Kyoto University: "No One Knows How Fukushima Could Be Wound Down" As the Corium May Be Melting Through the Foundation', http://ex-skf.blogspot.com/2011/05/hiroaki-koide-of-kyoto-university-no.html; 'Expert: Despite Japanese Gov't Claims of Decreasing Radiation, Fukushima a "Ticking Time Bomb" ', April 13, 2011, http://www.democracynow.org
31. 'EU promises 110 mn Euros for new Chernobyl sarcophagus', *FOCUS News Agency*, April 18, 2011, http://www.focus-fen.net; 'Chernobyl Nuclear Power Plant sarcophagus', *Wikipedia*, http://en.wikipedia.org; Andrew Osborne, 'Cracks in decaying shell of Chernobyl reactor threaten second disaster', April 28, 2005, http://www.independent.co.uk; Benjamin Bidder, 'The Overwhelming Challenge of Containing Chernobyl', *Spiegel Online*, April 26, 2011, http://www.spiegel.de
32. Benjamin Bidder, 'The Overwhelming Challenge of Containing Chernobyl', ibid.
33. 'Exclusive Arnie Gundersen interview: The dangers of Fukushima are worse and longer-lived than we think', op. cit.
34. 'Part II: Arnie Gundersen interview: Protecting yourself if the situation worsens', June 3, 2011, http://www.chrismartenson.com
35. Karl Grossman, 'Downplaying Deadly Dangers in Japan and at Home, After Fukushima, Media Still Buying Media Spin', April 30, 2011, http://karlgrossman.blogspot.com
36. Greg Mitchell, 'For 64th Anniversary: The Great Hiroshima Cover-Up – And the Nuclear Fallout for All of Us Today', Aug 6, 2009, http://www.huffingtonpost.com
37. Karl Grossman, 'Downplaying Deadly Dangers in Japan and at Home, After Fukushima, Media Still Buying Media Spin', op. cit.
38. Hans Blix cited in*: Anumukti*, Vol. 9, Nos. 5-6, p. 27: published by Dr Surendra Gadekar, Sampoorna Kranti Vidyalaya, Vedchhi, India-394641; available online at http://www.nonuclear.in/anumukti; also cited in: Vladimir Radyuhin, 'Twenty-five years after Chernobyl', *The Hindu*, April 26, 2011, http://www.hindu.com
39. Anil Kakodkar, 'What Fukushima tells us', *The Times of India*, Apr 16, 2011, http://m.timesofindia.com
40. See Chapter 3, Part V for more on this book.

41. INES: International Nuclear Events Scale, the scale created by the IAEA to categorise nuclear accidents. See Chapter 3, Endnote 107 for more on this.
42. Ryan Nakashima and Yuri Kageyama, 'Japan Equates Nuclear Crisis Severity to Chernobyl', April 12, 2011, http://www.aolnews.com
43. Alexander Higgins, 'US and IAEA knew Fukushima had Meltdown within 3.5 hours since March and hid it from the Public', May 24, 2011, http://blog.alexanderhiggins.com
44. 'Japanese withheld information on high radiation at Fukushima', April 4, 2011, http://www.adelaidenow.com.au
45. Justin McCurry, 'Japan under pressure to widen nuclear evacuation zone', March 31, 2011, http://www.guardian.co.uk
46. 'Radioactive strontium detected 62 km from Fukushima plant', *Japan Today*, June 9, 2011, http://www.japantoday.com
47. 'Unbelievable: TEPCO: Plutonium Found On 5 Locations Around Fukushima Plant A Week Ago', March 29, 2011, http://www.infiniteunknown.net
48. 'Gov't to scrap upper limit of radiation exposure for workers at Fukushima plant', May 30, 2011, http://mdn.mainichi.jp
49. John M. Glionna and Kenji Hall, 'Parent anger plays role in Japan's reversal of raised radiation limits at schools', *Los Angeles Times*, May 29, 2011, http://articles.latimes.com; Robert Alvarez, 'Radiation for Children's Day', *Counterpunch*, May 4, 2011, http://www.counterpunch.org; 'Japan Admits 3 Nuclear Meltdowns, More Radiation Leaked into Sea; U.S. Nuclear Waste Poses Deadly Risks', *Democracy Now*, June 10, 2011, http://www.democracynow.org
50. Christine Marran, *Contamination: From Minamata to Fukushima*, Institute for Policy Studies, May 16, 2011, http://www.ips-dc.org; 'IMF cuts Japan's 2011 growth forecast to 1.4 per cent; Kan assures safety of local produce', April 12, 2011, http://gantdaily.com
51. Christine Marran, ibid.; Jonathan Green, 'Please eat the vegetables, Japan tells radiation-wary nation', April 16, 2011, http://bionicbong.com
52. Toshifumi Kitamura, 'Japan, China, S. Korea leaders visit nuclear region', May 21, 2011, http://news.yahoo.com
53. 'Fukushima Radiation Update', *Sovereigns' Health Freedom Network Newsletter*, May 3, 2011, http://www.healthfreedom.info; 'No Nuclear Risk to North Pacific Fish, Officials Say', Apr 17, 2011, http://www.aolnews.com
54. 'US Government Responds to Fukushima by Trying to Raise Radiation

Limits, EPA Pulls 8 Of 18 Radiation Monitors Out Of CA, OR And WA', April 2, 2011, http://www.infiniteunknown.net; Richard Mauer, 'FDA claims no need to test Pacific fish for radioactivity', April 17, 2011, http://www.adn.com; Jeff Rense, 'FDA Won't Test Alaska Fish For Radiation—Don't Worry, Be Happy', April 20, 2011, http://newresearchfindingstwo.blogspot.com

55. Alexander Higgins, 'EU Secretly Authorizes Emergency Order Allowing Large Increase of Radiation in Food', April 3, 2011, http://blog.alexanderhiggins.com
56. 'Japan Gov. Agency: 40% of Chernobyl Radiation Released in the first week after March 11 at Fukushima', June 7, 2011, http://www.infiniteunknown.net
57. 'Japan Admits 3 Nuclear Meltdowns, More Radiation Leaked into Sea; U.S. Nuclear Waste Poses Deadly Risks', *Democracy Now*, op. cit.
58. According to Dr Surendra Gadekar, the Chernobyl accident was brought under control by May 9, 1986, that is, two weeks after the explosion. Other sources put it at 9-10 days.
59. 'Fukushima already Level 7 Chernobyl accident', March 24, 2011, http://www.nirs.org/reactorwatch
60. Dawn Woodworth, 'Fukushima ongoing Nuclear Crisis now has Nuclear Engineers urging IAEA to create "Level 8" ', June 3, 2011, http://dawnofthetruth.wordpress.com
61. 'Fukushima Groundwater Contamination Worst in Nuclear History', May 6, 2011, http://www.fairewinds.com
62. Ibid.
63. George Washington, 'No, the Amount of Radiation Released from the Japanese Nuclear Reactors is not "Safe" ', March 24, 2011, http://blog.alexanderhiggins.com; 'Radiation: There IS NO "SAFE" Level!!', http://forum.prisonplanet.com/index.php?topic=205339.0
64. See Chapter 3, Part II.
65. 'Epidemiologist, Dr Steven Wing, Discusses Global Radiation Exposures and Consequences with Gundersen', April 21, 2011, http://vimeo.com/22706805
66. Mark Willacy, 'Fukushima radiation traces spread across Asia', March 30, 2011, http://www.abc.net.au
67. 'Spinach in China Exposed to Radiation from Fukushima Nuclear Power Plant', April 7, 2011, http://www.worldnewsco.com
68. 'Low Levels of Fukushima Radiation Fallout Detected Worldwide', March 30, 2011, http://news.yahoo.com
69. Steven Hoffman, 'Fukushima in Our Food: Low Levels of Radiation

from Japan's Nuclear Meltdown Detected in Milk, Fruit and Vegetable Samples Tested from California Farms', June 1, 2011, http://www.compassnaturalmarketing.com

70. 'Part II: Arnie Gundersen interview: Protecting yourself if the situation worsens', op. cit.
71. Alexander Higgins, 'Radioactive Fukushima Plutonium and Strontium Bombarding US West Coast since March 18th', April 21, 2011, http://blog.alexanderhiggins.com; 'Fukushima Radiation Update', *Sovereigns' Health Freedom Network Newsletter*, op. cit.
72. 'Chief Nuclear Engineer Arnie Gundersen: Americium Found In New England', April 28, 2011, http://www.infiniteunknown.net
73. 'Nuclear crisis evacuees confused by policy shifts', *Japan Today*, April 12, 2011, http://www.japantoday.com; 'Earthquake and Tsunami Lead to Nuclear Crisis in Japan', March 14, 2011, http://environment.about.com; 'Why are evacuation zones circles?', April 11, 2011, http://seaandskyjp.wordpress.com
74. 'TEPCO Press Release: Americium, Curium Found with Plutonium at Fukushima', *Japan Fukushima Nuclear Fallout News*, April 28, 2011, http://www.facebook.com; 'Unbelievable: TEPCO: Plutonium found on 5 Locations around Fukushima Plant a Week Ago', March 29, 2011, http://www.infiniteunknown.net
75. Hiroko Tabuchi and Keith Bradsher, 'Japan Nuclear Disaster Put on Par With Chernobyl', *New York Times*, April 11, 2011, http://www.nytimes.com; 'Japan under pressure to widen nuclear evacuation zone', March 31, 2011, http://www.guardian.co.uk
76. 'Chronological factsheet on 2011 crisis at Fukushima Nuclear Power Plant, Updates: June 21 and June 17', National Information and Resource Service, USA, http://www.nirs.org
77. Alexander Higgins, 'Amid Protests and Doublespeak Japan widens Fukushima Evacuation Zone', April 11, 2011, http://blog.alexanderhiggins.com
78. 'Highly Radioactive Substances Detected in Tokyo', *Arirang TV*, May 15, 2011, http://www.arirang.co.kr; 'Highly radioactive sewage slag was found in Tokyo, and it's being sold as construction materials', *Action and Resistance*, May 25, 2011, http://actionandresistance.wordpress.com
79. Alexander Higgins, 'Chernobyl Evacuation Limit 5,000 uSv/yr, Yet 394,200 uSv/yr Detected In Fukushima School Zone', June 11, 2011, http://blog.alexanderhiggins.com
80. Chris Busby, *The health outcome of the Fukushima catastrophe: Initial*

analysis from risk model of the European Committee on Radiation Risk, March 30, 2011, http://llrc.org/fukushima/subtopic/fukushimariskcalc.pdf; 'Full meltdown in full swing? Japan maximum nuclear alert', *Russia Today*, March 31, 2011, http://info-wars.org

81. 'Wanted: The Right to Relocate - Growing Radiation Exposure and the July 19 Citizen-Government Talks in Fukushima', August 3, 2011, http://fukushima.greenaction-japan.org
82. 'Fukushima Much Worse than Chernobyl', March 31, 2011, http://alternetviews.wordpress.com
83. 'Pressure mounts on Japan to expand evacuation zone', March 31, 2011, http://news.xinhuanet.com
84. 'Nuclear crisis evacuees confused by policy shifts', *Japan Today*, op. cit.; Kanako Takahara, 'Evacuation zone to be widened', April 22, 2011, http://search.japantimes.co.jp; 'Radiation Hot Spots, Up To 200 KM From Fukushima Outside Evac Zone', June 18, 2011, http://www.myweathertech.com
85. 'Japan earthquake: Foreign evacuations increase', *BBC News*, March 18, 2011, http://www.bbc.co.uk; 'Singapore evacuating citizens within 100 km of Fukushima plant', *Kyodo News*, March 17, 2011, http://english.kyodonews.jp
86. 'Exclusive Arnie Gundersen interview: The dangers of Fukushima are worse and longer-lived than we think', op. cit.
87. 'Fukushima already Level 7 Chernobyl accident', March 24, 2011, http://www.nirs.org/reactorwatch
88. 'Fukushima meltdown – Caldicott says Japan may become uninhabitable – media silent', http://www.independentaustralia.net
89. Ryan Nakashima and Shino Yuasa, 'Radiation Leaks into Groundwater under Japan Nuclear Plant', *Huffpost World*, April 1, 2011, http://www.huffingtonpost.com
90. 'Fukushima Groundwater Contamination Worst in Nuclear History', May 6, 2011, http://www.fairewinds.com; 'Exclusive Arnie Gundersen interview: The dangers of Fukushima are worse and longer-lived than we think', op. cit.
91. Alexander Higgins, 'TEPCO Confirms 6,300 Becquerels Per Liter Strontium Radiation Contamination In Groundwater', June 13, 2011, http://blog.alexanderhiggins.com
92. Ken Belson, Hiroko Tabuchi, 'Japan finds Tainted Food up to 90 Miles from Nuclear Sites', *New York Times*, March 19, 2011, http://www.nytimes.com; 'Radioactivity scare hurts Japanese farmers', *ABC Rural*, March 30, 2011, http://www.abc.net.au; 'Radioactive substances

and their impact on health', *Reuters*, March 24, 2011, http://www.vancouversun.com
93. Danielle Demetriou, 'Japan: Green Tea Exports Banned due to High Levels of Radioactive Cesium', *The Telegraph*, June 3, 2011, http://www.telegraph.co.uk
94. 'Radioactive strontium detected 62 km from Fukushima plant', *Japan Today*, June 9, 2011, http://www.japantoday.com
95. 'High concentrations of plutonium from the soil of rice field 50 km or more distant', University of California, May 14, 2011, http://www.nuc.berkeley.edu; Kurt Nimmo, 'Media silent about Plutonium contamination of Japanese rice', *Infowars.com*, May 16, 2011, http://www.infowars.com
96. Statement by Dr Rima Laibow, Medical Director of the Natural Solutions Foundation, USA, cited in: 'Unbelievable: TEPCO: Plutonium found on 5 Locations around Fukushima Plant a week ago', March 29, 2011, http://www.infiniteunknown.net
97. Paul Joseph Watson, 'Fukushima Ocean contamination: worse than Chernobyl', *Infowars.com*, May 20, 2011, http://www.infowars.com; Ken Buesseler, 'Japan's irradiated waters: How worried should we be?', *CNN*, April 26, 2011, http://edition.cnn.com
98. 'Part II: Arnie Gundersen interview: Protecting yourself if the situation worsens', op. cit.
99. Ibid.
100. Robert Alvarez, 'Radiation for Children's Day', op. cit.
101. 'Health Effects, Radiation Protection', US EPA, http://www.epa.gov
102. Cited in Michel Fernex, 'The Chernobyl Catastrophe and Health Care', http://www.independentwho.info
103. Alexey V. Yablokov, Vassily B. Nesterenko, Alexey V. Nesterenko, *Chernobyl: Consequences of the Catastrophe for People and the Environment*, New York Academy of Sciences, 2009, p. 128, http://www.strahlentelex.de
104. Ibid., p. 42.
105. Ibid., pp. viii, 2.
106. Ibid., p. 128.
107. Ibid., p. 127.
108. 'Fukushima much worse than Chernobyl', *Alternet Views*, March 31, 2011, http://alternetviews.wordpress.com
109. Discussed in Chapter 3, Part II.
110. Alexander Higgins, 'From Japan To Seattle People Are Breathing In Virtually Undetectable Hot Radiation Particles', June 13, 2011, http:/

/blog.alexanderhiggins.com

111. Ibid.
112. Helen Caldicott, 'The Fukushima Nuclear Disaster in Perspective', *Global Research*, May 12, 2011, http://hamsayeh.net
113. Chris Busby, *The health outcome of the Fukushima catastrophe: Initial analysis from risk model of the European Committee on Radiation Risk*, op. cit.
114. 'WSJ Interview: Japan Senior Political Figure Ichiro Ozawa', May 26, 2011, http://www.infiniteunknown.net
115. Henry Sokolski, *Nuclear Power's Global Expansion: Weighing Its Costs and Risks*, Strategic Studies Institute, US Army War College, Dec 2010, p. 6, http://www.scribd.com
116. Judy Dempsey and Jack Ewing, 'Germany, in Reversal, Will Close Nuclear Plants by 2022', *New York Times*, May 30, 2011, http://www.nytimes.com; Matthew Day, 'Angela Merkel humiliated by Green Party in Baden-Wuerttemberg election', *The Telegraph*, March 28, 2011, http://www.telegraph.co.uk; 'International reaction to Fukushima I nuclear accidents', *Wikipedia*, http://en.wikipedia.org, accessed on June 11, 2011; 'Germany announces plans to phase out nuclear power', May 31, 2011, http://www.mohavedailynews.com
117. 'Swiss leaders OK N-power phase out', *Times of India*, June 9, 2011, http://articles.timesofindia.indiatimes.com; 'International reaction to Fukushima I nuclear accidents', *Wikipedia*, http://en.wikipedia.org, accessed on June 11, 2011.
118. Mathieu Gorse, 'Nuclear vote forces Italy to ponder renewables', June 14, 2011, http://www.google.com; 'Government Halts Nuclear Power Stations', April 20, 2011, http://www.corriere.it
119. 'Taiwan halts plans to build atomic reactors after Japan crisis', June 12, 2011, http://cnbusinessnews.com
120. 'Fukushima forces rethink on nuclear across Asia', *Nuclear Energy Insider*, April 28, 2011, http://analysis.nuclearenergyinsider.com; 'International reaction to Fukushima I nuclear accidents', *Wikipedia*, http://en.wikipedia.org, accessed on June 11, 2011.
121. Kaitlin Walter, 'Nuclear Commitment China's Nuclear Development Plan in Japan Aftermath', April 28, 2011, http://crisisjones.wordpress.com; 'Fukushima forces rethink on nuclear across Asia', ibid.
122. Holly Robin, 'Fukushima; the U.S. to Learn Nuclear Reactor Crisis', May 10, 2011, http://thebqb.com
123. Jessica Dailey, 'Japan's Prime Minister Calls for Complete Phase Out

of Nuclear Power', July 14, 2011, http://inhabitat.com; Chico Harlan, 'Japanese Prime Minister Naoto Kan calls for phase-out of nuclear power', *Washington Post*, July 13, 2011, http://www.washingtonpost.com

124. 'India to review nuclear safety systems: PM', March 14, 2011, http://www.deccanherald.com
125. Nitin Sethi, 'Mistimed? PM's N-meet fell on Chernobyl anniversary', April 28, 2011, http://articles.timesofindia.indiatimes.com
126. 'AERB to carry out comprehensive safety review of Indian nuclear power plants', *NetIndian News Network*, March 15, 2011, http://netindian.in/news; Ankur Paliwal, Ruhi Kandhari, 'What about nuclear transparency?', May 15, 2011, http://www.downtoearth.org.in
127. 'Japan N-blasts not accidents: Indian experts', *Press Trust of India*, March 15, 2011, http://ibnlive.in.com
128. 'Don't rush through Jaitapur N project: Top scientists, academics', http://www.hardnewsmedia.com
129. Dr A. Gopalakrishnan, 'Nuclear power: The missing safety audits', *DNA*, Apr 26, 2011, http://www.dnaindia.com
130. 'Japan N-blasts not accidents: Indian experts', *Press Trust of India*, op. cit.
131. Ibid.
132. 'PM firm on n-energy, directs safety upgrades', *The Assam Tribune*, June 1, 2011, http://www.assamtribune.com
133. 'Japan N-blasts not accidents: Indian experts', *Press Trust of India*, op. cit.
134. Ibid.
135. See Chapter 9, Part IV for more on this.
136. 'Japan N-blasts not accidents: Indian experts', *Press Trust of India*, op. cit.
137. See Chapter 9 for details of the accident.
138. 'Japan N-blasts not accidents: Indian experts', *Press Trust of India*, op. cit.
139. See Chapters 6 and 9 for more details.
140. Vaiju Naravane, 'French Nuclear Safety Authority cannot rule out moratorium on Flamanville reactor', *The Hindu*, April 1, 2011, http://www.hindu.com
141. 'Nukespeak - Daftspeak: Surendra Gadekar', http://www.anumukti.in
142. 'Maharashtra MLAs nap as scientists discuss risks of Japan nuclear plant', March 14, 2011, http://indiatoday.intoday.in; 'Can the authorities be trusted?', Greenpeace, http://www.greenpeace.org

143. 'Ex-AEC chief: Seismic history will be studied', *Times of India*, March 19, 2011, http://articles.timesofindia.indiatimes.com
144. See Chapter 9 for a more detailed discussion of this issue.
145. 'Update on Japan's Nuclear Reactors', *NPR*, March 18, 2011, http://www.npr.org
146. 'DOSE's Critique of Dr L.V. Krishnan's Post Tsunami Interview', Feb 7, 2005, http://www.kalpakkam.com
147. Zachary Shahan, 'Will the U.S. Close 23 Nuclear Reactors of Same Flawed Design as at Fukushima?', March 23, 2011, https://news.change.org
148. Latha Jishnu, 'Nuclear fault lines run deep', April 15, 2011, http://www.downtoearth.org.in
149. M.V. Ramana, Suchitra J.Y., 'Flaws in the pro-nuclear argument', *Infochange India*, Issue 5, 2006, http://infochangeindia.org
150. M.V. Ramana, 'Beyond our imagination: Fukushima and the problem of assessing risk', April 19, 2011, http://www.thebulletin.org
151. Dr Georgui Kastchiev, Prof. Wolfgang Kromp, Dipl.-Ing. Stephan Kurth, David Lochbaum, Dr Ed Lyman, Dipl.-Ing. Michael Sailer, and Mycle Schneider, *Residual Risk: An Account of Events in Nuclear Power Plants Since the Chernobyl Accident in 1986*, May 2007, http://www.greens-efa.org; See Chapter 3, Part V for more discussion on this report.
152. *Reactor Accidents*, http://www.nirs.org
153. Helen Caldicott, *Nuclear Power is not the answer to Global Warming or anything else*, Melbourne University Press, 2006, p. 83.
154. Figure as of Jan 19, 2011. Source: 'Nuclear power plants, world-wide', European Nuclear Society, http://www.euronuclear.org
155. Sabyasachi Bandopadhyay, 'Won't allow any nuclear plant in West Bengal, including Haripur: Gupta', Aug 18, 2011, http://www.expressindia.com

INDEX